PEARSON

# Android 开发秘籍（第2版）

The Android™ Developer's Cookbook

Building Applications with the Android SDK

Second Edition

[美] Ronan Schwarz  Phil Dutson  James Steele  Nelson To 著

钱昊 译

人民邮电出版社

北京

**图书在版编目（C I P）数据**

Android开发秘籍 ：第2版 / （美）施瓦茨
(Schwarz,R.) 等著 ；钱昊译. -- 2版. -- 北京 ：人民
邮电出版社，2014.8
书名原文：The Android developer’ s cookbook:
building applications with the Android SDK,second
Edition
ISBN 978-7-115-35517-1

Ⅰ. ①A… Ⅱ. ①施… ②钱… Ⅲ. ①移动终端—应用
程序—程序设计 Ⅳ. ①TN929.53

中国版本图书馆CIP数据核字(2014)第130822号

## 内 容 提 要

本书秉承“一个清晰可用的范例，胜过千言的文档”的原则，以一百多个范例为骨架，将知识、技巧和理念融入其中，从零开始，介绍了Android移动开发的方方面面。从Android及其设备的发展、Android项目的建立等入门内容到Activity、Intent、视图、线程、服务、用户界面布局、事件等基本要素，再到多媒体、硬件接口、网络、位置服务、应用内计费、消息推送等高级特性，最后还介绍了原生开发、测试与调试。本书致力于让读者充分理解和利用Android的各种特性，并十分强调设备与版本的兼容性、代码的复用性、项目的健壮性、方法的多样性等良好的开发理念。

本书的内容由浅入深，方便 Android 开发初学者上手；书中介绍的技巧彼此相关又相对独立，因此也适合有一定经验的开发者查阅参考。

◆ 著　　　　[美] Ronan Schwarz　Phil Dutson　James Steele
　　　　　　Nelson To
译　　　　钱　昊
责任编辑　杨海玲
责任印制　彭志环　焦志炜

◆ 人民邮电出版社出版发行　　北京市丰台区成寿寺路 11 号
邮编　100164　　电子邮件　315@ptpress.com.cn
网址　http://www.ptpress.com.cn
北京艺辉印刷有限公司印刷

◆ 开本：800×1000　1/16
印张：23
字数：521 千字　　　　2014 年 8 月第 2 版
印数：4 001 – 7 000 册　　2014 年 8 月北京第 1 次印刷
著作权合同登记号　图字：01-2013-7660 号

定价：59.00 元

**读者服务热线：(010)81055410　印装质量热线：(010)81055316**
**反盗版热线：(010)81055315**

# 版权声明

# 封面介绍

封面图片呈现了塞尔弗里奇百货公司[1]外墙的特写镜头，该建筑于 2003 年落成于英格兰伯明翰。其设计机构 Future Systems 宣称：其微光闪烁的设计营造了“精美的、如同蛇皮、或点缀着闪光金属片的帕科[2]服装一般的光彩熠熠的纹理”。它一落成，就立即被视为令人难忘的、有地标意义的建筑，成为重建的规模宏大的伯明翰 Bull Ring 购物商圈[3]的一部分。该区域作为众多商家落户的场所，已有超过 850 年的历史。塞尔弗里奇百货公司的建筑设计赢得了包括 Royal Fine Arts Commission、Retail Week、U.K.'s Concrete Society 等组织机构颁发的众多荣誉奖项。

---

① 塞尔弗里奇百货公司（Selfridge's department store）为英国一家有百年历史的著名百货公司，除本书所提到的伯明翰门店外，在伦敦、曼彻斯特等地尚有多家门店。——译者注

② 帕科（Paco Rabonne）是创建于 20 世纪 60 年代的法国高级时装品牌，以大胆地将各种新材料运用于服装上而著称。于 1966 年首次推出由金属和塑料片制成的服装，在当时引起轰动。——译者注

③ 伯明翰 Bull Ring（Birmingham's Bull Ring）为英格兰最著名的商业区之一。除上面提到的塞尔弗里奇百货公司外，还包含众多其他门店及露天市场等。——译者注

# 对本书的赞誉

《Android 开发秘籍（第 2 版）》包含了关于如何开发和营销一个成功的 Android 应用的各种技巧。每一个技巧都含有详细的解释和示例，讲解怎样正确地编写程序，使其能成为 Google Play Store[①]上别具一格的应用。以理解不同 Android 版本的基本特征、从而设计和构建出能适应这些不同的应用为开端，本书将教授诸多引领你走上成功之路的开发技巧。你将学会在 Android 系统的不同层面上进行工作，从使用硬件接口（如 NFC 和 USB），到能够指明如何有效地使用移动数据的网络接口，甚至还包括如何利用 Google 强大的计费接口（billing interface）。几位作者工作的杰出之处在于，他们为书中每个概念都配上了实用的、与现实生活息息相关的代码示例，这些示例可以被轻而易举地构建，并适用于各种各样的情况，使得本书成为一切 Android 开发者案头必备之书。

——David Brown，圣胡安（San Juan）学区信息数据管理员兼应用程序开发员

清晰易读、通俗易懂，但并非平淡无奇。这是我读过的 Android 开发领域书籍中最好的作品之一。如果你已经有了一定的基础，那么书中的技巧一定会引领你迈向大师的行列。

——Casey Doolittle，Icon Health and Fitness 首席 Java 开发者

《Android 开发秘籍（第 2 版）》提供了出类拔萃的 Android 开发基础知识。书中教授了诸如布局、Android 生命周期及基于多种多线程技术的响应性等内容，这些都是成为 Android 达人所必需的。

——Kendell Fabricius，自由 Android 开发者

每个人从本书中都能得到需要的东西。我自从 1.0 版起就从事 Android 编程，但仍从本书中学到了前所未见的东西。

——Douglas Jones，Fullpower Technologies 高级软件工程师

① 最初名为 Google Market，为 Google 官方提供的发布 Android 应用程序的平台。2013 年 3 月更名为 Google Play Store。——译者注

献给我挚爱的妻子 Susan 和 OpenIntents 社区：谢谢你们给予我的支持。

——Ronan

献给 Martin Simonnet 和 Niantic Project，为他们带给我的一切快乐致谢。

——Phil

深情地献给 Wei！

——Jim

献给我亲爱的妈妈！

——Nelson

# 前言

Android是当今成长最快的移动操作系统。同时，Google Play Store拥有的有效应用已超过80万人，意味着Android“生态系统”也在不断发展壮大。不同设备特性和不同无线运营商带来的多样性使Android几乎对所有人都有吸引力。

上网本（netbook）始终是一类能够自然而然地接受Android的平台，但Android的活力已经使其在平板电脑（tablet）、电视甚至汽车等诸多领域都茁壮成长。世界上许多顶级大公司，从银行，到快餐连锁企业，再到航空公司，已然共同营造了一个Android的世界，并提供相互兼容的各种服务。Android开发者获得了很多机遇，Android应用获得了比以往任何时候都多的受众这一事实，也增加了开发一个相关应用带给人们的满足感。

## 为什么要有一本Android秘籍

Android操作系统（后简称Android系统）简单易学，而且Google还提供了许多库（library），使功能丰富、复杂度高的应用的实现变得简单。唯一欠缺的，也正如Android开发社区中的许多人提到的那样，就是清楚明白的文档。Android是开源的这一事实，意味着任何人都能深入其中，用逆向工程[①]的方法得出一些文档。许多开发者电子公告栏（bulletin board）中都有运用这一方法演绎出的完美范例。尽管如此，如能写出一本可以一致地贯穿这一操作系统所有领域的著作，仍将是有意义的。

此外，一个清晰可用的范例，胜过千言的文档。开发者们遇到问题时，通常首选一种极限编程（extreme programming）方式，也就是找到一些功能与自己要实现的目标相近的工作代码范例，通过修改或扩展它们，来实现自己的需求。这些范例同时也起到编码风格的示范作用，帮助开发者塑造其他部分的代码。

本书通过许多自成一体的技巧，来满足读者不同的需求。而Android系统的各类基本概念则被融入到这些技巧之中。

## 目标读者

那些编写自己的Android应用程序的用户将从本书受益良多。书中大部分内容，都建立在

① 所谓逆向工程（reverse engineering），就是根据已有的实现结果，分析推导出其具体实现方法。文中所说的通过逆向工程方法得出文档，指的是通过分析开源代码，掌握其技术特点和实现思路，从而撰写出相关说明文档。——译者注

假定读者对 Java 和 Eclipse 有基本的了解的基础上，但这种了解也不是严格必需的。Java 是一种模块化程序设计语言，这使得绝大部分（如果不是全部）的技巧示例在经过较小的修改后可以被吸纳进读者自己的 Android 项目中。书中的各个主题所涵盖的范围及其创作动机，使得本书可以用于 Android 课程的补充读物。

# 使用书中的技巧

一般而言，本书中的代码技巧是相互独立的，而且囊括了对于开发运行一个可工作的应用程序所必需的所有信息。第 1 章和第 2 章给出了 Android 使用的概览，读者完全可以跳过它们，自由翻阅想读的章节。

本书首先是作为一本参考书编写的，主要以范例的形式传授知识，这些范例通过对富有趣味的技巧的实现，以求给读者带来最大的收益。每节的标题指明了该节中的技巧所介绍的主要技术。同时，那些为主要技巧提供支持的附加的技术也被包含在相应章节中。

在读过这本书后，开发者应该：

- 能够从零开始编写 Android 应用程序；
- 能够编写适用于多个 Android 版本的程序代码；
- 能够使用 Android 提供的各种各应用程序接口（Application Programming Interface，API）；
- 积累大量能够被迅速融入到应用程序中的代码片段；
- 领会 Android 中实现同一任务的不同方法，并能从每种方法中获益；
- 理解 Android 编程技术中那些独一无二的方面。

# 本书的结构

- 第 1 章“Android 概览”并不涉及代码，而是提供对 Android 各个方面的介绍。这是唯一一个不包含“技巧”的章节，但它提供了有用的背景材料。
- 第 2 章“应用程序基础：Activity 和 Intent”，给出了 4 种 Android 组件的概述，并解释了 Android 项目是如何组织的。Activity 作为构建应用的主要模块，是本节重点关注的对象。
- 第 3 章“线程、服务、接收器和警报”，介绍了线程（thread）、服务（service）、接收器（receiver）等后台任务，以及借助警报（alert）来实现的这些后台任务间的通知（notification）机制。
- 第 4 章“高级线程技术”，介绍了 AsyncTask 及装载器（loader）的使用。
- 第 5 章“用户界面布局”，涵盖了用户界面屏幕布局（Layout）和视图（View）的有关内容。
- 第 6 章“用户界面事件”介绍了由用户初始化的事件，例如各种触屏事件和手势。

- 第 7 章“高级用户界面技术”，涉及如何创建自定义视图、运用动画、提供可访问选项，以及以更大的屏幕工作等内容。
- 第 8 章“多媒体技术”，包括多媒体操作、视频和音频的录放等内容。
- 第 9 章“硬件接口”，介绍了 Android 设备上可用的硬件 API，以及如何使用它们。
- 第 10 章“网络”，探讨了 Android 设备与外界的交互，包括短信息（SMS）、Web 浏览和社交网络等。
- 第 11 章“数据存储方法”，涵盖了多种 Android 上可用的数据存储技术，包括 SQLite。
- 第 12 章“基于位置的服务”，聚焦于如何利用诸如 GPS 等不同方法访问位置信息以及如何使用 Google 地图（Google Maps）API 等服务。
- 第 13 章“应用内计费”，提供了一套利用 Google Play 服务将应用内计费机制引入应用程序的指令集合。
- 第 14 章“推送消息”，涉及如何使用 GCM 处理应用程序的推送消息。
- 第 15 章“原生 Android 开发”，探讨了原生开发用到的组件和结构。
- 第 16 章“测试和调试”，提供了可用于贯穿整个开发周期的测试和调试框架。

## 更多的参考资料

Android 相关的在线参考资料十分丰富，下面给出一些精华站点。

- Android 源代码：http://source.android.com/。
- Android 开发者主页：http://developer.android.com/。
- 开放源码目录：http://osdir.com/。
- 栈溢出论坛（Stack Overflow Discussion Threads）：http://stackoverflow.com/。
- Android 开发者讲坛（Talk Android Developer Forums）：www.talkandroid.com/android-forums/。

# 作者介绍

**Ronan "Zero" Schwarz** 是 OpenIntents 的创始人之一。OpenIntents 是一家以欧洲为基地、专门从事 Android 开发的开源软件公司。Ronan 拥有超过 15 年的编程经验，且涉足诸多领域，如增强现实（augmented reality）、Web、机器人学、商业系统等，以及包括 C、Java、Assembler 等在内的多种编程语言。他从 2007 年起就开始从事 Android 平台开发工作，在其他方面，还协助创建了 SplashPlay 和 Droidspray，它们分别入围第一届和第二届 Android 开发者挑战赛（Android Developer Challenge）的决赛圈，并名列前茅。

**Phil Dutson** 是 ICON Health and Fitness 的首席用户体验（UX）和移动开发员。他曾为 NordicTrak、ProForm、Freemotion、Sears、Costco、Sam's Club 及其他一些单位开发项目和提供解决方案。多年以来，他使用、修改和编写了许多移动设备（从他拥有的第一台 Palm Pilot 5000[①] 到现在手头的 iOS 和 Android 设备）上的程序。Phil 还著有《jQuery, JQueryUI and jQuery Mobile》、《Sams Teach Yourself jQuery Mobile in 24 Hours》和《Creating QR and Tag Codes》等书籍。

**James Steele** 15 年前，还是麻省理工学院（MIT）的一名物理学博士后时，他就投身于硅谷一家初创公司。15 年后的今天，他还在不断地推出创新成果，将消费者市场和移动市场上的若干研究项目转化为产品。他活跃在硅谷的多个技术创新团体中，并且是 Sensor Platforms 的工程副总裁。

**Nelson To** 在 Android Market 上发布的由他自己开发的应用数量已达到两位数。他还从事 Android 企业化应用工作，涉及 Think Computer 股份有限公司的 PayPhone、AOL 的 AIM、斯坦福大学（Stanford University）的 Education App 和罗技公司（Logitech）的 Google TV 等产品。此外，他还协助组织了硅谷的 Android Meetup 社区，并在美国湾区和中国讲授 Android 课程。

---

① Palm Pilot 5000 是 Palm Computing 公司于 1996 年 3 月推出的 Pilot 系列 PDA 掌上电脑中的一个机型。——译者注

# 目录

# 第1章 Android概览

自2007年底开放手持设备联盟（Open Handset Allience）宣告成立以来，Android系统已经走过了一段很长的道路[①]。嵌入式系统上的开源操作系统并不是一个新鲜事物，但Google雄心勃勃的支持，毫无疑问地将Android在几年之内就推到了这一领域的最前沿。

在许多国家诸多无线运营商五花八门的通信协议之上，都有至少一种Android手机与其相适应。而其他的嵌入式设备，诸如平板电脑、上网本、电视、机顶盒，甚至汽车当中，也都已经引入了Android系统。

本章讨论了Android的许多对开发者有意义的基本方面，提供了创建Android应用程序的基础知识，为本书后续部分的各个技巧做了铺垫。

## 1.1 Android的演化

Google在看到移动设备上因特网的使用和在线搜索量的迅猛发展后，于2005年并购了Android公司[②]，以致力于移动设备平台的开发工作。Apple公司在2007年推出了iPhone，其中体现出的若干理念可谓石破天惊，例如多点触控技术和开放式的应用程序市场。Android迅速地适应并吸纳了这些特性，但又提供了许多与之相异的特性，比如给予开发者更多的控制权限，以及对多任务的支持。此外，Android将一些企业级的需求考虑在内，如交易支持、远程擦除及对虚拟专用网络（Virtual Private Network，VPN）的支持，以追逐Research In Motion（RIM）公司在企业市场领域的脚步，后者曾以其出色的黑莓（BlackBerry）机型在这一领域独领风骚。

支持设备的多样性以及快速的适应能力使Android得以扩大其用户基础，但这样的成

---

① Android系统就是在开放手持设备联盟宣告成立的同时对外发布的。该联盟成立于2007年11月5日，由Google和其他34家手机制造商、软件开发商、电信运营商和芯片制造商等共同组成。——译者注

② Android公司创建于2003年10月，致力于开发名为Android的移动操作系统。2005年8月17日，Google收购了成立仅22个月的Android公司，并继续支持Android系统的开发。——译者注

长也为开发者带来了潜在的挑战。应用程序需要同时支持多种屏幕大小、分辨率、键盘、硬件传感器、操作系统版本、无线数据传输速率以及系统配置等。上述的任何一项都可能带来无法预测的行为。然而，在应用程序测试中遍历一切可能的环境又是一项无法完成的任务。

Android 因此被打造成为一套能确保跨平台体验尽可能一致的系统。通过将硬件的不同点加以抽象，Android 系统尽力将应用程序与特定于设备的改动隔绝开来，使得程序可以被灵活地按需调整。针对新的硬件平台及操作系统升级不会过时的应用程序也是一个考虑因素，需要开发者对这一系统方法有较好的认识。Android 提供的通用应用程序接口（API）以及如何确保设备和操作系统间的兼容性是贯穿本书的两条主线。

此外，同所有其他嵌入式平台一样，对应用程序进行广泛的测试是必需的。在这方面，Google 以多种形式为第三方开发者提供了的支持，比如 Eclipse 上的 Android 开发工具包（ADT）插件（它也可以作为独立的工具使用），其中包含实时日志功能、可以运行原生 ARM 代码的现实模拟器，以及由用户向 Google Play 应用程序开发者发送的现场错误报告等。

## 1.2　Android 的两面性

Android 身上有若干值得玩味的两面性。预先了解它们对于理解 Android 是什么或不是什么都是有益的。

Android 是一个嵌入式操作系统，其核心系统服务基于 Linux 内核开发，但 Android 本身并不是一套嵌入式 Linux。例如，一些标准 Linux 实用程序，像 X-Windows 和 GNU C 程序库都不为 Android 所支持。Android 应用程序是用 Java 框架编写的，但 Android 并非 Java，诸如 Swing 一类的标准 Java 库在 Android 上就不被支持。另外一些 Java 库，比如 Timer，也并非首选，而是可以被 Android 自己的库所替代。这些库是经过优化的，专门针对资源有限的嵌入式环境。

Android 系统是开源的，意味着开发者可以查看和使用所有系统源代码，包括射频协议栈（radio stack）。这些源代码对于需要查看活动的 Android 代码范例的人而言，是首选的资源之一。在文档匮乏时，它们也有助于人们搞清 Android 的某些用法。同时还意味着，开发者们可以像任何核心应用中所做的那样使用系统，并可将系统组件替换为自己的组件。然而，Android 设备确实包含一些开发者无权染指的专有软件（如 GPS 导航）。

## 1.3　运行 Android 的设备

全世界有种类数以百计的 Android 设备和为数众多的制造商，设备包括电话、平板电脑、电视、车载音响、运动器械及其他辅助设备。软件可通过 `android.os.Build` 来获取目标设备信息，例如：

```
if(android.os.Build.MODEL.equals("Nexus+One")) { ... }
```

所有 Android 支持的硬件由于操作系统的一些自然属性而具有一些共同点。Android 系统由下列镜像文件（image）组成。

- Bootloader（引导加载程序）：在启动时对引导镜像（boot image）的加载过程进行初始化。
- Boot image（引导镜像）：内核（kernel）和内存虚拟盘（RAMDisk）。
- System image（系统镜像）：Android 系统平台及应用程序。
- Data image（数据镜像）：在断电期间依然保存着的用户数据。
- Recovery image（恢复镜像）：重建或升级系统所用的文件。
- Radio image（射频镜像）：射频协议栈文件。

这些镜像被保存在非易失性闪存中，所以在设备断电期间仍然被保存。这块闪存被用作只读存储器（read-only memory，ROM），但它在必要时可以被改写（例如，对 Android 系统进行在线更新时）。

启动时，微处理器执行引导加载程序将内核和内存虚拟盘载入内存，以实现快速存取。接下来，微处理器执行一些指令，将系统镜像和数据镜像部分按需装载为内存页面。射频镜像常驻在基带处理器（baseband processor）中，后者则与射频硬件连接。

表 1-1 比较了一些早期的和新近的智能手机机型。通过比较可以看出，这些设备负责运算处理的硬件架构是相似的：一个微处理器单元（MPU）、同步动态随机存取存储器（SDRAM，简称 RAM），以及闪存（简称 ROM）。屏幕大小用像素（pixel）表示，而每英寸点数（dpi）这一指标则取决于屏幕的物理尺寸。例如，HTC Magic 机型屏幕对角线长为 3.2 英寸，分辨率则为 320×480 像素，相当于每英寸 180 像素，在 Android 的分类中属于中等像素密度的设备（平均水平为 160 dpi）。所有智能手机还都提供 CMOS 图像传感器摄像头、蓝牙（BT）、Wi-Fi（802.11）等，不过规格各异。

**表 1-1　具有代表性的几款 Android 智能手机***

| 型　号 | 移动处理器 | RAM/ROM | 显 示 屏 | 其 他 特 性 |
|---|---|---|---|---|
| Galaxy Nexus（2011 年 11 月） | 1.2 GHz 双核三星处理器 | 1024 MB /16 GB 或 32 GB | 超高清 AMOLED 720×1280 超高点密度（xhdpi） | GSM/UMTS/HSPA+/HSUPA/CDMA/1xEV-DO/带 A2DP 协议的 LTE 蓝牙 3.0、通过 Micro-USB 2.0 接口的移动高清连接（MHL）、802.11b/g/n、500 万像素摄像头、130 万像素前置摄像头、地理标记功能（Geotagging）、Wi-Fi 热点、AGPS、NFC |
| Droid RAZR MAXX（2012 年 5 月） | 1.2 GHz 双核 ARM Corex-A9 SoC 处理器 | 1024 MB/ 16 GB | 超级 AMOLED 540×960 QHD高点密度（hdpi） | GSM/CDMA/HSDPA/1xEV-DO/ 带 A2DP 和 EDR+LE 的 LTE 蓝牙 4.0、802.11b/g/n、HDMI、Wi-Fi 热点、800 万像素摄像头、130 万像素前置摄像头、地理标记功能、DLNA、AGPS、1080p 高清视频录制 |

续表

| 型　号 | 移动处理器 | RAM/ROM | 显　示　屏 | 其他特性 |
|---|---|---|---|---|
| Google Nexus 4（2012年11月） | 1.5 GHz 四核高通 Snapdragon 处理器 | 2 GB/8 GB 或 16 GB | TrueHD-IPS Plus LCD 786×1280 超高点密度 | GSM/UMTS/HSDPA、HSUPA、带 A2DP 和 LE 的 HSPA+蓝牙 4.0、802.11b/g/n、Wi-Fi 热点、Wi-Fi 直连、DLNA、HDMI、AGPS、SGPS、GLONASS、800 万像素摄像头、130 万像素前置摄像头、地理标记功能、1080p 高清视频录制 |
| Galaxy Note 2（2012年11月） | 1.6 GHz 三星 Exynos 4 Quad 4412 处理器 | 2 GB/32 GB | 超级 AMOLED 720×1280 | GSM/UMTS/HSDPA、HSUPA、HSPA+蓝牙 4.0、802.11a/b/g/n、GPS、AGPS、地理标记功能、800 万像素摄像头、190 万像素前置摄像头、1080p 高清视频录制、NFC |
| HTC One（2013年3月） | 1.7 GHz 高通 Snapdragon 600 APQ8064T 处理器 | 2 GB/32 GB 或 64 GB | 超级 LCD 3 1080×1920 | GSM/UMTS/HSDPA、HSUPA、HSPA+蓝牙 4.0、802.11a/b/g/n、GPS、AGPS、快速 GPS、地理标记功能、430 万像素摄像头、210 万像素前置摄像头、1080p 高清视频录制、NFC |

*数据来源于 http://en.wikipedia.org/wiki/Comparison_of_Android_devices 及 http://pdadb.net/。

新机型除容量和性能有所提升外，还有区别于旧机型的若干新特性。一些设备支持 4G，另一些则有调频收音机（FM）或额外的蜂窝式无线电台（cellular radio）、视频输出（通过 HDMI 或 micro-USB 接口），以及前置摄像头等。了解这些差别可以帮助开发者创建出色的应用。在内置硬件之外，许多 Android 设备还带有 Micro Secure Digital（microSD）卡槽。

microSD 卡可提供附加的存储空间，用于存储多媒体或额外的应用数据。不过在 Android 2.2 或更早的版本中，应用本身只能被存储在内部 ROM 里。

## 1.3.1 HTC 系列机型

HTC 是一家成立于 1997 年的台湾公司。HTC Dream（也被称为 G1，其中的 G 代表 Google）是运行 Android 系统的首款商用硬件，它发布于 2008 年 10 月。从那以后，HTC 已经推出了超过 20 种运行 Android 系统的手机，包括 Google 的 Nexus One、EVO 3D 和 One X$^{+}$。

Nexus One 是最先使用 1 GHz 微处理器的 Android 设备之一，这款微处理器为高通（Qualcom）公司的 Snapdragon 平台产品。Snapdragon 内含高通自己生产的核心，而非 ARM 核心；还包含解码 720p 高清视频的电路。随后的大多数智能手机也都采用了 1 GHz 微处理器。Nexus One 机型的其他特征包括：使用双麦克风以在通话时降低背景噪音，以及能根据不同的系统通知（notification）呈现不同颜色的背光轨迹球等。

HTC EVO 4G 于 2010 年 6 月发布，作为第一款支持 WiMAX（802.16e-2005）的商用机型，在当时引起了不小的轰动。HTC 还在 2011 年 8 月发布了 EVO 3D 机型，它与 EVO 4G 大体相似，但别具一格之处在于可以脱离 3D 眼镜来呈现 3D 效果。此外它还拥有两个后置摄像头，可以录制 720p 的 3D 视频。

## 1.3.2　摩托罗拉系列机型

摩托罗拉（Motorola）推出过将近十种不同标牌的 Android 机型。摩托罗拉 Droid X 拥有与 HTC Droid Incredible 机型近似的性能，包括高清视频摄制等。2011 年 Google 收购了摩托罗拉移动部门，力图在市场层面对 Android 加以提升，推动创新进程，并利用摩托罗拉移动的专利布局保护 Android 的“生态系统”。

摩托罗拉开发的 Droid RAZR MAXX 和 RAZR MAXX HD 两款手机具有超乎寻常的电池寿命，还兼具相对苗条的外形。

## 1.3.3　三星系列机型

三星（Samsung）是移动市场的一支强势力量，且如今已是 Android 设备的第一大制造商。2012 年的第 4 季度销售的全部 Android 设备中，三星所占份额高达 42%。如今市场上最流行的三星手机要属 Galaxy Note 2 和 Galaxy S3，此二者均支持蓝牙 4.0、近场通信（NFC）及三星独有的 S Beam 和 AllShare 等特性。

三星 Galaxy Nexus 是第一款基于 Android 4.2 的手机，并且是最先内嵌 NFC 模块的机型之一。三星还是最早推出试图弥合手机和平板电脑之间鸿沟的智能机型的厂商。Galaxy Note 和 Galaxy Note 2 的屏幕要大于 5 英寸，因此有人将这类手机称做“phablet[①]”。

## 1.3.4　平板电脑

在苹果公司推出 iPad 之后，人们期望 Android 制造商也能推出自己的平板电脑。平板电脑可被粗略地定义为拥有 4.8 英寸或更大尺寸的屏幕以及 Wi-Fi 连接的设备。由于许多平板电脑都支持 3G 无线服务，它们感觉更像是大屏幕的智能手机。

2009 年年底，爱可视（Archos）成为最早将 Android 平板电脑投向市场的厂商之一。最初的型号拥有对角线长 4.8 英寸的屏幕，并被称为 Archos 5。此后，Archos 系列又涌现出了屏幕尺寸在 7 英寸～10 英寸间的若干新机型。有的机型配备了真正的硬盘驱动器，有些则采用闪存来存储信息。三星也推出了屏幕尺寸在 7 英寸～10.1 英寸间的多款平板电脑。

亚马逊（Amazon）则推出了 Kindle Fire 系列，包括 4 种不同款式。这些平板电脑屏幕尺寸从 7 英寸～8.9 英寸不等，处理器有单核也有双核。它们均运行在一种 Android 的修改版系统之上，该系统可与 Amazon Appstore 及 Amazon MP3、Amazon Video 等连接。

Google 还与华硕（Asus）合作发布了 Nexus 7，这是一款运行在 Android 4.2.1 系统上 7 英寸平板电脑。没过多久，Google 又与三星合作制造了 Nexus 10。Nexus 10 是第一款分辨率高达

① phablet 为英文单词 phone（手机）和 tablet（平板电脑）的结合体。——译者注

2560×1600 的平板电脑，从而可与支持 retina 技术的 MacBook Pro 电脑以及较新的全尺寸 iPad 相媲美。表 1-2 给出了几种不同型号的 Android 平板电脑的对比。

表 1-2　具有代表性的几款 Android 平板电脑

| 型　号 | 移动处理器 | RAM/ROM | 显　示　屏 | 其 他 特 性 |
|---|---|---|---|---|
| 爱可视 80 G9（2011 年 9 月） | 1000 MHz TI OMAP 4430 | 512 MB/16 GB | 8 英寸 TFT LCD，1024×768 | BT2.1 + EDR、802.11b/g/n、90 万像素摄像头 |
| 爱可视 Gen10 101 XS（2012 年 9 月） | 1500 MHz TI OMAP 4470 | 1 GB/16 GB | 10.1 英寸 TFT LCD，1280×800 | 802.11b/g/n、蓝牙 4.0、可附加 QWERTY 型键盘、130 万像素摄像头、GPS |
| 三星 Galaxy Note 10.1（2012 年 2 月） | 1400 MHz 三星 Exynos | 2 GB/32 GB | 10.1 英寸 TFT LCD，1280×800 | 蓝牙 4.0、802.11a/b/g/n、GPS、地理标记功能、500 万像素摄像头、190 万像素前置摄像头 |
| Nexus 7 32GB（2012 年 11 月） | 1300 MHz 四核 Cortex-A9 T30L | 1 GB/32 GB | 7 英寸 IPS TFT LCD，1280×800 | GSM/UMTS/GPRS/EDGE/UMTS/HSDPA/HSUPDA/HSPA+、蓝牙 3、802.11a/b/g/ n、120 万像素摄像头、GPS、地理标记功能 |
| Nexus 10 32GB（2012 年 11 月） | 1700 MHz Exynos 5 Dual 5250 | 2 GB/32 GB | 10 英寸 PLS LCD，2560×1600 | 蓝牙 4、802.11b/g/n、500 万像素摄像头、190 万像素前置摄像头、GPS、地理标记功能 |

## 1.3.5　其他设备

既然 Android 是一种通用型嵌入式平台，那么除智能手机和平板电脑外，它也能应用于其他地方。第一辆基于 Android 的汽车荣威（Roewe）350 由上海汽车工业集团生产。Android 主要发挥 GPS 导航功能，同时也支持上网浏览。

萨博（Saab）公司也推出运行在 Android 平台上的、名为 IQon 的信息及娱乐系统。该系统能向驾驶员提供关于引擎工作负荷、时速、转矩等机械数据的实时反馈。它通过一个触屏式并支持 3G 或 4G 数据连接的 8 英寸嵌入式控制台来显示这些信息。而某些信息的获取需要通过给汽车的引擎控制单元（ECU）安装一个售后零件（aftermarket part）来实现。这种将 Android 直接植入 ECU 的创意是有趣而激动人心的。

Android 还被移植进了一些新颖刺激的平台，比如手表和 OUYA 主机。Pebble 手表是 Kickstarter 上的一个项目，力图制造可以与 Android 和 iOS 设备通信的手表。它可以通过使用某个 Android 设备的软件开发工具包（SDK）对其进行访问，并通过蓝牙通信显示呼叫者 ID、当前时间、接收到的短信息、邮件提醒等。OUYA 主机是将 Android 系统运用到极致的一个出色案例，它是一款专用于 Android 游戏的主机（与 PlayStation 及 Xbox 系列主机类似）。尽管这一新生事物在本书写作时还未向公众发布，但 OUYA 已经承诺每年将推出廉价而又前沿的硬件装备。

# 1.4 Android 设备间的硬件差异

每一款 Android 设备上可用的硬件各不相同，这在表 1-1 中已有所反映。总体而言，多数差异对开发者而言是透明的，这里不深入讨论。然而，了解其中某些硬件差异有助于我们编写设备无关的代码。在此我们将探讨屏幕、用户输入法和传感器这几方面。

## 1.4.1 屏幕

液晶显示（LCD）和发光二极管（LED）是目前显示器使用的两种技术。在 Android 手机上，则分别体现为薄膜晶体管（TFT）液晶屏和主动矩阵有机发光二极管（AMOLED）屏。

TFT 屏的一个优势是较长的寿命；而 AMOLED 的好处则在于无需背光，从而有更深的黑色和更低的能耗。

总的说来，Android 设备可以依照屏幕尺寸被分为小、中、大、超大，或按像素密度分为低、中、高、极高几类。注意，实际的像素密度可能多种多样，但总可以被归为低、中、高、极高中的某一种。表 1-3 给出了典型的屏幕尺寸、分辨率及一些与屏幕尺寸有关的名称。

表 1-3 Android 支持的设备屏幕情况

| 屏幕类型 | 低密度（～120ppi）ldpi | 中密度（～160ppi）mdpi | 高密度（～240ppi）hdpi | 极高密度（～3200dpi）xhdpi |
|---|---|---|---|---|
| 小屏幕（426× 320dp） | QVGA (240×320) ldpi | | 480×640 hdpi | |
| 中屏幕（470× 320dp） | WQVGA400 (240×400) ldpi、WQVGA432 (240×432) ldpi | HVGA (320×480) mdpi | WVGA800 (480× 800) hdpi、WVGA (480×854) hdpi | 640×960 xhdpi |
| 大屏幕（640× 480dp） | | 600×1024 mdpi | | |
| 超大屏幕（960× 20dp） | 1024×600 ldpi | WXGA (1280×800) mdpi、1024×768 mdpi、1280×768 mdpi | 1536×1152 hdpi、1920×1152 hdpi、1920×1200 hdpi | 2048×1536 xhdpi、2560×1536 xhdpi、2560×1600 xhdpi |

## 1.4.2 用户输入法

触摸屏使得用户可以与视觉显示进行互动。目前有三种类型的触屏技术。

- 电阻屏：在玻璃屏幕的上层覆盖两层电阻材料层。当手指、手写笔或其他物体施压时，两层电阻材料发生接触，而触点可以被确定。电阻屏性价比高，但透光率只能达到 75%，并且最近才实现多点触摸。

- 电容屏：玻璃屏上覆盖有一层带电材料。当手指或其他导体与该层接触，一些电荷被释放，引起电容变化，从而可检测出触点所在。电容屏透光率可达 90%，但精确度不如电阻屏。
- 表面声波：这是一种靠发送和接收超声波来定位的先进技术。当手指或其他物体接触屏幕时，引起的声波被吸收和测量，从而确定触点。这是一种最为经久耐用的解决方案，但更适合诸如银行自动柜员机一类的大尺寸的屏幕。

目前绝大多数的 Android 设备要么采用电阻屏，要么采用电容屏技术，并且均支持多点触摸。此外，Android 设备还可能采用下列某种替代性的屏幕访问方法。

- 十字键（D-Pad）：一种上下左右型的控制杆。
- 轨迹球（Trackball）：一种使用滚动球定位的指针设备，与鼠标类似。
- 触摸板（Trackpad）：用一块特殊的方形表面来定位的指针设备。

## 1.4.3 传感器

在某种程度上，智能手机正在成为一个传感器的集合体，带给用户丰富的体验。除每部手机都必备的麦克风外，首先被引入手机的附加传感器要属摄像头。不同的手机摄像头在性能上大相径庭，这已成为人们选择手机时的一个重要考虑因素。类似多样性也体现在其他的附加传感器上。

多数智能手机至少包含以下三种基本的传感器：用于测量重力的三轴加速度计（accelerometer）、用于测量周围磁场的三轴磁力计（magnetometer），以及用于测量周边温度的温度传感器。例如，HTC Dream（G1）包含下列传感器（可以通过调用 `getSensorList()` 函数列出它们，这将在第 9 章中进一步讲述）。

- AK8976A 三轴加速度计。
- AK8976A 三轴磁场传感器。
- AK8976A 方向传感器。
- AK8976A 温度传感器。

AK8976A 是出自旭化成微系统公司（Asahi Kasei Microsystems，AKM）的传感器包，整合了压阻式加速度计、霍尔效应磁力计及温度传感器。这些传感器都提供 8 位精度的数据。方向传感器则是一个虚拟传感器，实际是组合使用加速度计和磁力计来测定方向。

作为对比，我们看看摩托罗拉 Droid 手机采用的传感器。

- LIS331DLH 三轴加速度计。
- AK8973 三轴磁场传感器。
- AK8973 温度传感器。
- SFH7743 近距离传感器。
- 方向传感器。
- LM3530 光传感器。

LIS331DLH 是来自意法半导体公司（ST Microelectronics）的 12 位电容式加速度计，它提供更为精确的数据，并且采样频率可达 1 kHz。AK8973 是一个 AKM 传感器包，包含 8 位霍尔效应磁力计及温度传感器。

此外，Droid 还包含另两个传感器。SFH7743 是光电半导体的短程近距离传感器，可在约 40 mm 距离范围内有物体（比如耳朵）存在时关闭屏幕。LM3530 是美国国家半导体公司（National Semiconductor）生产的带有可编程光传感器的 LED 驱动器，可以检测周围光线并据此将屏幕背光和 LED 闪光灯调节到适当亮度。

关于 Android 设备上可用的传感器我们再举一例，即 HTC EVO 4G，它含有下列传感器。

- BMA150 三轴加速度计。
- AK8973 三轴磁场传感器。
- AK8973 方向传感器。
- CM3602 近距离传感器。
- CM3602 光传感器。

BMA150 是博世传感器公司（Bosch Sensortec）生产的 10 位加速度计，采样率可达 1.5 kHz。CM3602 则是 Capella 微系统公司生产的短程近距离传感器和环境光传感器的二合一产品。

总体而言，理解不同的 Android 设备拥有不同的底层硬件是很有必要的，这些不同会导致性能和传感器精度的差异。

## 1.5 Android 的特性

Android 的具体特性以及如何利用它们乃是贯穿全书的一大主题。从更广泛的层面讲，Android 的一些关键特性是其主要卖点和与众不同之处。认识 Android 的这些强项并充分利用它们将会带来益处。

### 1.5.1 多线程应用微件

Android 系统并不限制处理器某一时间只能处理单个应用。应用程序以及单个应用程序内不同线程的优先级由系统管理。这样的好处在于，在用户的前台进程占用设备时，后台任务依然可以保持运行。例如，当用户在玩游戏时，另一个后台进程可以检查股票价格并在需要时发出警告。

窗口微件（App Widget）是可以嵌入到其他应用程序（比如主屏幕）中的迷你应用程序。它们可以在其他应用程序运行期间处理事件（比如开启一个音乐流或更新外界温度信息）。

多线程的好处在于丰富了用户体验。然而，你必须小心谨慎，以防耗能的应用将电池电量榨干。多线程特性将在第 3 章中进一步讨论。

### 1.5.2 触摸、手势和多点触摸

触摸屏对于手持设备而言是一种直观的用户界面。如果善加利用，可以使用户轻松上手。当手指接触屏幕后，拖曳和投掷是与图形交互的自然而然的方式。多点触控提供了一种同时追踪多个手指触摸轨迹的办法，常用于缩放或旋转视图。

许多触摸事件对开发者是透明可用的，而无需实现其具体的行为。可以根据需要自定义手势。在应用程序间设法保持触摸事件用法的一致性是很重要的。触摸事件将在第6章中深入探讨。

### 1.5.3　硬键盘和软键盘

手持设备使用户发生两极分化的特性之一就是到底应该使用物理键盘（或称为硬键盘）还是软件键盘（或称为软键盘）。硬键盘提供的真实触感和明确的键位设置对于一些人而言会使得输入更迅速，但另一些人则更喜欢软键盘的简洁设计和易用性。

Android 设备种类繁多，两种键盘类型都有。这给开发者带来的麻烦就是要兼容二者。软键盘的一个劣势就是屏幕的一部分要专门用于输入，任何用户界面布局都需要将这一点纳入考虑范围并进行测试。

## 1.6　Android 开发

本书着眼于 Android 开发的最主要方面：编写 Android 代码。然而，略谈一谈开发的其他方面，如设计和发布，也是适宜的。

### 1.6.1　良好的应用设计

一个出色的应用程序应具备三个要素：好想法、好代码、好设计。通常，最后一点最不为人们所重视，因为多数开发者独立工作，且自身并不具备图形设计人员的素质。Google 应该已经意识到了这一点，因此创建了一套设计指南，包括图标设计、窗口微件设计、Activity 和任务设计，以及菜单设计等。这些可以在 http://developer.android.com/guide/practices/ui_guidelines/中找到。Google 还做了更进一步的工作，创建了一个专门演示设计原则及其在 Android 应用上的实现方法的网站，网址为 http://developer.android.com/design/index.html。

良好的设计无论怎样强调都不过分。它可以让应用显得与众不同，提高用户的接受度，并让用户赞不绝口。市场上许多最为成功的应用都是开发者与图形设计者通力合作的产物。应把应用开发时间的相当一部分投入到考虑它的最佳设计方案上来。

### 1.6.2　保持向前兼容[①]

新的 Android 版本通常在 API 层面上是逐渐增强且向前兼容的。事实上，一个设备，只有当其通过 Android API 的兼容性测试后，才能被称为 Android 设备。然而，如果一个应用对底层

① 向前兼容（forward compatibility），在应用程序自身层面，可以理解为使应用程序较新版本产生的输出能被较早的版本所兼容（可能需要通过忽略早期版本中未实现的功能来达到此目的）；而在应用程序与它所基于的操作系统的关系层面，则可理解为应用程序可以在未来发布的更新版系统上顺利运行。它与向后兼容（backward compatility）是一对截然相反的概念。后者在应用程序自身层面，意为应用程序的较新版本能够接受较早版本产生的结果；在应用程序与系统关系层面，则意味应用程序可以运行在任何操作系统的早期版本之上。具体到本节上下文，这里的向前兼容意为 Android 应用应能适应未来发布的 Android 系统更新。概念易混，特此辨析。——译者注

系统进行了改动，就无法确保兼容性。为确保未来的 Android 更新被安装到设备上时，应用程序能够向前兼容，Google 提出了以下规则。

- 不要使用内部的或不被支持的 API。
- 不要未经询问用户就直接修改设置。未来的系统版本可能会出于安全考虑对设置操作加以限制。例如，应用曾经可以自行打开 GPS 或数据漫游开关，但如今已经不被允许了。
- 不要在布局上做得太过火。虽然不常见，但太复杂的布局（层数超过 10 或总数超过 30）的确可能导致程序崩溃。
- 不要对硬件做不恰当的假定。不是所有的 Android 设备都包含全套可能支持的硬件。应确保对所需硬件进行检查，并在硬件不存在时处理相应的例外。
- 确保设备方向变化不会干扰应用程序运行，或者引发不可预测的行为。屏幕方向是可以锁定的，这在第 2 章中会提到。

注意 Android 并不保证向后兼容，因此最好如第 2 章中将要提到的那样，声明应用所支持的最低 SDK 版本，这样设备可以载入适当的兼容性设定。如何在旧程序上使用其他新特性这一问题，在本书中将会多次提及。

### 1.6.3 确保健壮性

与兼容性同等重要，在应用程序设计和测试中还应考虑健壮性（robustness）。下面给出确保健壮性的几条建议：

- 优先使用 Android 库而非 Java 库。Android 库专为嵌入式设备而设计，并且覆盖了应用程序的多种需求。在某些情况下，如使用第三方插件或应用程序框架时，可能要用到 Java 库。但在二者均可用的情况下，用 Android 库更佳。
- 注意内存分配。要对变量进行初始化。尽量对对象进行重用而不是再分配，这会提升应用程序运行速度并避免对垃圾收集（garbage collection）机制的过度使用。可以用 Dalvik 调试监视服务器（DDMS）工具对内存分配进行跟踪，这会在第 16 章中详述。
- 使用 LogCat 工具，并检查由其产生的警告或错误。这同样将在第 16 章中探讨。
- 调试要彻底，尽量涵盖不同的环境和设备。

## 1.7 软件开发工具包（SDK）

Android SDK 由平台、工具和示例代码以及开发 Android 应用所需的文档组成。它被构建成为 Java 开发工具包的附加组件，并包含一个面向 Eclipse 集成开发环境（IDE）的集成插件。

### 1.7.1 安装和升级

网上许多地方都有针对如何安装 Android SDK 的按部就班式的指导。例如，在 Google 的网站 http://developer.android.com/sdk/上就有关于安装过程的全套链接。如今 Google 已将 SDK 的所有必要组成部分捆绑为一个可以方便下载的 ADT Bundle[①]。该包包含已安装好 ADT 插件的 Eclipse、Android SDK 工具、Android 平台工具、最新的 Android 平台以及用在模拟器上的最新的 Android 系统镜像。ADT Bundle 适用于 Windows、Mac 和 Linux 系统。

这个包是一个预先配置好的 zip 压缩文件，我们需要做的全部事情就是将包解压，并启动 Eclipse 程序。启动后，程序将询问建立工作区（workspace）的路径，选定后，将出现一个界面，可以帮助我们建立新工程或者学习关于 Android 开发的更多知识。

对于那些不愿下载整个包，而宁愿只安装他们需要部分的开发者而言，下面的概要给出的一般过程阐明了最常见的安装步骤。应当在开发环境的主机上执行这些步骤。

（1）安装 Java 开发包（例如，要使用 Android 2.1 或更高版本，需安装 JDK 6.0；JDK 5.0 则是 Android 开发所要求的最低版本）。

（2）安装 Eclipse 经典版（例如 4.2.1 版）。在 Windows 下，只需将下载包解压到某个正确路径就可使用。

（3）安装 Android SDK 初始包（starter package），比如 r21 版。在 Windows 下，同样只需将下载包解压到一定的路径即可使用。

（4）启动 Eclipse 并选择 **Help→Install New Software...**②菜单项，然后键入 **https://dl-ssl.google.com/android/eclipse/**，以安装 Android DDMS 和 Android 开发工具（ADT）。

（5）在 Eclipse 中选择 **Window→Preferences...**（在 Mac 系统下，则选择 **Eclipse→Preferences**），然后选择 **Android** 菜单项。找到你解压 Android SDK 的路径并选择 Apply。

（6）在 Eclipse 中选择 **Window→Android SDK and AVD Manager→Available Packages** 菜单项，并选择需要安装的 API（例如 Android SDK 文档、SDK 平台、Google API 以及 API 17）。

（7）仍然是在 **Android SDK and AVD Manager** 菜单中，创建一个 Android 虚拟设备运行模拟器，或者安装 USB 驱动在连接到电脑上的真实手机上运行应用。

（8）在 Eclipse 中选择 **Run→Run Configurations...**为每个 Android 应用程序创建一套新的运行配置（或者与此类似地，创建一套调试配置）。Android JUnit 测试也可以在此处进行配置。

至此，开发环境应该配置成便于 Android 应用的开发，并让其在模拟器或真实的 Android 设备上运行了。对 SDK 进行版本更新也是件容易事，只要在 Eclipse 中选择 **Help→Software Updates...**，然后挑选适当版本即可。

## 1.7.2　软件特性和 API 级别

Android 系统会定期推出新特性，或是进行功能强化（如改善运行效率），又或是修复 bug。操作系统改进的一个主要动力是新设备上硬件性能的提升。事实上，大多数操作系统版本都是

① Android Development Tools Bundle，可译为 Android 开发工具包。——译者注

② 截至翻译完成时，ADT Bundle 并无官方中文版，因此对与开发环境交互时用到的各种命令和选项暂不翻译，保持英文原文。一部分感觉有必要的地方用括号注明中文含义。——译者注

伴随着新硬件的推出而发布的（比如 Eclair 版[①]的发布是伴随着 Droid 设备的推出）。

一些旧的 Android 设备无法支持新版本系统的要求，因此不能随着新操作系统的发布而升级。这就带来了体验各不相同的用户群体。开发者必须检查设备的性能，或者至少把所需的硬件特性告知用户。这项工作只需通过检查一个简单的数字——API 版本号——即可完成。附录 D 给出了 Android 不同版本及其变动的列表。

如今 Android 的发布符合一个大致的时间规律，每 6～9 个月会更新一次。尽管可以进行无线（over-the-air）升级，但这种方式有一定难度，因此不太被采用。而硬件制造商也希望保持一定的稳定性，这意味着市面上的主流产品并不一定会马上进行升级。但是，每当一个新版本发布，其带来的新特性对开发者而言是值得一试的。

### 1.7.3 用模拟器或 Android 设备进行调试

模拟器会在开发用的电脑上开启一个看上去与真实 Android 手机类似的窗口，并执行真实的 ARM 指令。注意，模拟器的初始化启动过程比较缓慢，即便在高配置的电脑上也是如此。尽管有办法对模拟器进行配置，使其模拟真实 Android 设备的多个方面，如电话呼入、有限的数据传输率、屏幕方向的改变等，但另一些特性（诸如传感器、音频和视频）还是有别于真实设备的。近期模拟器增加了利用宿主机 GPU 的能力，这有助于提升模拟器上产生视觉效果和图形变换的速度。对于那些开发者无法获取的设备，用模拟器对它们进行基本功能验证不失为一种好办法。例如，可以对平板电脑的屏幕尺寸进行模拟，而无需真正购买一台。

要注意，必须事先创建一个目标虚拟设备，才能正确地运行模拟器。Eclipse 提供了一种很好的方式来管理 Android 虚拟设备（AVD）。表 1-4 给出了模拟器的各种功能所对应的快捷键的速查清单。

**表 1-4 Android 系统的模拟器控制**

| 键 盘 键 位 | 模拟器功能 |
|---|---|
| Escape | 后退（Back）键 |
| Home | Home 键 |
| F2、PageUp | 菜单（Menu）键 |
| Shift+F2、PageDown | 开始（Start）键 |
| F3 | 呼叫/拨号键 |
| F4 | 挂断/结束呼叫键 |
| F5 | 搜索（Search）键 |
| F7 | 电源（Power）键 |
| Ctrl+F3、Ctrl+小键盘 5 | 照相（Camera）键 |
| Ctrl+F5、小键盘+ | 音量增键 |

① 2009 年 5 月以后发布的 Android 版本，除数字版本号外，还另有一个别名。且到目前为止，基本都用某种甜点的名字来命名。这里的 Eclair 即是一种法式甜点，可译为“闪电泡芙”。关于 Android 各个系统版本的详细情况可参见附录 D。——译者注

续表

| 键 盘 键 位 | 模拟器功能 |
|---|---|
| Ctrl+F6、小键盘- | 音量减键 |
| 小键盘 5 | 方向键（DPAD）中键 |
| 小键盘 4、小键盘 6 | 方向键左、方向键右 |
| 小键盘 8、小键盘 2 | 方向键上、方向键下 |
| F8 | 开启/关闭移动网络 |
| F9 | 开启/关闭代码分析（仅当设置了`-trace`标记时有效） |
| Alt+ENTER | 开启/关闭全屏模式 |
| Ctrl+T | 切换轨迹球模式 |
| Ctrl+F11、小键盘 7 | 将屏幕方向旋转为上一个布局 |
| Ctrl+F12、小键盘 9 | 将屏幕方向旋转为下一个布局 |

一般而言，首次测试最好使用 Android 手机来完成。这样可确保功能的完整性，还可以检验模拟器无法全完模拟的实时问题。要想把 Android 设备作为开发者平台，只需将其用自带的 USB 线与电脑连接，并确保 USB 驱动被成功检测出来（Mac 系统中检测过程会自动完成；Windows 系统的驱动包含在 SDK 里；Linux 用户则要参考 Google 提供的相关页面）。

为启用开发者功能，要对 Android 设备上的某些设置进行调整。在主屏幕下分别选择 **MENU**（菜单）→**Settings**（设置）→**Applications**（应用程序）→**Unknown sources**（未知来源）和 **MENU**（菜单）→**Settings**（设置）→**Applications**（应用程序）→**Development**（开发）→**USB debugging**（USB 调试）[①]，从而允许通过 USB 线安装应用程序。有关 Android 调试的更多详细内容参见第 16 章。

## 1.7.4　使用 Android 调试桥

通常通过命令行访问 Android 设备比较方便，只要用 USB 线将设备连接到电脑即可。SDK 自带的 Android 调试桥（Android Debug Bridge，ADB）可用于访问 Android 设备。例如，想要在运行 Linux 的计算机上登入到 Android 设备，可键入如下命令：

```
> adb shell
```

这样一来，就可在设备上使用很多 UNIX 命令。使用 `exit` 命令可退出 shell。也可以把一条命令加在 shell 命令之后来直接执行，而无需进入和退出 shell：

```
> adb shell mkdir /sdcard/app_bkup/
```

要想从设备上复制文件，可使用 `pull` 命令，如果需要的话，在复制同时还可以重命名文件：

① 因 Android 版本差异，上述两个功能选项可能并不在作者给出的位置。比如，在译者的 4.0.4 版系统中，前者位于 Settings（设置）→Security（安全）→Unknown sources（未知来源），后者则位于 Settings→Developer options（开发人员选项）→USB debugging（USB 调试）。——译者注

```
> adb pull /system/app/VoiceSearchWithKeyboard.apk VSwithKeyboard.apk
```

若想把文件复制到设备，则可使用 push 命令：

```
> adb push VSwithKeyboard.apk /sdcard/app_bkup/
```

要从设备上移除某个应用程序，例如 com.dummy.game，可以键入如下命令：

```
> adb uninstall com.dummy.game
```

这些都是最为常用的命令，另外还有其他很多命令，其中一些将在第 16 章介绍。

### 1.7.5 签名和发布

若想应用程序被 Google Play 接受，要对其进行签名。要做到这一点，首先需要生成一个私钥，并对其进行妥善保存。然后要在发布模式下用私钥对应用程序进行打包。对应用程序进行升级时，要使用相同的私钥进行签名，以保证升级对用户是透明的。

Eclipse 可以自动完成这些工作。只需右击要签名的项目，并选择 **Export...→Export Android Application** 开始打包工作。可以使用一个密码来生成私钥，该私钥会被保存起来，以备将来的应用程序和升级使用。接下来，通过菜单继续完成 APK 文件的创建。APK 文件是 Android 项目在发布模式下打包、并用私钥签了名的版本，它已可以被发布到 Google Play 上。

## 1.8 Google Play

完成了应用程序的设计、开发、测试和签名工作后，就可以在 Google Play 上对其进行部署。要使用 Google Play，首先要创建一个 Google Checkout 账户。该账户不仅用来支付 25 美元的开发者初始注册费，也用于供开发者获取付费应用的收益。创建的应用能亮相于众目睽睽之下，往往会令开发者激动不已。在上传后的数小时之内，应用程序就可能被来自全世界的用户数百次地浏览、下载、打分和评价。这里给出发布应用时需要考虑的若干事宜，谨供参考。

### 1.8.1 最终用户许可协议

在全球大部分地区，任何以有形形式发布的原创内容均会自动受到《伯尔尼公约》的版权保护。但通常作者还是会为发布内容添加带有发布日期的版权声明，如©2013。为 Android 应用添加版权符号的方法会在第 5 章中介绍。

更进一步地，可以使用最终用户许可协议（EULA）为开发者发布的软件提供保护，它是开发者（或开发商）与用户之间的一个合同。大多数 EULA 包含“许可证授权”、“版权”、“免责条款”等部分。为应用程序，特别是付费程序，加上 EULA 是一种常规做法。第 11 章中将会介绍为 Android 应用添加 EULA 的方法。

## 1.8.2 提升应用的曝光度

用户一般通过三种途径查找应用程序，对这些方法善加利用有助于吸引更多的眼球。

用户可能会在 Play 商店中通过浏览“最新应用”(Just In app) 寻找新发布的应用程序，这取决于用户的 Play 商店的版本。应当为应用程序选择一个一目了然的名字，并将其置于适当的分类之下，比如“游戏”或“通信”。描述用语应当简洁明了，才能吸引更多的浏览量。“游戏”分类下充斥了大量的应用，因此要分出子类别。如果应用很有趣，但没有获胜目标或评分机制，可以考虑将其归入“娱乐”类别。即便做到以上这些，由于每月都有超过 10 000 的应用程序被上传到 Android 市场，你上传的应用程序还是会在一两天内被挤出“最新应用”行列。

Google 有个专门的委员会负责检查新发布的 Android 应用程序，从中选出一些放在 Play 商店“应用”部分的几个显著位置。想要应用程序被选中，应确保它支持所有可能的屏幕分辨率和 dpi 级别，拥有清晰易懂的描述，并带有程序运行效果的图片及视频，还要不包含可能被用户认为侵犯隐私的服务（比如读取系统日志、无故发送短消息，或者在不必要的情况下使用精确定位而不是粗略定位）。

用户查找应用的第二种方式是通过关键词检索。我们可以推测用户可能选用的关键字，将它们包含在应用的标题或描述中。不同用户可能使用不同的语言，因此包含适当的国际通用关键字会有益处。

用户在 Play 商店中查找应用的最后一种可能途径是“热门程序排行”(Top app)，这里包含了那些获得了最高评分和最大下载量的应用。想要让你的应用跻身这一行列，需要付出时间和努力来升级应用和修正 bug。这就引出了决定应用受欢迎程度的最后一个因素：健壮性。要确保应用不包含重大的 bug，耗电量较低，且可以傻瓜式地退出。没有比“该应用要把我的电池耗光!”、“我无法卸载这个应用”这样的评价更能吓跑潜在用户的了。“热门”应用又被分为“免费”、“付费”和“时尚”几种。

还有一点需要注意：开发者和用户之间几乎所有的互动都要经过 Google Play。提供开发者联系方式或支持网站通常是多余的，移动市场的用户很少会用到它们。

## 1.8.3 让应用脱颖而出

有时开发者创建了一个应用程序，却发现 Android 市场上已经发布了类似的软件。应当把这种情况当成机遇而并非挫折。可以通过更良好的设计、界面及运行状况迅速赢得用户群体。一般而言，原创是好的，但不是必须的。只要注意避免使用侵犯版权的内容即可。

## 1.8.4 为应用收费

每当新的应用程序或应用更新被上载到 Android 市场时，开发者都需要选择是免费提供它还是对其收费。下面列出主要的选择。

- 免费提供应用，让所有访问 Google Play 的人都能看到和安装它。
- 免费提供应用，但在其中包含广告。有时，开发者会为应用拉取赞助；更多时候开发者会和第三方整合商合作。广告商通常按广告的点击量付费，按广告印象（浏览量）

付费的情况则较少。图 1-1 显示了一条来自 AdMob（如今已成为 Google 的一部分）的横幅广告。这类广告需要应用程序拥有访问 Internet 及获取设备定位的权限。应考虑使用粗略定位而不是精确定位，以免损失某些潜在用户。

图 1-1　来自 AdMob 的横幅广告

- 为应用收费。Google 负责处理包括收费在内的交易全过程，但会在过程中抽取 30%的费用，因此需要开发者在 Google Wallet 上建立一个商用账户。没有开通 Google Checkout 服务的国家无法查看或安装这些付费应用。一些开发者因此转而到第三方应用商店发布应用。
- 发布功能受限的免费版，但对完整版收费。这给了用户试用应用的机会，如果他们感觉良好，就会比较愿意购买完整版。对于某些应用而言，这种模式是上上之选（比如让一个游戏包含 10 个免费关卡），但并不是所有应用都适合这种方式。
- 在应用中售卖虚拟商品，或者采用应用内购买方式。售卖虚拟商品这一方式比较常见于“花钱买胜利”型的应用。这类应用可以免费获取，但需要用户花钱来提升装备质量、升级能力，甚至跳过游戏的某一部分。对于 Facebook 上的应用，这是一种重要经营方式，而在移动市场上也正在变得流行起来。

免费应用程序一般能获得较大的浏览量，即使那些最为莫名其妙和稀奇古怪的应用也能在发布到 Play 商店后的一个月内，被上千人浏览和下载。有的开发者明确指出自己的应用“绝对毫无用处”，但居然还能获得过万的下载量和四星级的评价。一些有用的免费应用下载量能达到 5 万，而特别有用的免费软件下载量可达 10 万以上。对于大多数开发者，这样的受欢迎程度相当令人满意。

移动广告目前尚处于幼年时期，通常不能吸引足够的用户点击从而让应用获利不菲。目前，最佳的获利方式是在 Play 商店上收费或者在应用内收费。只要应用对一定人群有用，描述清晰，并拥有一定数量的好评，就会有用户购买它。如果应用获得成功，或许还可以对其提价。

### 1.8.5　管理评价和更新

如果应用获得成功，多数独立开发者会继续发布更新版本，且会在修改时参考用户的反馈意见。这样做会吸引更多人下载应用，因为用户喜欢积极给予响应的开发者。随着下载量的增加，应用的口碑也随之提升。

一般来说，约 200 个用户中会有 1 人给应用打分，而这些人中又只有一小部分会发表评论。如果有人愿意花时间撰写评论，那么他的意见通常值得注意，特别是当评论有建设性时，例如“在 HTC Hero 上无法运行”或“不错的应用，如果能够……就更好了”。

响应用户评论而进行的升级会受到用户欣赏，从而吸引更多用户。在任何情况下，都应明确给出发布升级的理由。有的用户一天会收到两位数以上的可用更新提醒，如果看不到有说服力的理由，他们可能不会选择升级。

### 1.8.6　Google Play 以外的其他选择

有一些独立的 Android 应用商店存在。它们也许访问起来不如 Google Play 方便，但可能为开发者提供其他好处，例如更好的应用曝光度，更多的应用收费点，以及不抽取费用等。另外，一些 Android 设备制造商为他们的设备定制了自己的应用商店。比如，想让中国和拉丁美洲市场上的摩托罗拉 Android 手机看到你的应用，可以通过摩托罗拉应用商店（http://developer.motorola.com/shop4apps）来实现。

还有若干第三方应用商店，下面列出了其中一些。请注意，有的第三方商店会发布非法、盗版或损坏了的软件。如果你已下定决心使用第三方商店，请务必首先对其做一定的调查工作以确保它值得信任。

- 百度应用商店（中国）：http://as.baidu.com/。
- 亚马逊应用商店（Amazon Apps）：www.amazon.com/appstore。
- Opera 移动应用商店（Opera Mobile Apps Store）：http://apps.opera.com/en_us/。
- SlideMe：http://slideme.org/。
- Getjar：www.getjar.com/。
- AppBrain：www.appbrain.com/。

# 第2章

# 应用程序基础：Activity和Intent

每个 Android 应用程序都由单个的 Android 项目代表。本章给出了 Android 项目的目录结构，包括对于应用程序基本构建模块的简介，这为学习本书后面的技巧提供了有用的背景知识。本章的后半部分则转而介绍 Activity 以及负责调用 Activity 的 Intent。

## 2.1 Android 应用程序概览

Android 应用程序包含的功能五花八门，比如编辑文本、播放音乐、启动闹钟或是打开通讯录等。这些功能可以被分类对应到 4 类 Android 组件之中，如表 2-1 所示，每一类都对应一个 Java 基本类。

表 2-1 Android 应用程序可能包含的组件

| 功　能 | Java 基本类 | 例　子 |
|---|---|---|
| 用户持续聚焦的事情 | `Activity` | 编辑文本、玩游戏 |
| 后台进程 | `Service` | 播放音乐，更新天气图标 |
| 接收消息 | `BroadcastReceiver` | 根据事件触发警报 |
| 存储和检索数据 | `ContentProvider` | 打开通讯录 |

每个应用程序都由一个或多个这样的组件组成。当要用到某个组件时，Android 操作系统就会将其初始化。其他应用程序在指定的权限内也可以使用它们。

随着在操作系统中展现多种功能（有些功能甚至与预期的应用程序无关，如呼入电话），每个组件经历了生命周期的创建（create）、聚焦（focus）、失去焦点（defocus）和销毁（destroy）过程。对于优雅的操作（比如保存变量或者恢复用户界面元素）可以重写其默认行为，使交互对用户更加友好。

除了 `ContenProvider` 组件，每个组件都需要一个叫做 `Intent` 的异步消息来激活。`Intent` 可包含一组（`Bundle`）描述该组件的辅助数据。这也提供了一种在组件之间传递消息的方法。

本章最后将使用最常见的组件 `Activity` 演示前面提到的概念。由于 Activity 总是和具体的用户交互相关，所以每个 Activity 在创建时会自动创建一个新窗口。当然还会提到一些关于 UI 的概要介绍。至于 `Service` 和 `BroadcastReceiver` 这两个组件我们将会在第 3 章中讲解，而 `ContentProvider` 则会在第 11 章中阐述。

## 技巧 1：创建项目和 Activity

创建 Android 项目或其组件最直截了当的方式，当属使用 Eclipse 集成开发环境（IDE）。这样做可以确保所有辅助文件被正确安装。用 Eclipse 新建 Android 项目的步骤具体如下。

（1）在 Eclipse 下，选择 **File→New→Android Application Project**，这样会出现新建 Android 项目的界面。

（2）填写项目名称，比如 **SimpleActivityExample**。

（3）填写应用程序的名称，比如 **Example of Basic Activity**。

（4）填写包名称，例如 **com.cookbook.simpleactivity**。

（5）选择 SDK 的最低要求，这将成为可以运行应用的最低 Android 版本。建议最低选择 API Level 8 或 Android 2.2。

（6）从给出的选项中选择构建程序的目标 SDK 版本，此处应选择测试应用时会用到的最高 Android 版本。

（7）选择用于编译应用的 SDK 版本，这里应选择可用的最新版本，或者要用到的库所需的最低版本。

（8）选择应用程序的基本主题（theme）。以后还可以按需修改主题，不过在此早些指定一个没有坏处。

（9）接下来，设置另一些项目默认参数。选中 **Create custom launcher icon** 一项，以替换当前的默认图标。要在后面的步骤中创建主 Activity，请确保你勾选了 **Create Activity** 项。

（10）在 Configure Launcher Icon 界面，可以在文本、一小组剪贴画或磁盘中的图片里面选定应用的图标。系统会基于选定的图像，分别创建符合 4 种标准分辨率的图标。

（11）要在同一步骤中创建主 Activity，应确保勾选了 **Create Activity** 项，并选择 **BlankActivity** 选项。Fragment 的使用将在后面的技巧中予以介绍。

（12）填写 Activity 和布局的名字，或者维持默认值。要使用默认导航模式中的某一个，需要 SDK 版本为 14 或更高才行，因为这有赖于 ActionBar 的支持。

（13）点击 **Finish** 按钮完成这一实例工程的创建。

所有的 Activity 均继承了抽象类 `Activity` 或它的一个子类。每个 Activity 的入口为 `onCreate()` 方法。通常为初始化 Activity，都会对该方法进行重写，以完成设置 UI、创建按钮监听器（listener）、初始化参数，或是启动线程一类的工作。

如果主 Activity 并未随项目一起创建，或者需要另行创建一个 Activity，那么可以采取以下步骤。

（1）创建一个继承 `Activity` 类的新类。在 Eclipse 下可以这样做：在项目上右击，选择 **New→Class**，然后指定 `android.app.Activity` 作为父类。

（2）重写 `onCreate()` 函数。在 Eclipse 下可这样做：在类文件上右击，选择 **Source→Override/ImplementMethods...**，然后选中 `onCreate()` 方法。同大多数重写函数一样，重写 `onCreate()` 方法需要调用父类的该方法，如若不然，可能在运行时抛出异常。在此，应当首先调用 `super.onCreate()` 方法，以确保 Activity 的正确初始化，如代码清单 2-1 所示。

**代码清单 2-1　src/com/cookbook/simple_activity/SimpleActivity.java**

```
package com.cookbook.simple_activity;

import android.app.Activity;
import android.os.Bundle;

public class SimpleActivity extends Activity {

    @Override
    public void onCreate(Bundle savedInstanceState) {
        super.onCreate(savedInstanceState);
        setContentView(R.layout.main);
    }
}
```

（3）如果要用到 UI，需在 **res/layout/**目录下的一个 XML 文件中指定布局。此处该文件称为 **main.xml**，如代码清单 2-2 所示。

**代码清单 2-2　res/layout/main.xml**

```
<?xml version="1.0" encoding="utf-8"?>
<LinearLayout xmlns:android="http://schemas.android.com/apk/res/android"
    android:orientation="vertical"
    android:layout_width="match_parent"
    android:layout_height="match_parent"
    >
<TextView
    android:layout_width="match_parent"
    android:layout_height="wrap_content"
    android:text="@string/hello"
    />
</LinearLayout>
```

（4）使用 `setContentView()` 函数设置 Activity 的布局，并将 XML 布局文件对应的资源 ID 作为参数传递给它。此处该参数为 `R.layout.main`，已显示在代码清单 2-1 中。

（5）在 **AndroidManifest.xml** 文件中声明 Activity 的属性，详细内容会在稍后的代码清单 2-5 中涵盖。

注意，字符串类型资源均在 **res/values** 文件夹下的 **strings.xml** 中定义，如代码清单 2-3 所示。该文件为需要被修改或重用的字符串提供了一个集结地。

**代码清单 2-3　res/values/strings.xml**

```
<?xml version="1.0" encoding="utf-8"?>
<resources>
    <string name="hello">Hello World, SimpleActivity!</string>
    <string name="app_name">SimpleActivity</string>
</resources>
```

下面要对项目的结构及其他一些自动生成的内容进行详细探究。

### 2.1.1　项目目录结构及自动生成的内容

项目结构是用户生成的文件与自动生成文件的混合体。图 2-1 给出了 Eclipse Package Explorer（Eclipse 包浏览器）中显示的项目结构的一个示例。

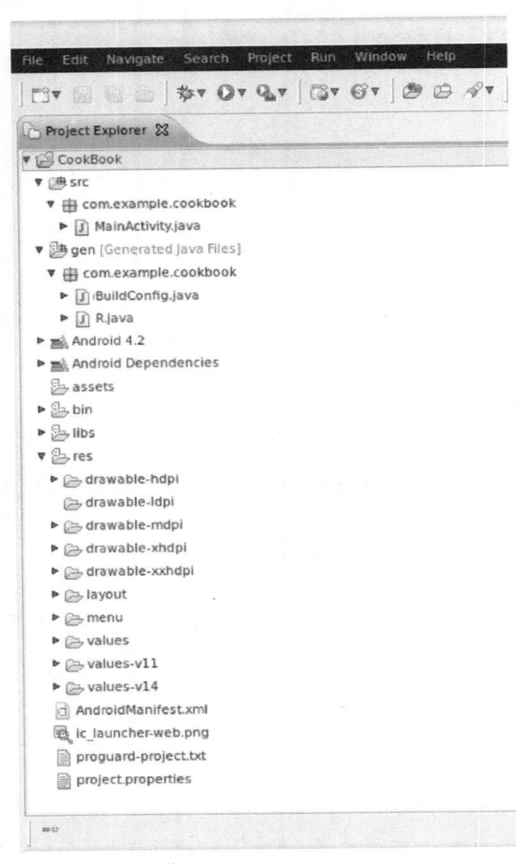


图 2-1　Eclipse IDE 中能看到的 Android 项目目录结构

用户生成的文件包括以下这些。

- **src/**目录包含了用户为应用编写或导入的 Java 包。每个包可以包含多个**.java** 文件，分别代表不同的类。

- **res/layout/**下包含了为每一屏界面指定布局的 XML 文件。
- **res/values/**下含有被其他文件所引用的 XML 文件。
- **res/values-v11/**用于存放 Honeycomb①及更高版本设备使用的 XML 文件。
- **res/values-v14/**用于存放 Ice Cream Sandwich②及以上版本设备使用的 XML 文件。
- **res/drawable-xhdpi/**、**res/drawable-hdpi/**、**res/drawable-mdpi/**和 **res/drawable-ldpi/**分别包含应用在极高、高、中、低每英寸点数分辨率下所用的图片。
- **assets/**下存有应用要用到的非媒体文件。
- **AndroidManifest.xml** 用于向 Android 系统指定该项目的特性。

每个名为 **res/values-XX** 的文件夹下均生成有 **styles.xml**。这是因为 Android 基本主题从 Honeycomb 版本起变为了 Holo，这导致应用的主题拥有不同的父主题。

自动生成的文件则有以下这些。

- **gen/**之下包含了自动产生的代码，包括生成的类文件 **R.java**。
- **project.properties** 里含有项目设定。尽管是自动生成的，它也应当被纳入版本控制之下。

应用程序的资源包括描述布局的 XML 文件、定义包括如字符串和 UI 元素标签等的各种值的 XML 文件，以及其他支持文件，比如图片和声音。编译时，对资源的引用都会收集到一个名为 `R.java` 的自动生成的包装类（wrapper class）中。Android Asset 打包工具会自动生成该文件。代码清单 2-4 给出了技巧 1 的项目产生的 `R.java` 的内容。

**代码清单 2-4 gen/com/cookbook/simple_activity/R.java**

```
/* AUTO-GENERATED FILE. DO NOT MODIFY.
 *
 * This class was automatically generated by the
 * aapt tool from the resource data it found. It
 * should not be modified by hand.
 */

package com.cookbook.simple_activity;

public final class R {
     public static final class attr {
     }
     public static final class drawable {
          public static final int icon=0x7f020000;
     }
     public static final class layout {
          public static final int main=0x7f030000;
     }
     public static final class string {
          public static final int app _ name=0x7f040001;
          public static final int hello=0x7f040000;
     }
}
```

① Honeycomb 为 Android 3.0～3.2 版的别名，这里特指对应 API Level 11 的 3.0 版本。——译者注

② Ice Cream Sandwich 为 Android 4.0 版的别名，对应 API Level 14。——译者注

在这里，每个资源都被映射到一个独一无二的整型（integer）值。通过这种方式，R.java 类提供了一种在 Java 代码中引用外部资源的办法。例如，要在 Java 中引用 **main.xml**，就使用整型值 R.layout.main。而要在 XML 文件里引用该资源，则应使用“@layout/main”字符串。

表 2-2 展示了在 Java 或 XML 文件中引用资源的方法。注意，要定义一个 ID 名为 home_button 的新按钮，需要在其标识字符串前面加上一个加号，就像这样：“@+id/home_button”。关于资源更完整的细节请见第 5 章，此处介绍的内容对于学习本章的技巧已经足够。

**表 2-2　Java 和 XML 文件中的资源**

| 资　源 | 在 Java 中的引用方式 | 在 XML 中的引用方式 |
|---|---|---|
| **res/layout/main.xml** | R.layout.main | @layout/main |
| **res/drawable-hdpi/icon.png** | R.drawable.icon | @drawable/icon |
| @+id/home_button | R.id.home_button | @id/home_button |
| <string name="hello"> | R.string.hello | @string/hello |

## 2.1.2　Android 包和 manifest 文件

Android 项目，有时也被称为 Android 包，乃是 Java 包的集合。不同的 Android 包可以拥有相同的 Java 包名，但 Android 设备上安装的每个 Android 包的名字都应当是独一无二的。

为让操作系统访问这些包，每个应用程序必须在一个名为 **AndroidManifest.xml** 文件中声明它所有可用的组件。此外，该文件还包含运行应用所需的权限和行为。代码清单 2-5 给出了技巧 1 对应的 **AndroidManifest.xml** 的内容。

**代码清单 2-5　AndroidManifest.xml**

```
<?xml version="1.0" encoding="utf-8"?>
<manifest xmlns:android="http://schemas.android.com/apk/res/android"
          package="com.cookbook.simple_activity"
          android:versionCode="1"
          android:versionName="1.0">
    <application android:icon="@drawable/icon"
                 android:label="@string/app_name">
        <activity android:name=".SimpleActivity"
                  android:label="@string/app_name">
            <intent-filter>
              <action android:name="android.intent.action.MAIN" />
              <category android:name="android.intent.category.LAUNCHER" />
            </intent-filter>
        </activity>
    </application>
    <uses-sdk android:minSdkVersion="8" />
</manifest>
```

第一行对于 Android 中的所有 XML 文件都是标准的和必须的，用于指定编码。manifest 元素定义了 Android 包的名字和版本。versionCode 是一个整型值，可在程序中对其求值，以确定版本的高低关系。versionName 以人们可读懂的格式表示了声明的主版本号和子版本号。

application 元素定义了用户会在 Android 设备菜单中看到的应用程序的图标和标签。标签是一个字符串，应当足够简短，以便在用户设备中正确显示在图标下方。通常，该名字最长为两个单词，每个单词由 10 个连续的英文字符构成。

activity 元素定义了应用启动时调用的主 Activity，以及 Activity 活动时标题栏（title bar）中显示的名称。在此需要给 Java 包指定名字，本例中为 com.cookbook.simple_activity.SimpleActivity。由于 Java 包名通常与 Android 包名相同，常常使用简化形式.SimpleActivity。然而，最好记得 Android 包和 Java 包还是有区别的。

intent-filter 元素向 Android 系统说明组件的功能，为此，它可以包含多个 action、category 或 data 元素。该元素在本书的许多技巧中都会用到。

uses-sdk 元素定义了运行应用所需的 API 级别。一般来说，API 级别会像下面那样指定：

```
<uses-sdk android:minSdkVersion="integer"
          android:targetSdkVersion="integer"
          android:maxSdkVersion="integer" />
```

由于 Android 系统按向前兼容原则构建，强烈不推荐使用 maxSdkVersion 参数（在 Android 2.0.1 及以后版本中它已经被剔除掉了）。然而 Google Play 仍然在使用它作为一个过滤参数，在运行大于这一参数的 SDK 版本的设备上，该应用程序就会被显示为不可下载。

targetSdkVersion 并非必须指定，但指定它可以让拥有相同 SDK 版本的设备关闭兼容性设置，这可能会提升操作速度。应该始终指定 minSdkVersion 的值，以确保一个应用运行在不支持其所需特性的平台上时不会崩溃。在指定该值时应总是选择可能的最低 API 级别。

**AndroidManifest.xml** 文件还可以包含运行该应用所需的权限设定。后面的章节会给出关于所提供选项的更完整介绍，以上所讲对于本章的技巧来说已经够用了。

## 技巧 2：重命名应用程序的某些部分

有时候 Android 项目的某部分需要被重命名。也许有某个文件被（比如从本书中）手动复制进了项目。也许应用程序在开发期间改了名，也需要在文件系统树上反映出来。这项工作由自动化工具协助我们完成，同时确保那些交叉引用也被自动更新。例如，在 Eclipse IDE 里，有多种办法实现对应用程序某一部分的重命名。

要重命名 Android 项目，步骤如下。

（1）在项目上右击，选择 **Refactor→Move** 将其移动到文件系统中的一个新目录中。

（2）在项目上右击，选择 **Refactor→Rename** 重命名项目。

要重命名 Android 包，步骤如下。

（1）在包上右击，选择 **Refactor→Rename** 重命名包。

（2）编辑 **AndroidManifest.xml** 文件，确保新的包名在其中得到反映。

要重命名 Android 类（比如像 Activity、Service、BroadCastReceiver 和 ContentProvider 这样的主要组件），步骤如下。

（1）在**.java** 文件上右击，选择 **Refactor→Rename** 重命名类。

（2）编辑 **AndroidManifest.xml** 文件，确保 `android:name` 之下的名字得到更新。

注意，重命名其他文件，诸如 XML 文件，通常需要手动更改 Java 代码中相应的引用。

## 技巧 3：使用库项目

库项目（library project）允许资源和代码在其他应用程序中被复用，它们也被用作 UI 库以使较老的设备能支持新的特性。库项目在 SDK 工具版本 14 中被首次引入。库项目与一般的 Android 项目很类似，同样包含源码、资源文件夹和 manifest 文件。二者主要的区别在于库项目不能独立运行，且无法被编译成**.apk** 文件。创建库项目的步骤如下。

（1）在 Eclipse 下，选择 **File→New→Android Application Project**，这样会出现新 Android 项目的创建界面。

（2）填写项目名称，比如 **SimpleLibraryExample**。

（3）填写应用程序的名称，比如 **Example of Basic Activity**。

（4）填写包名称，例如 **com.cookbook.simplelibrary**。

（5）从给出的选项中选择构建程序的目标 SDK 版本，这些选项基于开发机上已安装的 SDK 版本而提供。

（6）取消勾选 **Create custom launcher icon** 一项，因为库不需要图标。

（7）勾选 **Mark this project as a Library** 一项。

（8）要在库中包含 Activity，应选中 **Create Activity** 项。此处创建的 Activity 会在主项目中用到，目前勾选上 **Create Activity** 并选择 **BlankActivity** 即可。

（9）填写活动名称，或者维持默认。

（10）把布局名改为 **lib_activity_main**。由于所有资源最终都会被编译进单个 `R` 类文件中，最好给库的所有资源起一个统一的前缀，以避免命名冲突。

（11）点击 **Finish** 按钮完成库项目的创建。

要使用该库，需要一个主项目。依照技巧 1 的方法为工作区添加一个新项目，只需要进行一点微小的改动，即不需为该项目创建 Activity，取而代之地，使用之前创建的库项目中的 Activity。

要在主项目中引用库项目，操作如下。

（1）在 Eclipse 里右击项目并选择 **Properties→Android**。

（2）向下滚动到库部分，并点击 **Add**。

（3）会弹出一个对话框，显示工作区内所有可用的库项目，从中选择 **SimpleLibrary** 并按下 **OK** 按钮。

现在项目名称和引用路径已显示在 Properties 页中，会有一个绿箭头表明引用通过了检查。如果找不到引用路径，则会显示一个红叉。

添加了库项目后，建议对工作区做一次干净而完全的构建操作，以确保我们所做的变更能按预期工作。

在内部，库引用被保存在 **project.properties** 文件中，该文件同时为 Eclipse 和 Ant 所用。该文件内容应该像下面这样：

```
target=android-16
android.library.reference.1=../SimpleLibrary
```

代码清单 2-6 将库中的 Activity 添加到 **AndroidManifest.xml** 文件里，并将其设为默认启动的 Activity。

**代码清单 2-6　主项目的 AndroidManifest.xml**

```
<manifest xmlns:android="http://schemas.android.com/apk/res/android"
    package="com.example.simpleproject"
    android:versionCode="1"
    android:versionName="1.0">

    <uses-sdk android:minSdkVersion="8" android:targetSdkVersion="15" />

    <application android:label="@string/app_name"
        android:icon="@drawable/ic_launcher"
        android:theme="@style/AppTheme">

        <activity
            android:name="com.cookbook.simplelibrary.LibMainActivity"
            android:label="@string/title_activity_activity_main" >
            <intent-filter>
                <action android:name="android.intent.action.MAIN" />

                <category android:name="android.intent.category.LAUNCHER" />
            </intent-filter>

        </activity>
    </application>
</manifest>
```

在此注意，应用程序的包名与 Activity 的名字截然不同，由于 Activity 是从库项目中调用的，为其设定的名称必须是完全合格的包名和类名。

现在项目可以在 Eclipse 中运行了，库中的 Activity 会在前台显示。

库项目有很多用途，从为主项目设置不同主题、但运行在相同基础代码之上的 white-label[①] 应用，到使用像 `ActionBarSherlock` 这样的 UI 库为老 Android 设备带来新的外观和感觉。

## 2.2 Activity 的生命周期

应用程序中的每个 Activity 都要经历自己的生命周期。Activity 于 `onCreate()` 函数执行时被创建，该创建过程仅会执行一次。退出 Activity 时，则会执行 `onDestroy()` 函数。在此二者之间，各种各样的事件可以使 Activity 进入各种不同状态，如图 2-2 所示。下一个技巧将对这些函数一一举例。

① 所谓 white-label 是这样一类产品和服务，它们由一家公司（制造商）生产，但其他公司（市场商）可以对其进行重塑，让它们以自己的面貌面世。——译者注

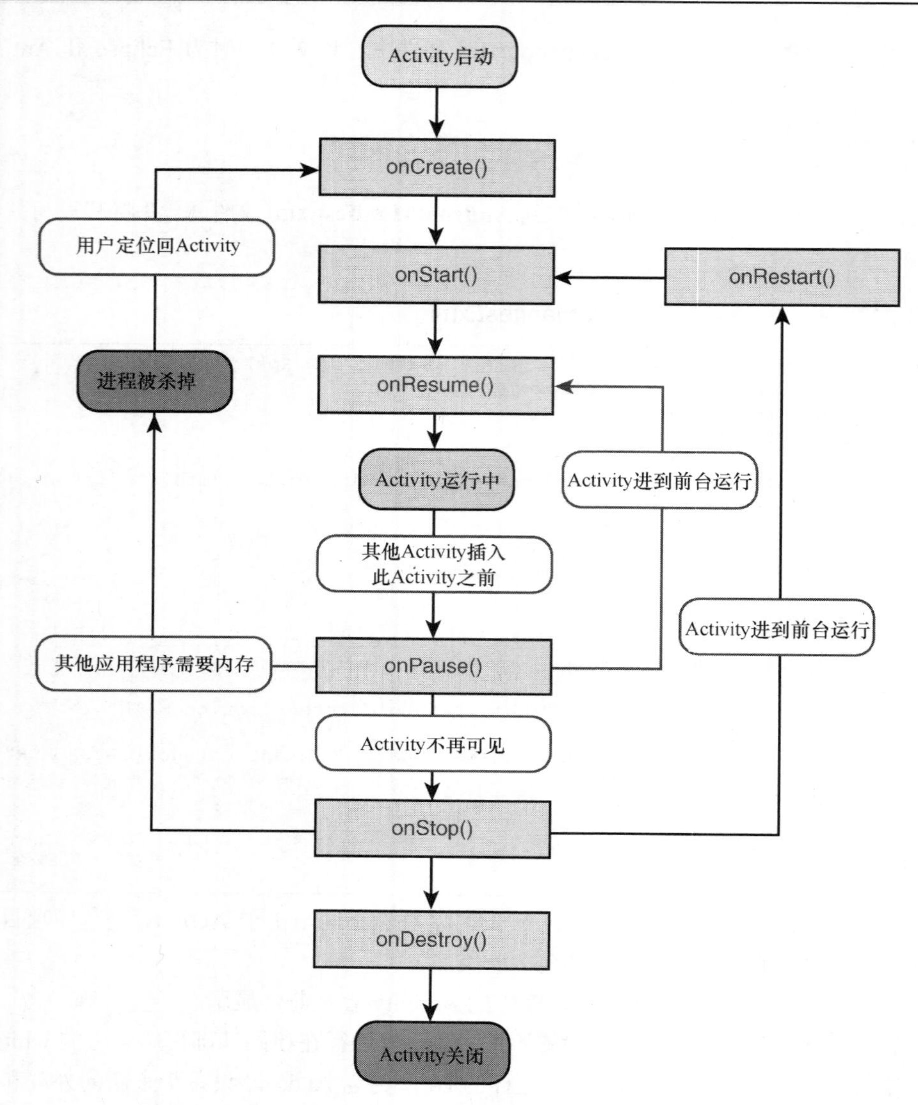


图 2-2　Activity 的生命周期，引自 http://developer.android.com/

## 技巧 4：使用 Activity 生命周期函数

本技巧提供了一种在 Activity 工作时查看其生命周期的简单方法。为清楚起见，对每个被重写的函数都予以显式声明，并在其中加入一条 Toast 命令，这样在执行某个函数时，在屏幕上能有所反映（关于 Toast 微件的更多细节将在第 3 章中给出）。代码清单 2-7 给出了 Activity 的内容。在 Android 设备上运行它，并尝试各种情况。特别注意以下几点。

- 改变屏幕方向将会销毁并重建 Activity。

- 按下 Home 键会使 Activity 暂停，但不会销毁它。
- 点击应用程序图标可能会启动一个 Activity 的新实例，甚至在旧的 Activity 并未销毁的情况下也会如此。
- 让屏幕休眠将会暂停 Activity，而在唤醒时会恢复它（与接电话时的状况类似）。

**代码清单 2-7 src/com/cookbook/activity_lifecycle/ActivityLifecycle.java**

```java
package com.cookbook.activity_lifecycle;

import android.app.Activity;
import android.os.Bundle;
import android.widget.Toast;

public class ActivityLifecycle extends Activity {

    @Override
    public void onCreate(Bundle savedInstanceState) {
        super.onCreate(savedInstanceState);
        setContentView(R.layout.main);
        Toast.makeText(this, "onCreate", Toast.LENGTH_SHORT).show();
    }

    @Override
    protected void onStart() {
        super.onStart();
        Toast.makeText(this, "onStart", Toast.LENGTH_SHORT).show();
    }

    @Override
    protected void onResume() {
        super.onResume();
        Toast.makeText(this, "onResume", Toast.LENGTH_SHORT).show();
    }

    @Override
    protected void onRestart() {
        super.onRestart();
        Toast.makeText(this, "onRestart", Toast.LENGTH_SHORT).show();
    }

    @Override
    protected void onPause() {
        Toast.makeText(this, "onPause", Toast.LENGTH_SHORT).show();
        super.onPause();
    }

    @Override
    protected void onStop() {
        Toast.makeText(this, "onStop", Toast.LENGTH_SHORT).show();
        super.onStop();
    }

    @Override
    protected void onDestroy() {
        Toast.makeText(this, "onDestroy", Toast.LENGTH_SHORT).show();
        super.onDestroy();
    }
}
```

由此可见，不少用户操作会导致 Activity 暂停或杀死，甚至可能启动应用程序的多个版本。在开始新内容之前，有必要提一下另外两个可以控制这种行为的简单技巧。

## 技巧 5：强制采用单任务模式

当跳出一个应用程序然后再次启动它时，可能在设备上产生同一 Activity 的多个实例。最终多余的实例会被杀死以释放内存，但期间可能引发莫名其妙的情况。为避免这种情况发生，开发者可以在 **AndroidManifest.xml** 文件中对每个 Activity 的此类行为进行控制。

要确保 Activity 只有一个实例在设备上运行，为拥有 MAIN 和 LAUNCHER 两个 Intent 过滤器的 Activity 元素内添加如下代码：

```
android:launchMode="singleInstance"
```

这样就使任务中的每个 Activity 始终只有一个实例。此外，任何子 Activity 都会作为一个单独的任务来启动。为进一步确保应用中的所有 Activity 都运行在同一个任务中，使用如下代码：

```
android:launchMode="singleTask"
```

这样就使多个 Activity 可以作为同一个任务来轻松地共享信息。

此外，有时我们会希望无论用户以何种方式进入 Activity 时，都能保持任务状态。例如，如果用户离开了应用，一段时间后又重新启动它，默认的做法通常是将任务重置为初始状态。为保证任务在用户返回时总是被还原到关闭之前的状态，可为任务的根 Activity 的 activity 元素指定如下属性：

```
android:alwaysRetainTaskState="true"
```

## 技巧 6：强制规定屏幕方向

每个拥有加速度计的 Android 设备都可以判定哪个方向是向下。当设备从纵向（portrait）模式切换到横向（landscape）模式[①]时，默认的动作是让应用程序的视图也随之旋转。然而，在技巧 4 中我们看到，Activity 会随着屏幕方向的改变而销毁和重启，一旦发生这种事，Activity 的当前状态可能丢失，以致干扰用户体验。

处理屏幕变向的办法之一是在改变前保存状态信息，并在改变后予以恢复。更为简单且有用的一种办法是强制屏幕方向保持恒定。可以为 **AndroidManifest.xml** 文件中的每一个 Activity 指定 screenOrientation 属性。例如，要让 Activity 始终保持纵向模式，可以在 Activity 标签内添加如下属性：

```
android:screenOrientation="portrait"
```

类似地，若要保持横向模式，则使用下面的代码：

---

① 在显示方向的英文术语中，纵向被称为 portrait（意为肖像），而横向被称为 landscape（意为风景），这大概是肖像图多呈纵向（高大于宽），而风景图多呈横向（宽大于高）的缘故。——译者注

```
android:screenOrientation="landscape"
```

然而，光有这样的代码，在硬键盘滑出时仍会引发 Activity 的销毁和重启。此时，可以采用第三种办法，告诉 Android 系统：应用程序会处理屏幕变向和键盘滑出事件。可以通过在 activity 元素中增添如下属性来实现这一点：

```
android:configChanges="orientation|keyboardHidden"
```

该属性可以单独使用，也可与 screenOrientation 属性一起使用，从而向应用程序规定我们想要的行为。

## 技巧 7：保存和恢复 Activity 信息

当 Activity 将被杀死时，会调用 onSaveInstanceState()函数。可重写该函数以保存有关信息。当 Activity 被重建时，会调用 onRestoreInstanceState()函数，重写它可以恢复保存的信息。这样可以在应用经历生命周期变化时，让用户获得无缝的体验。注意，多数 UI 状态不需要我们亲自处理，因为默认状况下系统会关照它们。

onSaveInstanceState()有别于 onPause()，例如，如果另一组件被启动并置于现有 Activity 的前端，就会调用 onPause()函数。随后，如果当操作系统要回收资源时该 Activity 仍处于暂停状态，就在结束它之前调用 onSaveInstanceState()。

代码清单 2-8 给出一个保存和恢复实例状态的例子，该状态包含一个字符串和一个 float 型数组。

**代码清单 2-8 onSaveInstanceState()和 onRestoreInstanceState()的示例**

```
float[] localFloatArray = {3.14f, 2.718f, 0.577f};
String localUserName = "Euler";

@Override
protected void onSaveInstanceState(Bundle outState) {
        super.onSaveInstanceState(outState);
        //Save the relevant information
        outState.putString("name", localUserName);
        outState.putFloatArray("array", localFloatArray);
}

@Override
public void onRestoreInstanceState(Bundle savedInstanceState) {
        super.onRestoreInstanceState(savedInstanceState);
        //Restore the relevant information
        localUserName = savedInstanceState.getString("name");
        localFloatArray = savedInstanceState.getFloatArray("array");
}
```

注意，onCreate()方法同样包含 Bundle savedInstanceState 对象。当 Activity 关闭后又被重新初始化时，在 onSaveInstanceState()中被保存的 bundle 也会被传递到 onCreate()方法中。在任何情况下，被保存的 bundle 都要传递给 onSaveInstanceState()函数，所以用它来恢复状态更为自然。

## 技巧 8：使用 Fragment

Fragment 是新近被加入 Android 基本构建模块行列的新鲜事物，它们是 Activity 下属的小部件，用于为视图与功能分组。对 Fragment 的一个形象类比是把它们想成小的积木块，可以堆在一起从而对更大的积木块进行填充。对小组块的需求源于平板电脑和电视屏幕的引入。

Fragment 使得视图可以被捆绑在一起，并按需要被混合并匹配到一个或两个（甚至更多）Activity 中去。对 Fragment 的经典应用是将屏幕从带有一个列表及一个详细视图的横向模式切换到带有一个单列表及一个详细视图的纵向模式。事实上，该模式已变得如此流行，以至于现在可以通过 **Create New Project** 对话框直接创建这一模式的应用骨架。

创建的步骤与前面技巧里描述过的类似。

（1）在 Eclipse 下，选择 **File→New→Android Application Project**。

（2）填写项目名称，比如 **SimpleFragmentExample**。

（3）填写应用程序的名称，比如 **Example of Basic Fragments**。

（4）填写包名称，例如 **com.cookbook.simplefragments**。

（5）将 SDK 的最低要求选为 API Level 11 或 Android Honeycomb。只有机器上安装了额外的支持库的情况下，Fragment 才能在更低的 API 版本下使用。

（6）在 **Create Activity** 界面中选择 **MasterDetailFlow** 作为起始点。

（7）为用于演示的项起名，例如叫做 **fruits**。

（8）点击 **Finfish** 按钮完成创建。

进一步探索这一范例功用的工作就留给读者完成。在此我们强调与 Fragment 有关的几点重要事项。

Fragment 有它们自己的生命周期，该周期依赖于宿主 Activity。由于 Fragment 可以在 Activity 生命周期的任意时间点被添加、显示、隐藏和移除，它们比其他组件要更短命一些。与 Activity 类似，Fragment 拥有 `onPause()`、`onResume()`、`onDestroy()` 和 `onCreate()` 方法。

但需要注意的是，对于 Fragment，`onCreate(Bundle)` 方法是第二个被调用的方法，第一个调用是 `onAttach(Activity)`，它产生信号，表明已存在与宿主 Activity 的连接。可以在 `onAttach` 中调用 Activity 上的方法，但此时并不能保证 Activity 已经将自己初始化完毕。只有当调用了 `onActivityCreated()` 方法之后，Activity 才算是通过了它自己的 `onCreate()` 方法。

鉴于 Fragment 可以在很晚时才被实例化和添加，我们不应依赖 Activity 在 `onAttach()` 中的状态。用于初始化视图并开始大部分工作的乃是 `onCreateView(LayoutInflater, ViewGroup, Bundle)` 方法。如果 Fragment 被是重建的，那么其中的 `Bundle` 类为事先保存的实例状态。

Fragment 还使用 bundle 来序列化参数。Fragment 所需的每一种属于可打包类型的外部信息，都可以通过调用 setArguments() 方法从宿主 Activity 获取。并且总能在 Fragment 中调用 `getArguments()` 方法读取它们。这使得从 Activity 的起始 Intent 来的信息能够被直接传递到 Fragment 中显示。

# 2.3 多个 Activity

就算是最简单的应用程序也会拥有不止一项功能，因此我们经常要应对多个 Activity。例如，一款游戏可能含有两个 Activity，其一为高分排行榜，另一为游戏画面。一个记事本可以有三个 Activity：浏览笔记列表、阅读选定笔记、编辑选定的或新建的笔记。

**AndroidManifest.xml** 文件中定义的主 Activity 会随应用程序启动而启动。该 Activity 可以开启另外的 Activity，通常由触发事件引起。这第二个 Activity 被激活时，主 Activity 会暂停。当第二个 Activity 结束时，主 Activity 会被重新调回前台恢复运行。

要想激活应用中的某个特定组件，可以用显式命名该组件的 Intent 来实现。而应用程序的所需可以通过 Intent 过滤器指定，这时可采用隐式 Intent。系统可随即确定最合适的某个或某组组件，不管它是属于另外的应用还是系统自带的。注意，与其他 Activity 不同，位于其他应用程序中的隐式 Intent 不需要在当前应用的 **AndroidManifest.xml** 文件中声明。

Android 尽可能选用隐式 Intent，它能为模块化功能提供强大的框架。当一个符合所需的隐式 Intent 过滤器要求的新组件开发完成，它就可以替代 Android 内部的 Intent。举个例子，假如一个用来显示电话联系人的新应用被装入到 Android 设备，当用户选择联系人时，Android 系统会找出符合浏览联系人这一 Intent 过滤器要求的所有可用的 Activity，并询问用户想使用哪一个。

## 技巧 9：使用按钮和文本视图

触发器事件有助于全面展示多 Activity 特性。为此我们引入一个按钮（button）按下动作，为给定的布局添加按钮并为其指派动作的步骤如下。

（1）为指定的布局 XML 文件添加一个按钮控件：

```
<Button android:id="@+id/trigger"
   android:layout_width="100dip" android:layout_height="100dip"
   android:text="Press this button" />
```

（2）声明一个指向布局文件中的按钮 ID 的按钮：

```
Button startButton = (Button) findViewById(R.id.trigger);
```

（3）为按钮点击事件指定一个监听器（listener）：

```
//Set up button listener
startButton.setOnClickListener(new View.OnClickListener() {
    //Insert onClick here
});
```

（4）重写监听器的 `onClick` 函数以执行要求的动作：

```
public void onClick(View view) {
    // Do something here
}
```

为展示动作的效果，改变屏幕上的文字不失为一招。定义文本域并通过编程对其进行改动的步骤如下：

（1）用一个 ID 为指定的布局 XML 文件添加文本域，该文本域可以有初始值（此处可用 **strings.xml** 中定义的 hello 字符串初始化它）。

```
<TextView android:id="@+id/hello_text"
    android:layout_width="match_parent"
    android:layout_height="wrap_content"
    android:text="@string/hello"
/>
```

（2）声明一个指向布局文件中 TextView ID 的 TextView（文本视图）：

```
private TextView tv = (TextView) findViewById(R.id.hello_text);
```

（3）如果需要变更文本，可使用 setText 函数：

```
tv.setText("new text string");
```

以上两项 UI 技术会在本章后续的一些技巧中用到。对于 UI 技术更详细的讲解请参见第 5 章。

## 技巧 10：通过事件启动另外一个 Activity

本技巧中 MenuScreen 是主 Activity，如代码清单 2-9 所示，它会开启名为 PlayGame 的 Activity。此处，触发器事件是作为按钮点击、用 Button 微件实现的。

当用户点击按钮，startGame() 函数会运行，并启动 PlayGame Activity。当用户点击 PlayGame Activity 中的按钮时，它会调用 finish() 函数将控制权交还给调用它的 Activity。下面是启动 Activity 的步骤。

（1）声明一个指向要启动的 Activity 的 Intent。

（2）在该 Intent 上调用 startActivity 方法。

（3）在 **AndroidManifest.xml** 中对这一额外的 Activity 加以声明。

**代码清单 2-9 src/com/cookbook/launch_activity/MenuScreen.java**

```
package com.cookbook.launch_activity;

import android.app.Activity;
import android.content.Intent;
import android.os.Bundle;
import android.view.View;
import android.widget.Button;
public class MenuScreen extends Activity {

    @Override
    public void onCreate(Bundle savedInstanceState) {
        super.onCreate(savedInstanceState);
        setContentView(R.layout.main);

        //Set up button listener
        Button startButton = (Button) findViewById(R.id.play_game);
        startButton.setOnClickListener(new View.OnClickListener() {
```

```
                public void onClick(View view) {
                    startGame();
                }
            });
        }

        private void startGame() {
            Intent launchGame = new Intent(this, PlayGame.class);
                startActivity(launchGame);
        }
    }
```

### 在匿名内部类里提供当前上下文环境

注意，通过点击按钮启动 Activity 时，还有一些东西需要考虑，如代码清单 2-9 显示的那样，Intent 需要一个上下文环境。然而，在 onClick 函数里使用 this 引用并不是个稳妥的解决办法。下面给出通过匿名内部类来提供当前上下文环境的几种不同方法。

- 使用 Context.this 代替 this。
- 使用 getApplicationContext() 代替 this。
- 显式地使用类名 MenuScreen.this。
- 调用一个在合适的上下文级别中声明的函数。在代码清单 2-8 的 startGame() 中使用的就是这个方法。

这些方法通常是可以互换的，可依照具体情况选择最好的方法。

代码清单 2-10 中给出的 PlayGame Activity 只不过是一个按钮，带有一个会调用 finish() 函数把控制权交还给主 Activity 的 onClick 监听器。可以按需给该 Activity 添加更多的功能，各个代码分支可以导致各自不同的 finish() 调用。

**代码清单 2-10　src/com/cookbook/launch_activity/PlayGame.java**

```
package com.cookbook.launch_activity;

import android.app.Activity;
import android.os.Bundle;
import android.view.View;
import android.widget.Button;

public class PlayGame extends Activity {

    public void onCreate(Bundle savedInstanceState) {
        super.onCreate(savedInstanceState);
        setContentView(R.layout.game);

        //Set up button listener
        Button startButton = (Button) findViewById(R.id.end_game);
        startButton.setOnClickListener(new View.OnClickListener() {
            public void onClick(View view) {
                finish();
            }
        });
    }
}
```

按钮必须像代码清单 2-11 所示的那样添加到 main 布局中，其 ID 应为 play_game，以与代码清单 2-9 中的设定匹配。此处，按钮的大小也以设备独立/无关像素（dip）[①]的方式声明，该方式会在第 5 章中进行更多讨论。

**代码清单 2-11　res/layout/main.xml**

```
<?xml version="1.0" encoding="utf-8"?>
<LinearLayout xmlns:android="http://schemas.android.com/apk/res/android"
    android:orientation="vertical"
    android:layout_width="match_parent"
    android:layout_height="match_parent"
    >
<TextView
    android:layout_width="match_parent"
    android:layout_height="wrap_content"
    android:text="@string/hello"
    />
<Button android:id="@+id/play_game"
    android:layout_width="100dip" android:layout_height="100dip"
      android:text="@string/play_game"
      />
</LinearLayout>
```

PlayGame Activity 引用它自己的按钮 ID——end_game，它位于布局资源 R.layout.game 中，R.layout.game 又对应名为 **game.xml** 的 XML 文件，如代码清单 2-12 所示。

**代码清单 2-12　res/layout/game.xml**

```
<?xml version="1.0" encoding="utf-8"?>
<LinearLayout xmlns:android="http://schemas.android.com/apk/res/android"
    android:orientation="vertical"
    android:layout_width="match_parent"
    android:layout_height="match_parent"
    >
<Button android:id="@+id/end_game"
      android:layout_width="100dip" android:layout_height="100dip"
      android:text="@string/end_game" android:layout_gravity="center"
      />
</LinearLayout>
```

尽管在各种情况下文本都可以显式地写在代码中，但更好的编码习惯是为每个字符串定义相应变量。本技巧里，名为 play_game 和 end_game 的两个字符串需要在字符串资源文件中分别定义，如代码清单 2-13 所示。

**代码清单 2-13　res/values/strings.xml**

```
<?xml version="1.0" encoding="utf-8"?>
<resources>
    <string name="hello">This is the Main Menu</string>
    <string name="app_name">LaunchActivity</string>
```

① 设备独立/无关像素（device-independent pixels），简写为 dip 或 dp，是 Android 为方便跨不同屏幕类型的设备的编程而推出的一种虚拟像素单位，用于定义应用的 UI，以密度无关的方式表达布局尺寸或位置。在运行时，Android 根据使用中的屏幕的实际密度，透明地处理任何所需 dip 单位的缩放。在第 5 章的表 5.1 中也有涉及。——译者注

```
    <string name="play_game">Play game?</string>
    <string name="end_game">Done?</string>
</resources>
```

最终，在 **AndroidManifest.xml** 文件里需要为 PlayGame 这个新类注册一个默认动作，如代码清单 2-14 所示。

**代码清单 2-14 AndroidManifest.xml**

```
<?xml version="1.0" encoding="utf-8"?>
<manifest xmlns:android="http://schemas.android.com/apk/res/android"
      android:versionCode="1"
      android:versionName="1.0" package="com.cookbook.launch_activity">
    <application android:icon="@drawable/icon"
                 android:label="@string/app_name">
        <activity android:name=".MenuScreen"
                  android:label="@string/app_name">
            <intent-filter>
              <action android:name="android.intent.action.MAIN" />
              <category android:name="android.intent.category.LAUNCHER" />
            </intent-filter>
        </activity>
        <activity android:name=".PlayGame"
                      android:label="@string/app_name">
                <intent-filter>
                    <action android:name="android.intent.action.VIEW" />
                    <category android:name="android.intent.category.DEFAULT" />
                </intent-filter>
            </activity>
    </application>
    <uses-sdk android:minSdkVersion="3" />
</manifest>
```

## 技巧 11：通过使用语音转文本功能启动一个 Activity

本技巧演示了如何调用一个 Activity 以获取其返回值，还演示了如何使用 Google 的 RecognizerIntent 中的语音转文本功能，并将转换结果输出到屏幕上。这里采用按钮点击作为触发事件，它会启动 RecognizerIntent Activity，后者对来自麦克风的声音进行语音识别，并将其转换为文本。转换结束时，文本会被传递回调用 RecognizerIntent 的 Activity。

返回时，首先会基于返回的数据调用 onActivityResult() 函数，然后会调用 onResume() 函数使 Activity 正常继续。调用的 Activity 可能会出现问题而不能正确返回，因此，在解析返回的数据之前，应当始终检查 resultCode 确保返回值为 RESULT_OK。

注意，一般来讲启动任何会返回数据的 Activity 都将导致同一个 onActivityResult() 函数被调用。因此，要使用一个请求代号来辨别是哪个 Activity 在返回数据。当被启动的 Activity 结束时，它会将控制权交还给调用它的 Activity，并使用相同的请求代码调用 onActivityResult()。

调用 Activity 获取返回值的步骤如下：

（1）用一个 Intent 调用 startActivityForResult() 函数，定义被启动的 Activity 及一个起识别作用的 requestCode 变量。

（2）重写 onActivityResult() 函数，检查返回结果的状况，检查所期望的 requestCode，并解析返回的数据。

下面是使用 RecognizerIntent 的步骤：

（1）声明一个动作为 ACTION_RECOGNIZE_SPEECH 的 Intent。

（2）为该 Intent 传递附加内容，至少 EXTRA_LANGUAGE_MODEL 是必需的，它可以被设置成 LANGUAGE_MODEL_FREE_FORM 或者 LANGUAGE_MODEL_WEB_SEARCH。

（3）返回的数据包中包含可能与原始文本匹配的字符串的列表。使用 data.getStringArrayListExtra 检索这一数据，它将在稍后以 ArrayList 的形式传送给用户。

返回的文本用一个 TextView 显示。主 Activity 在代码清单 2-15 中给出。

所需的支持文件还有 **main.xml** 和 **strings.xml**，其中需要定义一个按钮以及用于存放结果的 TextView，这可以借助技巧 10 中的代码清单 2-11 和 2-13 来实现。**AndroidManifest.xml** 文件中只需要声明主 Activity，这与前面的技巧 1 相同。RecognizerIntent Activity 是 Android 系统原生的 Activity，不需要显式声明即可使用。

**代码清单 2-15　src/com/cookbook/launch_for_result/RecognizerIntent Example.java**

```
package com.cookbook.launch_for_result;

import java.util.ArrayList;

import android.app.Activity;
import android.content.Intent;
import android.os.Bundle;
import android.speech.RecognizerIntent;
import android.view.View;
import android.widget.Button;
import android.widget.TextView;

public class RecognizerIntentExample extends Activity {
    private static final int RECOGNIZER_EXAMPLE = 1001;
    private TextView tv;

    protected void onCreate(Bundle savedInstanceState) {
        super.onCreate(savedInstanceState);
        setContentView(R.layout.main);

        tv = (TextView) findViewById(R.id.text_result);

        //Set up button listener
        Button startButton = (Button) findViewById(R.id.trigger);
        startButton.setOnClickListener(new View.OnClickListener() {
            public void onClick(View view) {
                // RecognizerIntent prompts for speech and returns text
                Intent intent =
                new Intent(RecognizerIntent. ACTION_RECOGNIZE_SPEECH);

                intent.putExtra(RecognizerIntent.EXTRA_LANGUAGE_MODEL,
                RecognizerIntent. LANGUAGE_MODEL_FREE_FORM);
                intent.putExtra(RecognizerIntent.EXTRA_PROMPT,
                "Say a word or phrase\nand it will show as text");
                startActivityForResult(intent, RECOGNIZER_EXAMPLE);
            }
```

```
        });
    }

    @Override
    protected void onActivityResult(int requestCode,
                                    int resultCode, Intent data) {
        //Use a switch statement for more than one request code check
        if (requestCode==RECOGNIZER_EXAMPLE && resultCode==RESULT_OK) {
                // Returned data is a list of matches to the speech input
                ArrayList<String> result =
                data.getStringArrayListExtra(RecognizerIntent.EXTRA_RESULTS);

            //Display on screen
            tv.setText(result.toString());
            }
        super.onActivityResult(requestCode, resultCode, data);
    }
}
```

## 技巧 12：实现选择列表

应用程序中常常需要提供给用户一个选择列表，供用户点选。这一功能利用 ListActivity 可以轻松地实现。ListActivity 是 Activity 的一个子类，它会根据用户选择触发事件。

下面是创建选择列表的步骤。

（1）创建一个扩展 ListActivity 而不是 Activity 的类。

```
public class ActivityExample extends ListActivity {
  //content here
}
```

（2）创建一个存储各个选项名称的字符串数组：

```
static final String[] ACTIVITY_CHOICES = new String[] {
                    "Action 1",
                    "Action 2",
                    "Action 3"
        };
```

（3）以 ArrayAdapter 为参数调用 setListAdapter()，为其指定选择列表及一个布局：

```
setListAdapter(new ArrayAdapter<String>(this,
        android.R.layout.simple_list_item_1, ACTIVITY_CHOICES));
getListView().setChoiceMode(ListView.CHOICE_MODE_SINGLE);
getListView().setTextFilterEnabled(true);
```

（4）启动 OnItemClickListener 以确定选中了哪个选项，并做出对应的动作：

```
getListView().setOnItemClickListener(new OnItemClickListener()
{
   @Override
   public void onItemClick(AdapterView<?> arg0, View arg1,
         int arg2, long arg3) {
      switch(arg2) {//Extend switch to as many as needed
      case 0:
```

```
                //code for action 1
                break;
            case 1:
                //code for action 2
                break;
            case 2:
                //code for action 3
                break;
            default: break;
            }
        }
});
```

这一技术在下一个技巧中也会用到。

## 技巧 13：使用隐式 Intent 创建 Activity

隐式 Intent 不需要指定要使用哪个组件。相反，它们通过过滤器指定所需的功能，而 Android 系统必须决定使用哪个组件是最佳选择。Intent 过滤器可以是动作（action）、数据（data）或者分类（category）。

最常用的 Intent 过滤器是动作，而其中最常用的要属 `ACTION_VIEW`。该模式需要指定一个统一资源标识符（URI），从而将数据显示给用户。它为给定的 URI 执行最合理的动作。比如，在下面的例子中，case 0、case 1、case 2 中的隐式 Intent 拥有相同的语法，却产生不同的结果。

下面是使用隐式 Intent 启动 Activity 的具体步骤。

（1）声明 Intent，同时指定合适的过滤器（如 `ACTION_VIEW`、`ACTION_WEB_SEARCH` 等）。

（2）为运行 Activity 所需的该 Intent 附加额外的信息。

（3）将该 Intent 传递给 `startActivity()` 方法。

**代码清单 2-16　src/com/cookbook/implicit_intents/ListActivityExample.java**

```
package com.cookbook.implicit_intents;

import android.app.ListActivity;
import android.app.SearchManager;
import android.content.Intent;
import android.net.Uri;
import android.os.Bundle;
import android.view.View;
import android.widget.AdapterView;
import android.widget.ArrayAdapter;
import android.widget.ListView;
import android.widget.AdapterView.OnItemClickListener;

public class ListActivityExample extends ListActivity {
    static final String[] ACTIVITY_CHOICES = new String[] {
        "Open Website Example",
        "Open Contacts",
        "Open Phone Dialer Example",
        "Search Google Example",
        "Start Voice Command"
    };
    final String searchTerms = "superman";
```

```
    protected void onCreate(Bundle savedInstanceState) {
        super.onCreate(savedInstanceState);

        setListAdapter(new ArrayAdapter<String>(this,
                android.R.layout.simple_list_item_1,
ACTIVITY_CHOICES));
        getListView().setChoiceMode(ListView.CHOICE_MODE_SINGLE);
        getListView().setTextFilterEnabled(true);
        getListView().setOnItemClickListener(new OnItemClickListener()
        {
            @Override
            public void onItemClick(AdapterView<?> arg0, View arg1,
                    int arg2, long arg3) {
                switch(arg2) {
                case 0: //opens web browser and navigates to given website
                    startActivity(new Intent(Intent.ACTION_VIEW,
                                         Uri.parse("http://www.android.com/")));
                    break;
                case 1: //opens contacts application to browse contacts
                    startActivity(new Intent(Intent.ACTION_VIEW,
                                         Uri.parse("content://contacts/people/")));
                    break;
                case 2: //opens phone dialer and fills in the given number
                    startActivity(new Intent(Intent.ACTION_VIEW,
                                         Uri.parse("tel:12125551212")));
                    break;
                case 3: //searches Google for the string
                    Intent intent= new Intent(Intent.ACTION_WEB_SEARCH);
                            intent.putExtra(SearchManager.QUERY, searchTerms);
                            startActivity(intent);
                    break;
                case 4: //starts the voice command
                    startActivity(new
                                        Intent(Intent.ACTION_VOICE_COMMAND));
                    break;
                default: break;
                }
            }
        });
    }
}
```

## 技巧 14：在 Activity 间传递基本数据类型

有时需要向某个启动的 Activity 传递数据，有时启动的 Activity 需要把其创建的数据传回给调用它的 Activity。例如，需要把游戏的最终得分返回给高分排行榜界面。以下是在 Activity 之间传递信息的几种不同方式。

- 在发起调用的 Activity 中声明相关变量（如 `public int finalScore`），并在启动的 Activity 中为其赋值（例如：`CallingActivity finalScore=score`）。
- 给 bundle 附加额外数据（在本技巧中有所体现）。
- 使用 Preference 属性存储数据，以备后面检索（将在第 6 章中介绍）。
- 使用 SQLite 数据库储存数据，以备后面检索（将在第 11 章中介绍）。

Bundle 是从字符串值到各种可打包（parcelable）类型的映射，可以通过向 Intent 添加额外数据创建它。本技巧显示了将数据从主 Activity 传递给启动的 Activity，在其中修改后再传递回来的全过程。

变量（本例中一个为 integer 型，另一个为 String 型）在 StartScreen Activity 中定义。在创建 Intent 调用 PlayGame 类时，通过 putExtra 方法把这两个变量附加给 Intent。当结果从启动的 Activity 中返回时，可借助 getExtras 方法读取变量值。以上调用过程如代码清单 2-17 所示。

**代码清单 2-17　src/com/cookbook/passing_data_activities/StartScreen.java**

```
package com.cookbook.passing_data_activities;

import android.app.Activity;
import android.content.Intent;
import android.os.Bundle;
import android.view.View;
import android.widget.Button;
import android.widget.TextView;

public class StartScreen extends Activity {
    private static final int PLAY_GAME = 1010;
    private TextView tv;
    private int meaningOfLife = 42;
    private String userName = "Douglas Adams";

    @Override
    public void onCreate(Bundle savedInstanceState) {
        super.onCreate(savedInstanceState);
        setContentView(R.layout.main);
        tv = (TextView) findViewById(R.id.startscreen_text);

        //Display initial values
        tv.setText(userName + ":" + meaningOfLife);

        //Set up button listener
        Button startButton = (Button) findViewById(R.id.play_game);
        startButton.setOnClickListener(new View.OnClickListener() {
            public void onClick(View view) {
                startGame();
            }
        });
            }

    @Override
    protected void onActivityResult(int requestCode,
            int resultCode, Intent data) {
        if (requestCode == PLAY_GAME && resultCode == RESULT_OK) {
            meaningOfLife = data.getExtras().getInt("returnInt");
            userName = data.getExtras().getString("returnStr");
            //Show it has changed
            tv.setText(userName + ":" + meaningOfLife);
        }
        super.onActivityResult(requestCode, resultCode, data);
    }

    private void startGame() {
        Intent launchGame = new Intent(this, PlayGame.class);

        //passing information to launched activity
        launchGame.putExtra("meaningOfLife", meaningOfLife);
```

```
        launchGame.putExtra("userName", userName);

        startActivityForResult(launchGame, PLAY_GAME);
    }
}
```

传入 PlayGame Activity 的变量可以用 getIntExtra 和 getStringExtra 读取。当该 Activity 结束并准备通过一个 Intent 返回时，可以用 putExtra 方法将数据传回给发起调用的 Activity。上述调用如清单 2-18 所示。

**代码清单 2-18 src/com/cookbook/passing_data_activities/PlayGame.java**

```
package com.cookbook.passing_data_activities;

import android.app.Activity;
import android.content.Intent;
import android.os.Bundle;
import android.view.View;
import android.widget.Button;
import android.widget.TextView;

public class PlayGame extends Activity {
    private TextView tv2;
    int answer;
    String author;

    public void onCreate(Bundle savedInstanceState) {
        super.onCreate(savedInstanceState);
        setContentView(R.layout.game);

        tv2 = (TextView) findViewById(R.id.game_text);

        //reading information passed to this activity
        //Get the intent that started this activity
        Intent i = getIntent();
        //returns -1 if not initialized by calling activity
        answer = i.getIntExtra("meaningOfLife", -1);
        //returns [] if not initialized by calling activity
        author = i.getStringExtra("userName");

        tv2.setText(author + ":" + answer);

        //Change values for an example of return
        answer = answer - 41;
        author = author + " Jr.";

        //Set up button listener
        Button startButton = (Button) findViewById(R.id.end_game);
        startButton.setOnClickListener(new View.OnClickListener() {
            public void onClick(View view) {
                //将信息返回给发起调用的activity
                Intent i = getIntent();
                i.putExtra("returnInt", answer);
                i.putExtra("returnStr", author);
                setResult(RESULT_OK, i);
                finish();
            }
        });
    }
}
```

# 第 3 章

# 线程、服务、接收器和警报

本章继续介绍 Android 应用程序的基本构建模块。首先介绍线程（thread）的显式定义，这是一种分离任务的方法。接着介绍服务（service）和广播接收器（broadcast receiver）。你将会从某些技巧中看到使用线程如何能给服务和广播接收器带来方便。随后，介绍应用程序微件（widget），它会用到接收器，从而引出对各种开发者可用的警报（alert）的探讨。

## 3.1 线程

每个应用程序在创建时都会默认运行单一的进程，其中包含所有的任务。为避免用户界面被挂起，那些耗时的任务，诸如网络下载或密集的计算，应当驻留于单独的后台线程中。如何适当地实现这种操作由开发者决定，但之后则是由 Android 系统基于开发者的实现来确定线程的优先级。

大多数应用程序都能通过使用线程来提升性能。如果用户界面的挂起在软件设计阶段没有被发觉，到了测试阶段就会迅速体现出来，因为 Android 系统会在用户界面挂起时弹出警告，如图 3-1 所示。

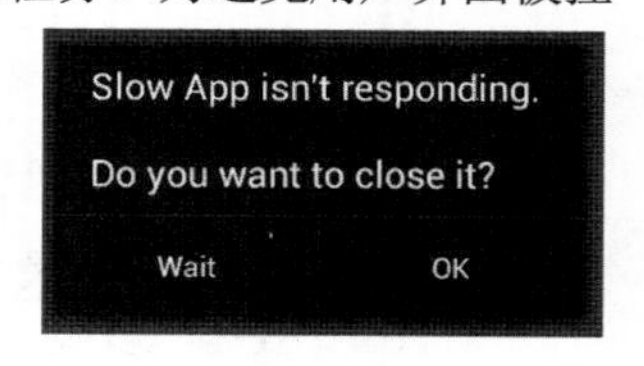


图 3-1　线程挂起时显示的消息示例

### 技巧 15：启动一个辅助线程

本技巧中，当屏幕上的一个按钮被按下时，就会播放一段铃声。这个例子为线程如何被用于处理耗时的操作提供了一个简单的例证。下面的代码调用 play_music() 函数时并未指定单独的进程，这将会导致播放音乐时阻塞应用程序：

```
Button startButton = (Button) findViewById(R.id.trigger);
startButton.setOnClickListener(new View.OnClickListener() {
    public void onClick(View view){
         // BAD USAGE: function call too time-consuming
         //   function causes main thread to hang
         play_music();
    }
});
```

这就意味着在音乐播放完之前，任何用户请求，比如返回主界面或多次按下屏幕上的按钮这样的操作将都不会被提交给系统。没有响应的用户界面甚至可能导致 Android 系统显示像图 3-1 所示那样的错误信息。

要想解决此问题，可以启动一个辅助进程来调用 `play_music()` 函数，步骤如下。

（1）创建一个新线程来容纳 `Runnable` 对象：

```
Thread initBkgdThread = new Thread(
    //Insert runnable object here
);
```

（2）创建一个 `Runnable` 对象，重写 `run()` 方法来调用耗时的任务：

```
new Runnable() {
  public void run() {
    play_music();
  }
}
```

（3）启动新线程，然后该线程会执行前述任务。

```
initBkgdThread.start();
```

包含耗时任务的辅助线程的设置工作很快就能完成，所以主线程可以继续为其他事件服务了。

在给出完整的 Activity 的代码之前，要先讨论一下支持文件。媒体播放将在第 8 章中更全面地涉及，但这里要说明，本例中用到的铃声音乐是用铃声文本传输语言（ringtone text transfer language，RTTTL）指定的一连串音符来实现的。例如，下面的 RTTTL 代码描述了一个紧邻中央 C 之下的 A（频率为 220 Hz）的四分音符。将该代码放在 **res/raw** 目录下的一个单行文本文件中，从而将其注册为名为 `R.raw.a4` 的资源。

```
a4:d=4,o=5,b=250:a4
```

接着，在 Activity 里调用媒体播放器来播放这一铃声音符：

```
m_mediaPlayer = MediaPlayer.create(this, R.raw.a4);
                m_mediaPlayer.start();
```

本技巧用到了 4 个不同的音符，分别存储在 4 个 RTTTL 文件里：**g4.rtttl**、**a4.rtttl**、**b4.rtttl** 以及 **c5.rtttl**。它们不过是前面的 `R.raw.a4` 的翻版，只需把其中的 `a4` 按照具体需求改成代表相应音符的符号就可以了，但它也可以被扩展为其他的音符或格式。

补充一点，`MediaPlayer` 会启动自己的后台线程来播放媒体。因此，对于单个较长文件来说，可以有办法避免使用显式的线程，这会在第 8 章中提及。当有多个文件需要被迅速播放时（比如本例），这个办法就不灵了，不过我们还是应该了解，线程并不总是必要的。

使音乐开始播放的触发器是按下按钮的操作。`Button` 微件需要在主布局文件（这里名为 **main.xml**）中指定，并用 `trigger` 这个名字来标识，如代码清单 3-1 所示。

**代码清单 3-1 res/layout/main.xml**

```
<?xml version="1.0" encoding="utf-8"?>
<LinearLayout xmlns:android="http://schemas.android.com/apk/res/android"
```

```
    android:orientation="vertical"
    android:layout_width="match_parent"
    android:layout_height="match_parent"
    >
  <Button android:id="@+id/trigger"
    android:layout_width="100dip" android:layout_height="100dip"
    android:text="Press Me"
  />
</LinearLayout>
```

启动单独的线程的一个副作用是，即使主 Activity 已经中断，该线程仍会继续执行。这一点可以通过实现后台线程，并在音乐播放期间返回到主屏幕这样的操作中看出来。音乐会持续播放，直到播完为止。如果不希望这样，可以在主 Activity 的 `onPause()` 函数中设置一个标记（本例中叫做 `paused`），让 `play_music()` 函数检查它，当主线程暂停时就停止播放音乐。

将前述的全部项目组合成完整的 `PressAndPlay` Activity，在代码清单 3-2 中给出。

**代码清单 3-2 src/com/cookbook/launch_thread/PressAndPlay.java**

```
package com.cookbook.launch_thread;

import android.app.Activity;
import android.media.MediaPlayer;
import android.os.Bundle;
import android.view.View;
import android.widget.Button;

public class PressAndPlay extends Activity {

    @Override
    public void onCreate(Bundle savedInstanceState) {
        super.onCreate(savedInstanceState);
        setContentView(R.layout.main);

        Button startButton = (Button) findViewById(R.id.trigger);
        startButton.setOnClickListener(new View.OnClickListener() {
            public void onClick(View view){

                //Stand-alone play_music() function call causes
                //main thread to hang. Instead, create
                //separate thread for time-consuming task.
                Thread initBkgdThread = new Thread(new Runnable() {
                    public void run() {
                        play_music();
                    }
                });
                initBkgdThread.start();
            }
        });
    }

    int[] notes = {R.raw.c5, R.raw.b4, R.raw.a4, R.raw.g4};
    int NOTE_DURATION = 400; //millisec
    MediaPlayer m_mediaPlayer;
    private void play_music() {
        for(int ii=0; ii<12; ii++) {
            //Check to ensure main activity not paused
            if(!paused) {
                if(m_mediaPlayer != null) {m_mediaPlayer.release();}
```

```
                m_mediaPlayer = MediaPlayer.create(this, notes[ii%4]);
                m_mediaPlayer.start();
                try {
                    Thread.sleep(NOTE_DURATION);
                } catch (InterruptedException e) {
                    e.printStackTrace();
                }
            }
        }
    }

    boolean paused = false;
    @Override
    protected void onPause() {
        paused = true;
        super.onPause();
    }
    @Override
    protected void onResume() {
        super.onResume();
        paused = false;
    }
}
```

注意，Thread.sleep()方法会让线程大致在指定的时长（以毫秒计）内暂停，用以实现音符的延响。

还要注意生命周期方法的使用惯例：针对特定 Activity 的附加逻辑应当在其父方法中用大括号括起来。这是一种能确保命令被正确执行的良好编程习惯。因此，将内部的暂停标记在 Activity 真正暂停之前设置为真，而 Activity 在内部暂停标记被设为假之前就已经完全恢复运行了。

## 技巧 16：创建实现 Runnable 接口的 Activity

这个技巧展示了一个用来为高计算密度函数求值（例如为图像做边缘检测）的 Activity。此处，一个名为 detectEdges()的虚拟函数被用来模拟真实的图像处理算法。

如果在 onCreate()方法中自行调用 detextEdges()方法，就会将主线程挂起。在未完成计算之前将不会显示用户界面布局。因此，需要为这一费时的功能创建并启动一个独立的线程。因为这个 Activity 的主要目的就是处理这一费时的操作，让该 Activity 自己实现 Runnable 接口就是一件很自然的事。如代码清单 3-3 中所示，后台线程在 onCreate()方法中被声明。后台线程启动时，会调用 Activity 的 run()方法，而该方法则按照我们的功能意图被重写。

按钮的实现与上一条技巧中的完全相同。按下按钮后会发现，在后台任务 detectEdges()运行的同时，用户界面仍然是有响应的。

**代码清单 3-3 src/com/cookbook/runnable_activity/EdgeDetection.java**

```
package com.cookbook.runnable_activity;

import android.app.Activity;
import android.os.Bundle;
import android.view.View;
import android.widget.Button;
```

```
import android.widget.TextView;

public class EdgeDetection extends Activity implements Runnable {
    int numberOfTimesPressed=0;

    @Override
    public void onCreate(Bundle savedInstanceState) {
         super.onCreate(savedInstanceState);
         setContentView(R.layout.main);
         final TextView tv = (TextView) findViewById(R.id.text);
         //In-place function call causes main thread to hang:
         /* detectEdges(); */
         //Instead, create background thread for time-consuming task
         Thread thread = new Thread(EdgeDetection.this);
         thread.start();

         Button startButton = (Button) findViewById(R.id.trigger);
         startButton.setOnClickListener(new View.OnClickListener() {
             public void onClick(View view){

                 tv.setText("Pressed button" + ++numberOfTimesPressed
                         + " times\nAnd computation loop at "
                         + "(" + xi + ", " + yi + ") pixels");

             }
         });
    }

    @Override
    public void run() {
         detectEdges();
    }

    //Edge Detection
    int xi, yi;
    private double detectEdges() {
         int x_pixels = 4000;
         int y_pixels = 3000;
         double image_transform=0;

         //Double loop over pixels for image processing
         //Meaningless hyperbolic cosine emulates time-consuming task
         for(xi=0; xi<x_pixels; xi++) {
              for(yi=0; yi<y_pixels; yi++) {
                   image_transform = Math.cosh(xi*yi/x_pixels/y_pixels);
              }
         }
         return image_transform;
    }
}
```

## 技巧 17：设置线程的优先级

Android 系统会处理线程的优先级。默认情况下，一个新线程，如 myThread，其优先级被定为 5。开发者可以在 myThread.start()执行前，通过调用 myThread.setPriority(priority)来为线程设定另外的优先级。优先级不能高于 Thread.MAX_PRIORITY（该值为 10）或者低于 Thread.MIN_PRIORITY（该值为 1）。

Android 还提供了另一种设定线程优先级的方法。通过 android.os.Process.setThread

Priority()可以请求一个基于“良好的”Linux 值（介于-20～20）的优先级。这两种方法最后都会映射到同一个底层系统调用，但 android.os.Process.setThreadPriority()可以设定的粒度更细。

## 技巧 18：取消线程

有时，当一个组件完成或被杀死时，开发者希望由它产生的线程也同样被杀死。例如在某个 Activity 中定义了线程：

```
private volatile Thread myThread;
```

myThread.stop()方法已经被弃用，因为它会将应用程序置于不可预知的状态。取而代之的是下面的方法，比如在父组件的 onStop()方法中加入：

```
//Use to stop the thread myThread
if(myThread != null) {
   Thread dummy = myThread;
   myThread = null;
   dummy.interrupt();
}
```

在应用程序层面上还有另外一种办法来完成相同的工作：使用 setDaemon(true)方法将所有生成的线程都声明为守护线程，如后面例子中所做的那样。这就可确保所有与应用程序关联的线程在应用程序的主线程终结时，也都随之被杀死。

```
//Use when initially starting a thread
myThread.setDaemon(true);
myThread.start();
```

最后，总是有办法在 run()方法中使用 while(stillRunning)循环，然后通过从外部设定 strillRunning=false 来杀死线程。然而，这样做可能无法有效地控制线程结束的时间。

## 技巧 19：在两个应用程序间共享线程

前几个技巧激发了我们在单个应用程序中使用多线程的积极性。相反的情形有时也会有用处，即让一个线程被多个应用程序所使用。例如，如果两个应用程序需要互相通信，它们可以通过用绑定器（binder）共享线程来达到目的，从而避免使用更为复杂的进程间通信（IPC）协议。实现的具体步骤如下。

- 出于安全考虑，首先要保证每个应用程序在打包发布时都使用相同的密钥进行签名。
- 确保每个应用程序都运行在相同的用户 ID 之下。这一点可通过在 **ActivityManifest.xml** 文件中声明相同的属性实现：android:sharedUserId="my.shared.userid"
- 将所有相关的 Activity 或组件声明为运行在相同的进程中。实现的办法是，在 **ActivityManifest.xml** 文件里为各个组件声明相同的属性：android:process="my.shared.process-name"

这样就保证了两个组件运行在相同的线程中，并且彼此间透明地共享信息。更复杂的情况，即权限不能被共享时的处理办法，将在第4章的技巧34中讲解。

## 3.2 线程间的消息机制：Handler

在多个线程（比如一个主应用程序线程和一个后台线程）同时运行之后，它们之间就需要一种相互通信的途径。下面是这类通信的例子。

- 主线程负责应对那些需要立即处理的信息，并给负责处理耗时任务的后台进程发送消息，对其进行更新。
- 一个大的计算任务完成，将结果传回给发起调用的线程。

这些都可以通过Handler来实现，它是一类在线程间传递消息的对象。每个Handler都被绑定在单一的线程上，为其传送消息，并执行它的命令。

### 技巧20：从主线程调度Runnable型的任务

这个技巧实现了在应用程序中会被屡屡用到的计时器。例如，游戏中会用它来追踪玩家通过一个关卡的时间。这为后台线程持续运行的同时处理用户交互，提供了一个简单的方法。

计时器运行在一个后台线程里，因此不会阻塞UI线程，但需要在时间变化后对UI进行更新。如代码清单3-4所示，ID为text的TextView一开始显示一条欢迎信息，而ID为trigger的按钮开始时值为"Press Me"。

**代码清单3-4 res/layout/main.xml**

```
<?xml version="1.0" encoding="utf-8"?>
<LinearLayout xmlns:android="http://schemas.android.com/apk/res/android"
     android:orientation="vertical"
     android:layout_width="match_parent"
     android:layout_height="match_parent"
     >
<TextView android:id="@+id/text"
     android:layout_width="match_parent"
     android:layout_height="wrap_content"
     android:text="@string/hello"
     />
<Button android:id="@+id/trigger"
     android:layout_width="100dip" android:layout_height="100dip"
     android:text="Press Me"
     />
</LinearLayout>
```

布局XML文件里的文本资源被用来与名为BackgroundTimer的Java Activity中的TextView变量相关联，所使用的初始化语句如下：

```
mTimeLabel = (TextView) findViewById(R.id.text);
mButtonLabel = (TextView) findViewById(R.id.trigger);
```

被Java识别后，文本可以在运行时被修改。当应用程序启动时，mUpdateTimeTask开启一个计时器，并将 mTimeLabel 的文本重写为几分几秒格式的新时间。按钮被按下之后，它的 onClick()方法会对 mButtonLabel 的值进行重写，重写的内容是按钮被按下的次数。

名为 mHandler 的 Handler 被创建并用于对 mUpdateTimeTask 这一 runnable 对象进行排队。它首先在 onCreate()方法中被调用，然后在 mUpdateTimeTask 中以每 200 ms 一次的频率递归调用自身，以持续地对时间进行更新。往往我们会使用这种方法确保每秒一次地更新时间，并且没有过度的任务调用开销。完整的 Activity 如代码清单 3-5 所示。

**代码清单 3-5　src/com/cookbook/background_timer/BackgroundTimer.java**

```
package com.cookbook.background_timer;

import android.app.Activity;
import android.os.Bundle;
import android.os.Handler;
import android.os.SystemClock;
import android.view.View;
import android.widget.Button;
import android.widget.TextView;

public class BackgroundTimer extends Activity {
    //Keep track of button presses, a main thread task
    private int buttonPress=0;
    TextView mButtonLabel;

    //counter of time since app started, a background task
    private long mStartTime = 0L;
    private TextView mTimeLabel;

    //handler to handle the message to the timer task
    private Handler mHandler = new Handler();

    @Override
    public void onCreate(Bundle savedInstanceState) {
        super.onCreate(savedInstanceState);
        setContentView(R.layout.main);
        if (mStartTime == 0L) {
            mStartTime = SystemClock.uptimeMillis();
            mHandler.removeCallbacks(mUpdateTimeTask);
            mHandler.postDelayed(mUpdateTimeTask, 100);
        }

        mTimeLabel = (TextView) findViewById(R.id.text);
        mButtonLabel = (TextView) findViewById(R.id.trigger);

        Button startButton = (Button) findViewById(R.id.trigger);
        startButton.setOnClickListener(new View.OnClickListener() {
            public void onClick(View view){
                mButtonLabel.setText("Pressed " + ++buttonPress
                                                  + " times");
            }
        });
    }

    private Runnable mUpdateTimeTask = new Runnable() {
        public void run() {
            final long start = mStartTime;
            long millis = SystemClock.uptimeMillis() - start;
            int seconds = (int) (millis / 1000);
```

```
                int minutes = seconds / 60;
                seconds     = seconds % 60;

                mTimeLabel.setText("" + minutes + ":"
                                      + String.format("%02d",seconds));
                mHandler.postDelayed(this, 200);
            }
        };

        @Override
        protected void onPause() {
             mHandler.removeCallbacks(mUpdateTimeTask);
             super.onPause();
        }

        @Override
        protected void onResume() {
             super.onResume();
             mHandler.postDelayed(mUpdateTimeTask, 100);
        }
    }
```

## 技巧21：使用倒数计时器

上个技巧是Handler和计时器的一个例子。在系统的内建类CountDownTimer中还提供了另外一种计时器，它将创建后台线程及对Handler的排队操作封装为一个便于使用的类调用。

倒数计时器要用到两个参数：倒计时完成前要经过的毫秒数，以及以毫秒计的处理onTick()回调的频率。onTick()方法用来更新倒计时文本。注意本技巧剩下的部分都与上个技巧相同。完整的代码如代码清单3-6所示。

**代码清单3-6　src/com/cookbook/countdown/CountDownTimerExample.java**

```
package com.cookbook.countdown;

import android.app.Activity;
import android.os.Bundle;
import android.os.CountDownTimer;
import android.view.View;
import android.widget.Button;
import android.widget.TextView;

public class CountDownTimerExample extends Activity {
     //Keep track of button presses, a main thread task
     private int buttonPress=0;
     TextView mButtonLabel;

     //countdown timer, a background task
     private TextView mTimeLabel;

     @Override
     public void onCreate(Bundle savedInstanceState) {
          super.onCreate(savedInstanceState);
          setContentView(R.layout.main);

          mTimeLabel = (TextView) findViewById(R.id.text);
          mButtonLabel = (TextView) findViewById(R.id.trigger);
```

```
        new CountDownTimer(30000, 1000) {
            public void onTick(long millisUntilFinished) {
                mTimeLabel.setText("seconds remaining: "
                        + millisUntilFinished / 1000);
            }
            public void onFinish() {
                mTimeLabel.setText("done!");
            }
        }.start();

        Button startButton = (Button) findViewById(R.id.trigger);
        startButton.setOnClickListener(new View.OnClickListener() {
            public void onClick(View view){
              mButtonLabel.setText("Pressed " + ++buttonPress + " times");
            }
        });
    }
}
```

## 技巧 22：处理耗时的初始化工作

当应用程序启动时，常要运行耗时的初始化工作，本技巧对此给出了一种解决办法。程序运行之初，布局上会持续显示一个特定的“Loading...”画面，该画面在 **loading.xml** 中被指定。本例中它不过是条简单的文本信息，但也可以被替换为公司的 logo 或介绍性的动画。

**代码清单 3-7 res/layout/loading.xml**

```
<?xml version="1.0" encoding="utf-8"?>
<LinearLayout
  xmlns:android="http://schemas.android.com/apk/res/android"
  android:layout_width="wrap_content"
  android:layout_height="wrap_content">
  <TextView android:id="@+id/loading"
     android:layout_width="match_parent"
     android:layout_height="wrap_content"
     android:text="Loading..."
     />
</LinearLayout>
```

在显示该布局的同时，需要花一段时间才能完成的 `initializeArrays()` 函数也在后台线程中被启动，以避免用户界面被挂起。初始化过程使用静态变量，来确保屏幕切换或当 Activity 的另一个实例启动时，不需要对数据进行重复计算。

初始化过程完成时，会给 `mHandler` 发送一条消息。鉴于所需要的信息仅仅是发送消息这一动作本身，所以我们用 `mHandler.sendEmptyMessage(0)` 来发送一条空消息。

收到消息后，UI 线程会执行 `handleMessage()` 方法。该方法被重写，以实现初始化过程开始后 Activity 能继续运行这一功能，本例中具体而言就是设置 **main.xml** 布局文件所指定的主屏幕。完整的 Activity 如代码清单 3-8 所示。

**代码清单 3-8 src/com/cookbook/handle_message/HandleMessage.java**

```
package com.cookbook.handle_message;

import android.app.Activity;
```

```
import android.os.Bundle;
import android.os.Handler;
import android.os.Message;

public class HandleMessage extends Activity implements Runnable {

    @Override
    public void onCreate(Bundle savedInstanceState) {
        super.onCreate(savedInstanceState);
        setContentView(R.layout.loading);

        Thread thread = new Thread(this);
        thread.start();
    }

    private Handler mHandler = new Handler() {
        public void handleMessage(Message msg) {
            setContentView(R.layout.main);
        }
    };

    public void run(){
        initializeArrays();
        mHandler.sendEmptyMessage(0);
    }

    final static int NUM_SAMPS = 1000;
    static double[][] correlation;
    void initializeArrays() {
        if(correlation!=null) return;

        correlation = new double[NUM_SAMPS][NUM_SAMPS];
        //calculation
        for(int k=0; k<NUM_SAMPS; k++) {
            for(int m=0; m<NUM_SAMPS; m++) {
                correlation[k][m] = Math.cos(2*Math.PI*(k+m)/1000);
            }
        }
    }
}
```

# 3.3 警报

警报在应用程序的主用户界面之外为用户提供了快速的消息。警报可以放在像 Toast 或 AlertDialog 这样叠加的窗口里，也可以位于屏幕顶部的通知栏中。Toast 警报仅用一行代码就可在屏幕上产生一条输出消息，而并不需要修改布局文件。因此，它对于调试工作也算是一种便利的工具，与 C 语言程序中的 printf 语句等价。

## 技巧 23：利用 Toast 在屏幕上显示一条简单的信息

在上一章里我们已经介绍过 Toast 方法的一种简洁形式：

```
Toast.makeText(this, "text", Toast.LENGTH_SHORT).show();
```

它也可以写成多行命令的形式：

```
Toast tst = Toast.makeText(this, "text", Toast.LENGTH_SHORT);
        tst.show();
```

当文本需要多次输出时，这个形式就比较有用，因为第一行代码中的实例可以被反复调用。

另两个用到多行形式的 Toast 命令的情形是：当需要改变文本位置的时候，或者需要添加图像的时候。要改变文本位置，或者要使 Toast 在屏幕上居中，可在调用 show() 方法之前使用 setGravity 方法：

```
tst.setGravity(Gravity.CENTER, tst.getXOffset() / 2,
                               tst.getYOffset() / 2);
```

要给 Toast 添加图像，可使用下面的代码：

```
Toast tst = Toast.makeText(this, "text", Toast.LENGTH_LONG);
ImageView view = new ImageView(this);
view.setImageResource(R.drawable.my_figure);
tst.setView(view);
tst.show();
```

## 技巧 24：使用 AlertDialog 对话框

如果要给用户提供一条警报，且所需的动作按钮不超过三个，就可以利用 AlertDialog 类来实现，例如：

- “您的最终得分为 90/100，重玩本关卡，进入下一关卡，或返回主菜单。”
- “图像已损坏。请另选图像或取消操作。”

本技巧实现了前一种情形，并展示了如何依据用户点选的按钮来执行相应的操作。代码展示在代码清单 3-9 中。

使用 create() 方法来初始化 AlertDialog；用 setMessage() 方法来指定文本；用 setButton() 方法来指定按钮上的文字以及按钮对应的动作；最后，用 show() 方法让对话框显示在屏幕上。请注意，清单中几个 onClick() 回调函数中的逻辑仅仅给出了如何指定按钮动作的一个例子。

**代码清单 3-9　用 AlertDialog 来提供动作选项**

```
AlertDialog dialog = new AlertDialog.Builder(this).create();

dialog.setMessage("Your final score: " + mScore + "/" + PERFECT_SCORE);
dialog.setButton(DialogInterface.BUTTON_POSITIVE, "Try this level again",
        new DialogInterface.OnClickListener() {
            public void onClick(DialogInterface dialog, int which) {
                mScore = 0;
                start_level();
            }
        });
dialog.setButton(DialogInterface.BUTTON_NEGATIVE, "Advance to next level",
        new DialogInterface.OnClickListener() {
            public void onClick(DialogInterface dialog, int which) {
                mLevel++;
                start_level();
            }
        });
dialog.setButton(DialogInterface.BUTTON_NEUTRAL, "Back to the main menu",
        new DialogInterface.OnClickListener() {
            public void onClick(DialogInterface dialog, int which) {
                mLevel = 0;
```

```
                finish();
            }
        });
    dialog.show();
```

以上代码会生成如图3-2所示的弹出式对话框。注意其中的按钮会按照BUTTON_POSITIVE、BUTTON_NEUTRAL、BUTTON_NEGATIVE的顺序显示。如果需要只含一个或两个选项的对话框，就不要指定全部三种选项。

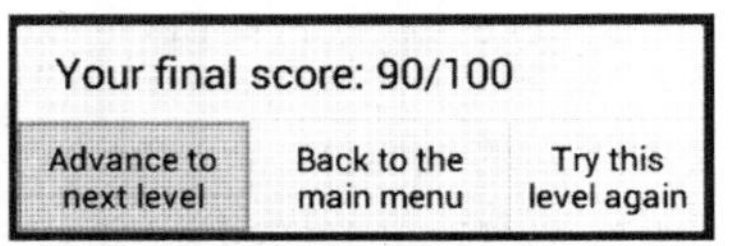


图3-2　带三个选项的警报对话框

## 技巧25：在状态栏中显示通知

设备屏幕上部的状态栏可以显示待决定的通知，用户可随时阅读它们。一般而言，因为Activity通常能与用户交互，所以相比之下，服务会较多地用到状态栏通知。为了最佳的用户体验，通知应当尽可能简洁，这是开发者应遵循的规则。

下面是创建状态栏通知的步骤。

（1）声明一条通知，并为其指定在状态栏上的显示方式：

```
String ns = Context.NOTIFICATION_SERVICE;
mNManager = (NotificationManager) getSystemService(ns);
final Notification msg = new Notification(R.drawable.icon,
        "New event of importance",
        System.currentTimeMillis());
```

（2）定义当状态栏展开时通知的外观，以及通知被点击后的动作（该动作通过PendingIntent类来定义）：

```
Context context = getApplicationContext();
CharSequence contentTitle = "ShowNotification Example";
CharSequence contentText = "Browse Android Cookbook Site";
Intent msgIntent = new Intent(Intent.ACTION_VIEW,
                    Uri.parse("http://www.pearson.com"));
PendingIntent intent =
            PendingIntent.getActivity(ShowNotification.this,
                    0, msgIntent,
                    Intent.FLAG_ACTIVITY_NEW_TASK);
```

（3）添加其他配置信息，比如当通知被选择后，是否要让LED灯闪烁，或播放声音，或自动将通知消除。这里给出后两种操作的代码：

```
msg.defaults |= Notification.DEFAULT_SOUND;
msg.flags |= Notification.FLAG_AUTO_CANCEL;
```

（4）为系统设定通知事件的相关信息：

```
msg.setLatestEventInfo(context, contentTitle, contentText, intent);
```

（5）当相应事件在系统中发生时，通过唯一的标识符触发通知：

```
mNManager.notify(NOTIFY_ID, msg);
```

（6）通知完毕后，如需要可使用同一个标识符来清除通知。

如果通知信息被改变，应当更新通知，而不是发送另一条通知。可以通过更新第 2 步中的有关信息并再次调用 `setLatestEventInfo` 来实现这一点。代码清单 3-10 给出了一个显示通知的 Activity 范例。

**代码清单 3-10 src/com/cookbook/show_notification/ShowNotification.java**

```
package com.cookbook.show_notification;

import android.app.Activity;
import android.app.Notification;
import android.app.NotificationManager;
import android.app.PendingIntent;
import android.content.Context;
import android.content.Intent;
import android.net.Uri;
import android.os.Bundle;
import android.view.View;
import android.view.View.OnClickListener;
import android.widget.Button;

public class ShowNotification extends Activity {

    private NotificationManager mNManager;
    private static final int NOTIFY_ID=1100;

    /** called when the activity is first created */
    @Override
    public void onCreate(Bundle savedInstanceState) {
        super.onCreate(savedInstanceState);
        setContentView(R.layout.main);

        String ns = Context.NOTIFICATION_SERVICE;
        mNManager = (NotificationManager) getSystemService(ns);
        final Notification msg = new Notification(R.drawable.icon,
                "New event of importance",
                System.currentTimeMillis());

        Button start = (Button)findViewById(R.id.start);
        Button cancel = (Button)findViewById(R.id.cancel);

        start.setOnClickListener(new OnClickListener() {
            public void onClick(View v) {
                Context context = getApplicationContext();
                CharSequence contentTitle = "ShowNotification Example";
                CharSequence contentText = "Browse Android Cookbook Site";
                Intent msgIntent = new Intent(Intent.ACTION_VIEW,
                        Uri.parse("http://www.pearson.com"));
                PendingIntent intent =
                    PendingIntent.getActivity(ShowNotification.this,
                            0, msgIntent,
                            Intent.FLAG_ACTIVITY_NEW_TASK);

                msg.defaults |= Notification.DEFAULT_SOUND;
                msg.flags |= Notification.FLAG_AUTO_CANCEL;

                msg.setLatestEventInfo(context,
                        contentTitle, contentText, intent);
                mNManager.notify(NOTIFY_ID, msg);
```

```
            }
        });

        cancel.setOnClickListener(new OnClickListener() {
            public void onClick(View v) {
                mNManager.cancel(NOTIFY_ID);
            }
        });
    }
}
```

Android 4.1 引入了新的通知风格，并提供了基于一种构建器模式[①]的 API 来创建它们。要使用这些通知风格，推荐使用 `NotificationCompat` API。要使用该 API，需要将 **android-support-v4.jar** 添加到项目的 **"/libs/"** 文件夹中。4 种新的风格分别是：大文本、大图片、收件箱风格以及可带有进度条的通知。这几种通知依旧需要拥有小图标、标题和内容文本。大文本和大图片风格十分近似（就是在内容里添加文本或图片对象）。代码清单 3-11 给出大图片风格的示例。

**代码清单 3-11　一个大图片风格的通知**

```
        Button startBigPic = (Button)findViewById(R.id.startBigPic);
        Button stopBigPic = (Button)findViewById(R.id.stopBigPic);

        startBigPic.setOnClickListener(new OnClickListener() {
            public void onClick(View v) {
                Context context = getApplicationContext();
                CharSequence contentTitle = "Show Big Notification Example";
                CharSequence contentText = "Browse Android Cookbook Site";

                Intent msgIntent = new Intent(Intent.ACTION_VIEW,
                        Uri.parse("http://www.pearson.com"));

                PendingIntent intent =
                    PendingIntent.getActivity(ShowNotification.this,
                            0, msgIntent,
                            Intent.FLAG_ACTIVITY_NEW_TASK);

                NotificationCompat.Builder builder =
                    new NotificationCompat.Builder(context);
                builder.setSmallIcon(R.drawable.icon);
                builder.setContentTitle(contentTitle);
                builder.setContentText(contentText);
                builder.setContentIntent(intent);

                NotificationCompat.BigPictureStyle pictureStyle = new
                    NotificationCompat.BigPictureStyle();
               Bitmap bigPicture= BitmapFactory.decodeResource(getResources(),
                    R.drawable.bigpicture);
                pictureStyle.bigPicture(bigPicture);

                builder.setStyle(pictureStyle);

                mNManager.notify(NOTIFY_ID+1,builder.build());
            }
        });
        stopBigPic.setOnClickListener(new OnClickListener() {
            public void onClick(View v) {
                mNManager.cancel(NOTIFY_ID+1);
```

① 构建器模式（builder pattern），或译为生成器模式，是一种经典的设计模式。——译者注

```
            }
        });
```

其中，Intent、标题和内容都与代码清单 3-10 中的通知相同。这里获取了一个新的构建器实例，并通过调用 builder.setSmallIcon(..)和 builder.setContextXX(..)方法来为文本、标题和 Intent 设置必需的信息。需要提供给 NotificationCompact.BigPictureStyle 类一个位图对象，该位图可以通过 BitmapFactory.decodeResource (getResources(), R.drawable.bigpicture)语句从 drawable 文件夹中获取。对 builder.setStyle (pictureStyle);的调用可确保图像在显示通知的同时被显示出来。

收件箱风格会显示与电子邮箱的收件箱类似的、由若干行文本组成的列表。显示通知的过程与上例类似，但有两点不同：一是要把 NotificationCompat.InboxStyle 的一个实例提供给构建器；二是要将各行文本添加到 InboxStyle 对象上，这点通过基于 CharSequence 参数调用 inboxStyle.addline(..)来实现。具体代码如代码清单 3-12 所示。

**代码清单 3-12　收件箱风格的通知**

```
    Button startInbox = (Button)findViewById(R.id.startInbox);
     Button stopInbox = (Button)findViewById(R.id.stopInbox);

     startInbox.setOnClickListener(new OnClickListener() {
          public void onClick(View v) {

               Context context = getApplicationContext();
               CharSequence contentTitle = "Show Big Notification Example";
               CharSequence contentText = "Browse Android Cookbook Site";

               Intent msgIntent = new Intent(Intent.ACTION_VIEW,
                        Uri.parse("http://www.pearson.com"));
               PendingIntent intent =
                    PendingIntent.getActivity(ShowNotification.this,
                            0, msgIntent,
                            Intent.FLAG_ACTIVITY_NEW_TASK);

               NotificationCompat.Builder builder =
                    new NotificationCompat.Builder(context);
               builder.setSmallIcon(R.drawable.icon);
               builder.setContentTitle(contentTitle);
               builder.setContentText(contentText);

               NotificationCompat.InboxStyle inboxStyle =
                   new NotificationCompat.InboxStyle();

               for(int i=0;i<4;i++){
                    inboxStyle.addLine("subevent #"+i);
               }

               builder.setStyle(inboxStyle);
               builder.setContentIntent(intent);

               mNManager.notify(NOTIFY_ID+2,builder.build());
          }
    });

    stopInbox.setOnClickListener(new OnClickListener() {
        public void onClick(View v) {
            mNManager.cancel(NOTIFY_ID+2);
        }
    });
```

## 3.4 服务

服务是运行在后台且不与用户发生任何交互的 Android 组件。它可以被任意其他组件开启或停止。当其处于运行状态时，任何组件都可与它进行绑定。服务也可以自行停止。下面给出几个概念性场景，帮助理解什么是服务。

- Activity 让用户可以选择一组音乐文件，然后启动一个服务来播放这些文件。播放期间，一个新 Activity 被启动并绑定到这个服务之上，允许用户变更乐曲或停止播放。
- Activity 启动一个服务向网站上传一组图片。一个新的 Activity 被启动并绑定到服务上，以确定当前正在上传的文件，并把该图片显示到屏幕上。
- 广播接收器收到一条消息，消息表明用户拍摄了一张图片。随即启动一个服务以将该图片上传到某网站上。然后广播接收器转入非活动状态，并最终被杀死以回收内存，但服务会继续运行，直至上传完成。最后，服务自动停止。

通常情况下服务的生命周期如图 3-3 所示。

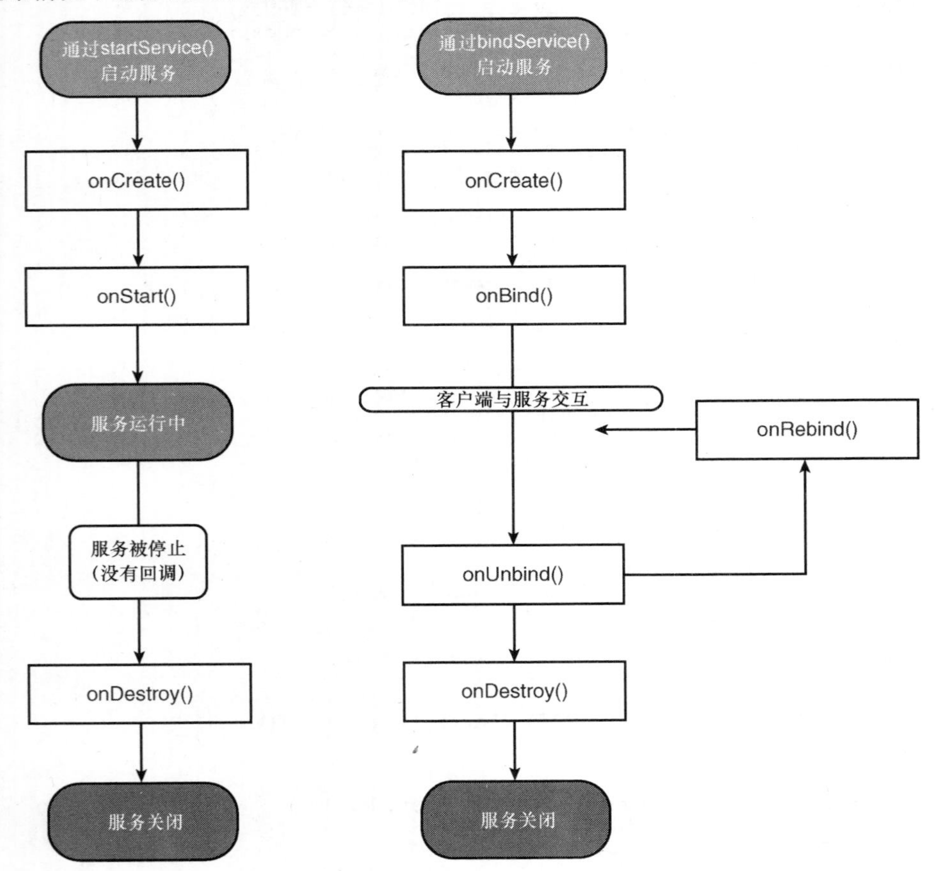


图 3-3　服务的生命周期（http://developer.android.com/）

再对第 3 个场景进行一点补充说明：当组件被杀死时，运行在组件中的后台任务也同样会被杀死。因此，那些在组件停止后还要继续运行的任务需要通过启动服务来完成。这确保了操作系统能意识到仍有进程运行主动的工作。

所有的服务都继承了 Service 这一抽象类或者它的某个子类。所有服务的入口都是 onCreate()方法，这与 Activity 类似。没有暂停服务这个概念，但可以通过调用 onDestory()方法来停止服务。

## 技巧 26：创建自足式服务

下面是创建关联于单个组件的自足式（self-contained）服务的步骤。

（1）创建一个扩展了 Service 的类。（在 Eclipse 里可以这样做：在项目上右击，选择 **New→Class**，并指定 android.app.Service 作为父类。）

（2）通过对 **AndroidManifest.xml** 文件进行如下改变，来对服务进行声明（如果在 Eclipse 里采取了上一步，则本步会被自动完成）：

```
<service android:name=".myService"></service>
```

（3）重写 onCreate()和 onDestory()方法。（在 Eclipse 里，可以在类文件上右击，选择 **Source→Override/Implement Methods...**，并选中 onCreate()和 onDestory()两个方法。）这两个方法会包含服务启动和停止时的功能。

（4）重写 onBind()方法，用于处理当服务被创建后，有新组件与之绑定的情况。

（5）通过一个外部触发器来激活服务。服务不能自启动，而需要被其他组件激活，或者用其他方法触发。比如，某组件可以创建一个 Intent，使用 startService()或 stopService()方法来启动或停止服务。

为进一步阐明前述过程，在代码清单 3-13 里给出了一个简单的服务的例子，该服务使用了本章首个技巧（技巧 15）里的 play_music()函数。要注意以下几点。

- 用一个 Toast 来显示服务启动和停止。
- onBind()方法虽然被重写，但并未被使用（可以根据需要扩展这个方法）。
- 仍需创建一个线程来播放音乐，以避免 UI 被阻塞。
- 当 Activity 被销毁（例如，通过改变屏幕方向）或被暂停（例如按下 Home 键），服务并不随之结束。这表明服务虽然被 Activity 开启，但却是作为一个独立实体在运行。

**代码清单 3-13 src/com/cookbook/simple_service/SimpleService.java**

```
package com.cookbook.simple_service;

import android.app.Service;
import android.content.Intent;
import android.media.MediaPlayer;
import android.os.IBinder;
import android.widget.Toast;

public class SimpleService extends Service {
```

```
    @Override
    public IBinder onBind(Intent arg0) {
          return null;
    }

    boolean paused = false;

    @Override
    public void onCreate() {
          super.onCreate();
          Toast.makeText(this,"Service created ...",
                              Toast.LENGTH_LONG).show();
          paused = false;
          Thread initBkgdThread = new Thread(new Runnable() {
              public void run() {
                   play_music();
              }
          });
          initBkgdThread.start();
    }

    @Override
    public void onDestroy() {
          super.onDestroy();
          Toast.makeText(this, "Service destroyed ...",
                              Toast.LENGTH_LONG).show();
          paused = true;
    }

    int[] notes = {R.raw.c5, R.raw.b4, R.raw.a4, R.raw.g4};
    int NOTE_DURATION = 400; //millisec
    MediaPlayer m_mediaPlayer;
    private void play_music() {
         for(int ii=0; ii<12; ii++) {
              //Check to ensure main activity not paused
              if(!paused) {
                   if(m_mediaPlayer != null) {m_mediaPlayer.release();}
                   m_mediaPlayer = MediaPlayer.create(this, notes[ii%4]);
                   m_mediaPlayer.start();
                   try {
                        Thread.sleep(NOTE_DURATION);
                   } catch (InterruptedException e) {
                        e.printStackTrace();
                   }
              }
         }
    }
}
```

Activity 和服务二者都在 **AndroidManifest.xml** 文件中声明，如代码清单 3-14 所示。

### 代码清单 3-14　AndroidManifest.xml

```
<?xml version="1.0" encoding="utf-8"?>
<manifest xmlns:android="http://schemas.android.com/apk/res/android"
          package="com.cookbook.simple_service"
          android:versionCode="1"
          android:versionName="1.0">
    <application android:icon="@drawable/icon"
                 android:label="@string/app_name">
        <activity android:name=".SimpleActivity"
```

```
                android:label="@string/app_name">
            <intent-filter>
                <action android:name="android.intent.action.MAIN" />
                <category android:name="android.intent.category.LAUNCHER" />
            </intent-filter>
        </activity>
        <service android:name=".SimpleService"></service>
    </application>
    <uses-sdk android:minSdkVersion="3" />
</manifest>
```

用于设置触发服务的开启及关闭过程的 UI 的示例 Activity 在代码清单 3-15 中给出，而与之关联的布局文件则如代码清单 3-16 所示，其中定义了两个按钮。

**代码清单 3-15 src/com/cookbook/simple_service/SimpleActivity.java**

```
package com.cookbook.simple_service;

import android.app.Activity;
import android.content.Intent;
import android.os.Bundle;
import android.view.View;
import android.widget.Button;

public class SimpleActivity extends Activity {
    @Override
    protected void onCreate(Bundle savedInstanceState) {
        super.onCreate(savedInstanceState);
        setContentView(R.layout.main);

        Button startButton = (Button) findViewById(R.id.Button01);
        startButton.setOnClickListener(new View.OnClickListener() {
            public void onClick(View view){
                startService(new Intent(SimpleActivity.this,
                                        SimpleService.class));
            }
        });

        Button stopButton = (Button)findViewById(R.id.Button02);
        stopButton.setOnClickListener(new View.OnClickListener() {
            public void onClick(View v){
                stopService(new Intent(SimpleActivity.this,
                                       SimpleService.class));
            }
        });
    }
}
```

**代码清单 3-16 res/layout/main.xml**

```
<?xml version="1.0" encoding="utf-8"?>
<LinearLayout xmlns:android="http://schemas.android.com/apk/res/android"
    android:orientation="vertical"
    android:layout_width="match_parent"
    android:layout_height="match_parent">
  <TextView
    android:layout_width="match_parent"
    android:layout_height="wrap_content"
    android:text="@string/hello"
```

```
    />
  <Button android:text="Do it" android:id="@+id/Button01"
    android:layout_width="wrap_content"
    android:layout_height="wrap_content"></Button>
  <Button android:text="Stop it" android:id="@+id/Button02"
    android:layout_width="wrap_content"
    android:layout_height="wrap_content"></Button>
</LinearLayout>
```

## 技巧 27：添加唤醒锁

当用户按下电源键，或设备在一定时间内未被使用时，屏幕会关闭，设备会进入待机状态。待机状态下，大多数进程会关闭或被取消，而处理器会进入休眠模式，以节省宝贵的电池电量。尊重这一习惯，不要通过阻止待机来消耗用户的电池是有益之举。话虽如此，但依然存在在屏幕关闭时需要保持程序运行的情况，音乐播放就是这样的典型例子。要想让应用在屏幕关闭时仍然运行，需要设置一个唤醒锁（WakeLock）。本技巧沿用前一技巧的服务，并为其添加唤醒锁，使得当设备的电源键被按下以后，音乐能继续播放。表 3-1 显示了可供使用的唤醒锁类型。

**表 3-1　可用的唤醒锁的比较**

| 唤醒锁类型 | CPU | 屏　幕 | 物理键盘 |
|---|---|---|---|
| PARTIAL_WAKE_LOCK | 开启 | 关闭 | 关闭 |
| SCREEN_DIM_WAKE_LOCK | 开启 | 变暗 | 关闭 |
| SCREEN_BRIGHT_WAKE_LOCK | 开启 | 明亮 | 关闭 |
| FULL_WAKE_LOCK | 开启 | 明亮 | 明亮 |

有两种主要的唤醒锁类型：一种是部分（partial）唤醒锁，允许屏幕关闭，但确保应用仍然在运行；另一种是完全（full）唤醒锁，即便电源键被按下，也会让屏幕和键盘保持开启。部分唤醒锁有两种子类型，可以提供对屏幕更细粒度的控制。本技巧采用部分唤醒锁，因为在音频播放时屏幕并不需要开启。要使用唤醒锁，需要一个特殊的唤醒锁许可，它被添加到代码清单 3-17 所示的 **AndroidManifest.xml** 文件中。

**代码清单 3-17　AndroidManifest.xml**

```
<?xml version="1.0" encoding="utf-8"?>
<manifest xmlns:android="http://schemas.android.com/apk/res/android"
          package="com.cookbook.simple_service"
          android:versionCode="1"
          android:versionName="1.0">

        <uses-permission android:name="android.permission.WAKE_LOCK"/>

        <application android:icon="@drawable/icon"
                  android:label="@string/app_name">

        <activity android:name=".SimpleActivity"
                  android:label="@string/app_name">
```

```
            <intent-filter>
                <action android:name="android.intent.action.MAIN" />
                <category android:name="android.intent.category.LAUNCHER" />
            </intent-filter>
        </activity>
        <service android:name=".SimpleService"></service>
    </application>
    <uses-sdk android:minSdkVersion="3" />
</manifest>
```

要获取一个唤醒锁，需要访问 PowerManager 类。该类是一个系统服务，可以通过调用 Context.getSystemService(Context.PowerService)来检索到它。要创建一个新唤醒锁实例，可执行下面的调用：

```
powerManager.mWakeLock =
    powerManager.newWakeLock(PowerManager.PARTIAL_WAKE_LOCK, LOG_TAG);
```

此处，所需的所有唤醒锁的属性都由通过 OR 操作符连接的第一个参数中的标志来指定；第二个参数则是赋予唤醒锁的名字标签，通常是服务或 Activity 的日志标签。任何时候需激活唤醒锁，只要调用 mWakelock.acquire()即可。

当不再需要唤醒锁时，应当显式地将其释放。具体方法是利用 mWakelock.isHeld()来检查它是否仍被保持。如果仍被保持，就调用 mWakelock.release();来释放它。唤醒锁被释放以后，设备就可以像通常那样在屏幕关闭后转入休眠了。要注意的一点是，少数情况下，当不止一个唤醒锁被使用时，唤醒锁释放的顺序应当与获取它们的顺序相反。上述内容体现在代码清单 3-18 中。

**代码清单 3-18　一个带唤醒锁的简单服务**

```
public class SimpleService extends Service {
    private static final String LOG_TAG = SimpleService.class.getSimpleName();

        @Override
    public IBinder onBind(Intent arg0) {
            return null;
    }

    boolean paused = false;
        private WakeLock mWakeLock=null;

    @Override
    public void onCreate() {
            super.onCreate();
            Toast.makeText(this,"Service created ...", Toast.LENGTH_LONG).show();
            setWakeLock();
            paused = false;
            Thread initBkgdThread = new Thread(new Runnable() {
                public void run() {
                    play_music();
                }
            });
            initBkgdThread.start();
    }

    private void setWakeLock() {
      PowerManager powerManager =
          (PowerManager)getSystemService(Context.POWER_SERVICE);
     mWakeLock=powerManager.newWakeLock(PowerManager.PARTIAL_WAKE_LOCK, LOG_TAG);
```

```
    mWakeLock.acquire();
    }

    @Override
public void onDestroy() {
        super.onDestroy();
        releaseWakeLock();
        Toast.makeText(this, "Service destroyed ...", Toast.LENGTH_LONG).show();
        paused = true;
}

private void releaseWakeLock() {
     if(mWakeLock!=null && mWakeLock.isHeld())
     {
     mWakeLock.release();
     }
     }

     int[] notes = {R.raw.c5, R.raw.b4, R.raw.a4, R.raw.g4};
int NOTE_DURATION = 400; //millisec
MediaPlayer m_mediaPlayer;
private void play_music() {
     for(int ii=0; ii<12; ii++) {
          //Check to ensure main activity not paused
          if(!paused) {
               if(m_mediaPlayer != null) {m_mediaPlayer.release();}
               m_mediaPlayer = MediaPlayer.create(this, notes[ii%4]);
               m_mediaPlayer.start();
               try {
                    Thread.sleep(NOTE_DURATION);
               } catch (InterruptedException e) {
                    e.printStackTrace();
               }
          }
     }
}
}
```

在代码清单3-18中，唤醒锁分别由两个函数来处理，彼此分工明确。setWakeLock()在服务的onCreate()方法里被调用，因此在服务启动的那一刻它就被设置了。在服务的onDestory()方法里则调用releaseWakeLock()方法，以确保如果服务收到了停止请求，系统资源能得以释放。

## 技巧28：使用前台服务

服务原本是运行在后台的，用于处理短期、低优先级的任务。这意味着当前台进程如一个Activity，需要更多的内存或计算能力时，服务将是首先被清理的对象。在多数情况下这也符合我们的期望。然而，有时会需要某个服务维持运行，并拥有高于其他后台任务的优先级。我们再一次以音乐播放为例。Android允许服务被标记为前台服务，但需要强制设置一个始终保持在通知栏的通知，以告知用户有个服务正处于高优先级。

激活前台模式十分简单，只需进行下面的调用：

```
startForeground(NOTIFICATION_ID, getForegroundNotification());
```

startForeground的参数一个是用来识别通知的ID，另一个是待显示通知的新实例。

可以通过调用stopForeground(true);停止前台模式，其中的布尔型标志用来告知现在可以移除通知。此外，两个调用分别放在两个方法中，从服务的 onCreate() 和 onDestroy() 方法调用。此处最大的困难是创建通知，如代码清单 3-19 所示。

**代码清单 3-19 前台服务类**

```
public class SimpleService extends Service {
    private static final int NOTIFICATION_ID = 1;

    @Override
    public IBinder onBind(Intent arg0) {
        return null;
    }

    boolean paused = false;

    @Override
    public void onCreate() {
        super.onCreate();
        enforceForegroundMode();
        Toast.makeText(this,"Service created ...", Toast.LENGTH_LONG).show();
        paused = false;
        Thread initBkgdThread = new Thread(new Runnable() {
            public void run() {
                play_music();
            }
        });
        initBkgdThread.start();
    }

    @Override
    public void onDestroy() {
        super.onDestroy();
        releaseForegroundMode();
        Toast.makeText(this, "Service destroyed ...", Toast.LENGTH_LONG).show();
        paused = true;
    }

    private final void enforceForegroundMode() {
        startForeground(NOTIFICATION_ID, getForegroundNotification());
    }

    private final void releaseForegroundMode(){
        stopForeground(true);
    }

    protected Notification getForegroundNotification(){
        // Set the icon, scrolling text, and timestamp

        Notification notification = new Notification(R.drawable.icon,
            "playback running",System.currentTimeMillis());

        // the PendingIntent to launch the activity if the user selects this notification
        Intent startIntent=new Intent(getApplicationContext(),SimpleActivity.class);
        startIntent.setFlags(Intent.FLA);
        PendingIntent contentIntent=PendingIntent.getActivity(this,0,startIntent,0);

        // Set the info for the views that show in the notification panel
        notification.setLatestEventInfo(
                this,
                "Playing Music",
                "Playback running",
                contentIntent
        );
```

```
            return notification;
        }
            int[] notes = {R.raw.c5, R.raw.b4, R.raw.a4, R.raw.g4};
        int NOTE_DURATION = 400; //millisec
        MediaPlayer m_mediaPlayer;
        private void play_music() {
            for(int ii=0; ii<12; ii++) {
                //Check to ensure main activity not paused
                if(!paused) {
                    if(m_mediaPlayer != null) {m_mediaPlayer.release();}
                    m_mediaPlayer = MediaPlayer.create(this, notes[ii%4]);
                    m_mediaPlayer.start();
                    try {
                        Thread.sleep(NOTE_DURATION);
                    } catch (InterruptedException e) {
                        e.printStackTrace();
                    }
                }
            }
        }
    }
```

从这个代码清单中可以看到，创建通知是对服务类进行的最大的修改。首先，创建一个新的通知实例并提供用于在通知栏显示的图标和提示文字。一个 Intent 被封装进 Pending Intent，这个 Pending Intent 用于当用户在全显示模式下点击了通知后启动相应的 Activity。在 `notification.setLastestEventInfo(..)` 中设置 Intent 的标题和简单的描述文字，然后这个通知实例返回并传送给 `startForeground(..)` 方法。

## 技巧 29：使用 IntentService

`IntentService` 是这样一类服务：它保持一个它收到的 Intent 的队列，并逐个执行它们。它对于许多后台任务（比如对服务器进行轮询以获取新信息，或下载大量的数据）而言是个理想的处理者。由于 Intent 十分灵活，可以在其额外信息中包含任何类型的可打包对象，所以在其中可以给出的配置信息量差不多是无限制的。这就允许向 `IntentService` 发送非常复杂的查询，然后让其做出反应。本技巧给出了一个很简单的例子，其中的服务负责接收文本信息，并将其显示为通知。为此，需要一个布局，其中具有一个可以输入信息的文本编辑域以及一个控制信息发送的按钮，如代码清单 3-20 所示。

**代码清单 3-20　main.xml**

```
<?xml version="1.0" encoding="utf-8"?>
<LinearLayout xmlns:android="http://schemas.android.com/apk/res/android"
    android:orientation="vertical"
    android:layout_width="match_parent"
    android:layout_height="match_parent"
    >
<TextView
    android:layout_width="match_parent"
    android:layout_height="wrap_content"
    android:text="@string/hello"
    />

<EditText
```

```
        android:id="@+id/editText1"
        android:layout_width="match_parent"
        android:layout_height="wrap_content"
        android:ems="10" >

        <requestFocus />
    </EditText>

    <Button
        android:id="@+id/Button01"
        android:layout_width="wrap_content"
        android:layout_height="wrap_content"
        android:text="send message" />

    </LinearLayout>
```

本技巧里的 Activity 与其他服务范例中的基本如出一辙。OnCreate 中的文本编辑框和按钮都已被注册，且为按钮设置一个 onClickListener。在 onClickListener 中，从编辑框中读取字符串，并作为一个名为 msg 的附加字符串传递给 Intent。接着，在 Intent 中调用 startService(..)。以上过程显示在代码清单 3-21 中。

**代码清单 3-21　主 Activity**

```
public class SimpleActivity extends Activity {
        EditText editText;

        @Override
    protected void onCreate(Bundle savedInstanceState) {
        super.onCreate(savedInstanceState);
        setContentView(R.layout.main);

       editText=(EditText) findViewById(R.id.editText1);

        Button sendButton = (Button) findViewById(R.id.Button01);
        sendButton.setOnClickListener(new View.OnClickListener() {
            public void onClick(View view){
                Intent intent =
                    new Intent(SimpleActivity.this, SimpleIntentService.class);
                intent.putExtra("msg",editText.getText().toString());
                startService(intent);
            }
        });
    }
}
```

全部“脏活累活”都由 IntentService 类来完成，它甚至还会在 IntentQueue 为空时自行停止。只有两项需要被实现：一个是通过调用 super("myname") 为队列命名的构造函数；另一个是 handleIntent 方法，在其中对 Intent 进行反编列（demarshal）并执行查询。实际上，不过是像前面技巧做过的那样创建一个通知，并通过 intent.getStringExtra("msg") 方法把发来的消息解压。得到的字符串则用于替换通知的描述文本。代码清单 3-22 给出了完整的 IntentService。

**代码清单 3-22　一个简单的 IntentService**

```
public class SimpleIntentService extends IntentService {

        Notification notification=null;
```

```
    private NotificationManager mNManager;

    public SimpleIntentService() {
            super("SimpleIntentService");
    }

    @Override
    protected void onHandleIntent(Intent intent) {
            createNotification();
            if(intent.hasExtra("msg")){
                    updateNotification(intent.getStringExtra("msg"));
                mNManager.notify(1, notification);
            }
    }

protected void createNotification(){
    if(mNManager==null){
        mNManager = (NotificationManager)
            getSystemService(Context.NOTIFICATION_SERVICE);
    }
    notification = new Notification(
            R.drawable.icon,
            "New event of importance",
            System.currentTimeMillis());
    // Set the icon, scrolling text, and timestamp
}

protected void updateNotification(final String text){
    // the PendingIntent to launch the activity if the user
    // selects this notification
    Intent startIntent=new Intent(getApplicationContext(),SimpleActivity.class);

    PendingIntent contentIntent =
        PendingIntent.getActivity(this, 0,startIntent, 0);

    // Set the info for the views that show in the notification panel
    notification.setLatestEventInfo(
            this,
            "Message received",
            text,
            contentIntent
    );

}

}
```

## 3.5　广播接收器

广播接收器（broadcast receiver）监听有关的广播消息，从而触发事件。以下是操作系统发出的一些广播事件的例子：

- 按下拍照按钮；
- 电池电量低；
- 安装了一个新应用程序。

用户生成的组件也可以发出广播，比如：

- 某个计算完成；
- 某个特定线程启动。

所有的广播接收器要么继承了 BroadcastReceiver 抽象类，要么继承了它的一个子类。广播接收器的生命周期比较简单，当消息到达接收器时，会调用 onReceive()方法，该方法完成后，BroadcastReceiver 实例转入非活动状态。

广播接收器通常会在 onReceive()方法中启动一个独立的组件或向用户发送通知，我们将在本章稍后讨论这件事。如果广播接收器需要做一些更费时的工作，它应该启动一个服务而不是产生一个线程，因为广播接收器在不活动时可能被系统杀死。

## 技巧 30：当按下拍照按钮时启动一个服务

这个技巧将展示如何基于广播事件（比如在拍照按钮被按下时）来启动服务。BroadcastReceiver 需要监听特定的事件，随之启动服务。BroadcastReceiver 自身会被另外的组件启动。（在这里，它被实现为一个独立的 Activity，名为 SimpleActivity。）

如代码清单 3-23 所示，该 Activity 建立一个 BroadcastReceiver，并构建一个带有拍照按钮事件过滤器的 Intent。为了演示得更明白，还添加了“包被添加”（package-added）这一消息过滤器。然后将启动 BroadcastReceiver，并通过 registerReceiver()方法将前述 Intent 过滤器传递给它。

**代码清单 3-23 src/com/cookbook/simple_receiver/SimpleActivity.java**

```java
package com.cookbook.simple_receiver;

import android.app.Activity;
import android.content.Intent;
import android.content.IntentFilter;
import android.os.Bundle;

public class SimpleActivity extends Activity {
    SimpleBroadcastReceiver intentReceiver =
        new SimpleBroadcastReceiver();

    /** called when the activity is first created */
    @Override
    public void onCreate(Bundle savedInstanceState) {
        super.onCreate(savedInstanceState);
        setContentView(R.layout.main);

        IntentFilter intentFilter =
            new IntentFilter(Intent.ACTION_CAMERA_BUTTON);
        intentFilter.addAction(Intent.ACTION_PACKAGE_ADDED);
        registerReceiver(intentReceiver, intentFilter);
    }

    @Override
    protected void onDestroy() {
        unregisterReceiver(intentReceiver);
        super.onDestroy();
    }
}
```

要注意，如果 Activity 被销毁，接收器也要随之被解除注册。这并不是必须要做的，但很有用。BroadcastReceiver 组件展示在代码清单 3-24 中。生命周期方法 onReceive()被重写，以对广播事件进行检查。如果它匹配了特定的事件（本例中为 ACTION_CAMERA_BUTTON 事件），一个服务将在原有上下文中被启动。

代码清单 3-24 src/com/cookbook/simple_receiver/SimpleBroadcastReceiver.java

```
package com.cookbook.simple_receiver;

import android.content.BroadcastReceiver;
import android.content.Context;
import android.content.Intent;

public class SimpleBroadcastReceiver extends BroadcastReceiver {
    @Override
    public void onReceive(Context rcvContext, Intent rcvIntent) {
        String action = rcvIntent.getAction();
        if (action.equals(Intent.ACTION_CAMERA_BUTTON)) {
            rcvContext.startService(new Intent(rcvContext,
                    SimpleService2.class));

        }
    }
}
```

在代码清单3-24的 SimpleBroadcastReceiver 中启动的服务展示在代码清单3-25中。该服务仅使用 Toast 来显示它自己是否被启动或是停止了。

代码清单 3-25 src/com/cookbook/simple_receiver/SimpleService2.java

```
package com.cookbook.simple_receiver;

import android.app.Service;
import android.content.Intent;
import android.os.IBinder;
import android.widget.Toast;

public class SimpleService2 extends Service {
    @Override
    public IBinder onBind(Intent arg0) {
        return null;
    }

    @Override
    public void onCreate() {
        super.onCreate();
        Toast.makeText(this,"Service created ...",
                Toast.LENGTH_LONG).show();
    }

    @Override
    public void onDestroy() {
        super.onDestroy();
        Toast.makeText(this, "Service destroyed ...",
                Toast.LENGTH_LONG).show();
    }
}
```

## 3.6 应用微件

应用微件乃是应用程序中看上去像图标一样的视图。它们实现了 BroadcastReceiver

的一个子类，用来更新视图。应用微件又被简称为微件，可以被嵌入到其他应用程序中。比如在主屏幕，通过长按（按下并保持一段时间）触屏的空白区域来进行嵌入。这样会显示一个菜单，允许选择一个微件安装到用户长按的区域上。Android 4 设备上还可以直接从启动器来安装微件。可以通过在微件上长按后将再其拖动到垃圾箱来删除它。总之，微件需要以下几样东西。

- 一个描述微件外观的视图。它被定义在 XML 布局资源文件中，包含文本、背景及其他布局参数。
- 一个应用微件提供器，负责接收广播事件、与微件连接并更新视图。
- 应用微件的详细信息，比如大小及更新频率。注意，手机的主屏幕被分割成 4×4 的单元格，平板电脑的主屏幕则被分割成 8×7 的单元格，因此微件通常会占用一个或多个单元格的大小。
- 可选地，可以定义一个应用微件配置 Activity 来为微件设置适宜的参数，该 Activity 于微件创建时启动。

## 技巧 31：创建应用微件

这个技巧创建了一个简单的应用微件，在主屏幕上显示一些文本。文本被设为每秒刷新一次，但注意，默认情况下 Android 系统强制最小刷新间隔为 30 分钟，这有助于防止粗制滥造的微件耗干电池。代码清单 3-26 实现了 `AppWidgetProvider`，它是 `BroadcastReceiver` 的一个子类。要重写的主要方法是 `onUpdate()` 函数，它在系统认为有必要更新微件时被调用。

**代码清单 3-26　src/com/cookbook/widget_example/SimpleWidgetProvider.java**

```
package com.cookbook.simple_widget;

import android.appwidget.AppWidgetManager;
import android.appwidget.AppWidgetProvider;
import android.content.Context;
import android.widget.RemoteViews;

public class SimpleWidgetProvider extends AppWidgetProvider {
    final static int APPWIDGET = 1001;
    @Override
    public void onUpdate(Context context,
            AppWidgetManager appWidgetManager, int[] appWidgetIds) {
        super.onUpdate(context, appWidgetManager, appWidgetIds);
        // Loop through all widgets to display an update
        final int N = appWidgetIds.length;
        for (int i=0; i<N; i++) {
            int appWidgetId = appWidgetIds[i];
            String titlePrefix = "Time since the widget was started:";
            updateAppWidget(context, appWidgetManager, appWidgetId,
                        titlePrefix);
        }
    }

    static void updateAppWidget(Context context, AppWidgetManager
            appWidgetManager, int appWidgetId, String titlePrefix) {
        Long millis = System.currentTimeMillis();
        int seconds = (int) (millis / 1000);
```

```
        int minutes = seconds / 60;
        seconds       seconds % 60;

        CharSequence text = titlePrefix;
        text += " " + minutes + ":" + String.format("%02d",seconds));

        // Construct the RemoteViews object
        RemoteViews views = new RemoteViews(context.getPackageName(),
                R.layout.widget_layout);
        views.setTextViewText(R.id.widget_example_text, text);

        // Tell the widget manager
        appWidgetManager.updateAppWidget(appWidgetId, views);
    }
}
```

描述微件详细信息的XML文件如代码清单3-27所示。它展示了微件在主屏幕上占据的空间大小，以及它每隔多少毫秒刷新一次（系统默认的最小值是30分钟）。

**代码清单3-27　src/res/xml/widget_info.xml**

```
<?xml version="1.0" encoding="utf-8"?>
<appwidget-provider xmlns:android="http://schemas.android.com/apk/res/android"
    android:minWidth="146dp"
    android:minHeight="72dp"
    android:updatePeriodMillis="1000"
    android:initialLayout="@layout/widget_layout">
</appwidget-provider>
```

描述微件外观的视图放在一个XML文件中，如代码清单3-28所示。

**代码清单3-28　src/res/layout/widget_layout.xml**

```
<?xml version="1.0" encoding="utf-8"?>
<TextView xmlns:android="http://schemas.android.com/apk/res/android"
    android:id="@+id/widget_example_text"
    android:layout_width="wrap_content"
    android:layout_height="wrap_content"
    android:textColor="#ff000000"
    android:background="#ffffffff"/>
```

# 第4章

# 高级线程技术

本章展示了一组 Android 框架提供的线程技术，能让线程的使用更容易和更安全。首先展示的是支持包所带有的装载器（loader）API。接着将呈现 AsyncTask API，这是原始 Java 线程的非常灵活和强大的替代品。然后讨论在 Android 上使用进程间通信的方法。

## 4.1 装载器

在应用程序生命周期中，用数据或缓存对屏幕进行初始化的阶段是线程典型的应用场景。如果让主线程完成此工作，就可能阻塞布局本身的装载，又让用户在若干秒内盯着空空如也的屏幕却无可奈何。为克服数据初始化的挑战，Android 提供了一个装载器 API。装载器是通过管理器启动的一个小型对象，它在后台执行查询工作，再通过一个通用回调接口将结果呈现出来。装载器主要有以下两种。

- CursorLoader：用于查询数据库和 ContentProvider。
- AsyncTaskLoader：用于上一条功能之外的所有其他情况。

### 技巧 32：使用 CursorLoader

本技巧使用 CursorLoader 在一个简单的 ListView 中显示手机上的全部联系人。装载这一查询所需的时长取决于手机中联系人的数量。这是一个简单的 Activity，包含一个 ListView，而不含其他 Fragment。装载器既可以随 Activity 使用，也可以随 Fragment 使用。每个类中需要重写的方法以及实施的步骤都是相同的。要访问支持 API，Activity 需要扩展 `FragmentActivity` 类。Activity 如代码清单 4-1 所示。

**代码清单 4-1　MainActivity.java**

```
public class MainActivity extends FragmentActivity
    implements LoaderCallbacks<Cursor>{
```

```
    private static final int LOADER_ID = 1;
    SimpleCursorAdapter mAdapter;
    ListView mListView;

    static final String[] CONTACTS_PROJECTION = new String[] {
        Contacts._ID,
        Contacts.DISPLAY_NAME
    };

    @Override
    protected void onCreate(Bundle savedInstanceState) {
        super.onCreate(savedInstanceState);
        setContentView(R.layout.activity_main);
        mListView=(ListView)findViewById(R.id.list);

        mAdapter=new SimpleCursorAdapter(
                getApplicationContext(), //context for layout inflation, etc.
                android.R.layout.simple_list_item_1, //the layout file
                null, //We don't have a cursor yet
                new String[]{Contacts.DISPLAY_NAME}, //the name of the data row
                //The ID of the layout for data is displayed
                new int[]{android.R.id.text1}
        );

        mListView.setAdapter(mAdapter);

        getSupportLoaderManager().initLoader(LOADER_ID,null,this);
    }

    @Override
    public Loader<Cursor> onCreateLoader(int loaderId, Bundle args) {
        return new CursorLoader(
                getApplicationContext(),
                Contacts.CONTENT_URI,
                CONTACTS_PROJECTION,
                null,
                null,
                Contacts.DISPLAY_NAME + "COLLATE LOCALIZED ASC"
        );
    }

    @Override
    public void onLoadFinished(Loader<Cursor> loader, Cursor cursor) {
        if(loader.getId()==LOADER_ID){
            mAdapter.swapCursor(cursor);
        }
    }

    @Override
    public void onLoaderReset(Loader<Cursor> loader) {
        if(loader.getId()==LOADER_ID){
            mAdapter.swapCursor(null);
        }
    }
}
```

要使用装载器，Activity 需要实现 LoaderCallbacks<T>接口，该接口有三个函数：onCreateLoader、onLoadFinished 和 onLoaderReset。

只要生成了初始化装载器的请求，onCreateLoader(int loaderId, Bundle args)

函数就会被调用。ID 可被用来区分不同的装载器，如果仅使用一个装载器，可将其设为 0。这里会返回一个新的 CursorLoader 类，并给它赋予针对一个 ContentProvider 查询的若干相同参数，主要包括一个内容 URI、一个投影图（projection map）、一个带参数的选择字符串以及一个序函数（order function）。

onLoadFinished(Loader<Cursor> loader, Cursor cursor)方法在装载器运行结束时被调用。装载器的 ID 可以由 loader.getId()检查，如果该值符合要求，就调用 mAdaptor.swapCursor()将结果游标（resulting cursor）设置为列表组件的数据源。如果适配器在此前还有另一个被激活的游标，该游标将会被关闭。

对 onLoaderReset(Loader<Cursor> loader)函数的调用将会关闭装载器产生的游标。此时须调用 mAdaptor.swapCursor(null)来确保游标不再在列表中使用。

当游标下属的数据发生变化时，装载器被设置成自动重新对其进行查询，这样显示视图就可以"与时俱进"。重新查询会导致 onLoadFinished 被再次调用。这显示了使用装载器对于初始化 Activity 或 Fragment 而言是一种简单但强大的方法。

## 4.2 AsyncTask

AsyncTask 是一种在单独的线程中执行某个方法的类，允许其容易地将工作转移到 UI 线程之外。AsyncTask 有趣的部分在于，只有 doInBackGround()方法运行在单独的线程上，允许 onPreExecute()和 onPostExecute()方法对启动它们的 Activity 或 Fragment 的视图或其他部分进行操作。AsyncTask 会被定为 Parameters、Progress Reporting 或 Result 类型。如果上述三种类型均未被选用，AsyncTask 会被定为 Void 型。AsyncTask 允许通过在 doInBackground()中调用 setProgress(int)将进度报告传回给 UI 线程。该操作会触发对 AsyncTask 的 onProgressChanged()函数的调用，该函数还可以在 UI 线程中再次被调用，并被用来更新进度对话框。

## 技巧 33：使用 AsyncTask

本技巧使用 AsyncTask 来装载远程图像，将其显示在一个 ImageView 里。从服务器装载图像是 AsyncTask 的常见应用场景，又因为在主线程里做这件事有可能阻塞 UI，所以应当避免。将 AsyncTask 实现为一个内部类，可以在下载完成后直接将图像置入 ImageView 中。完整的代码如代码清单 4-2 所示。

**代码清单 4-2 MainActivity**

```
public class MainActivity extends Activity {
    private static final int LOADER_ID = 1;
    ImageView mImageView;

    private static final String IMAGE_URL =
            "http://developer.android.com/images/brand/Android_Robot_100.png";

    @Override
```

```
    protected void onCreate(Bundle savedInstanceState) {
        super.onCreate(savedInstanceState);
        setContentView(R.layout.activity_main);
        mImageView=(ImageView)findViewById(R.id.image);
        new ImageLoaderTask(getApplicationContext()).execute(IMAGE_URL);
    }

    public class ImageLoaderTask extends AsyncTask<String, Void, String>{
        private Context context;

        public ImageLoaderTask(Context context){
            this.context=context;
        }

        @Override
        protected String doInBackground(String. . . params) {
            String path=context.getFilesDir()
                        +File.pathSeparator
                        +"temp_"
                        +System.currentTimeMillis()
                        +".png";

            HttpURLConnection connection=null;

            android.util.Log.v("TASK","opening url="+params[0]);

            try {
                final URL url=new URL(params[0]);
                connection=(HttpURLConnection) url.openConnection();
                InputStream in =
                    new BufferedInputStream(connection.getInputStream());

                OutputStream out= new FileOutputStream(path);
                int data = in.read();
                while (data != -1) {
                    out.write(data);
                    data = in.read();
                }
            } catch (IOException e) {
                android.util.Log.e("TASK","error loading image",e);
                e.printStackTrace();
                return null;
            }finally {
                if(connection!=null){
                    connection.disconnect();
                }
            }
            return path;
        }

        @Override
        protected void onPostExecute(String imagePath) {
            super.onPostExecute(imagePath);
            if(imagePath!=null){
                android.util.Log.v("TASK","loading image from temp
➥file"+imagePath);
                Bitmap bitmap=BitmapFactory.decodeFile(imagePath);
                mImageView.setImageBitmap(bitmap);
            }
        }
    }
}
```

onCreate()方法加载布局，将 ImageView 的实例保存为域变量，并启动 AsyncTask。

ImageLoaderTask 类接受一个用于表示要装载图像的 URL 的字符串型参数，并返回另一个包含存储图像的临时文件路径的字符串。此路径的构成为，文件夹是从上下文读取的应用程序的内部文件夹，文件名则是由一个以 **temp** 作为前缀，其后以毫秒表示的当前时间，并以**.png** 作为后缀的文件名。doInBackground 函数使用 URLConnection 以打开一个指向图像的 InputStream，逐字节地读入，并将其写入一个指向临时文件路径的 FileOutputStream。

onPostExecute 方法将获取的 doInBackground 结果作为参数。如果该结果非空，说明相应路径下存在图像。因为 onPostExecute 运行在 Activity 的 UI 线程上，图像可以被解码为位图，并被置入 Activity 的 ImageView 中。由于 onPreExecute()和 onPostExecute()方法始终运行在 UI 线程上，如果需要，可以用它们和宿主 Activity 或 Fragment 通信，以操纵视图。

## 4.3 Android 进程间通信

如果两个应用程序要共享资源，但无法获得权限，可以通过定义 IPC 消息来解决。要支持 IPC，需要一个接口作为应用程序之间的桥梁。这类接口由 Android 接口定义语言（Android Interface Definition Language，AIDL）来提供。

定义 AIDL 与 Java 接口类似。实际上，在 Eclipse 中可以轻而易举地实现：先创建一个新 Java 接口，在定义完毕后，将扩展名由**.java** 改成**.aidl** 即可。

AIDL 目前支持的数据类型有以下几种。

- Java 基本数据类型，包括 int、boolean 和 float。
- String。
- CharSequence。
- List。
- Map。
- 其他 AIDL 生成的接口。
- 实现了 Parcelable 协议并以值方式传递的自定义类。

### 技巧 34：实现远程过程调用

这个技巧实现了两个 Activity 之间的远程过程调用（RPC）。首先要定义 AIDL 接口，如代码清单 4-3 所示。

**代码清单 4-3 com.cookbook.advance.rpc 包中的 IAdditionalService.aidl**

```
package com.cookbook.advance.rpc;

// Declare the interface
interface IAdditionService {
    int factorial(in int value);
}
```

AIDL文件被创建之后，Eclipse会在项目生成时在**gen/**文件夹下建立名为**IadditionalService.java**的文件。该文件的内容不应被人为修改，它包含了在实现远程服务时要用到的一个存根（stub）类。

在第一个Activity——**rpcService**中，要将一个mBinder成员声明为来自IAdditionalService的存根，它也可以被称为IBinder。mBinder在onCreate()方法中被初始化，并被定义调用factorial()函数。当onBind()方法被调用时，它将mBinder返回给调用者。在onBind()函数执行过后，其他进程的Activity就可以连接到此服务上。如代码清单4-4所示。

**代码清单4-4　src/com/cookbook/advance/rpc/rpcService.java**

```
package com.cookbook.advance.rpc;

import android.app.Service;
import android.content.Intent;
import android.os.IBinder;
import android.os.RemoteException;

public class RPCService extends Service {

    IAdditionService.Stub mBinder;
    @Override
    public void onCreate() {
        super.onCreate();
        mBinder = new IAdditionService.Stub() {
            public int factorial(int value1) throws RemoteException {
                int result=1;
                for(int i=1; i<=value1; i++){
                    result*=i;
                }
                return result;
            }
        };
    }

    @Override
    public IBinder onBind(Intent intent) {
      return mBinder;
    }

    @Override
    public void onDestroy() {
      super.onDestroy();
    }
}
```

现在要指定运行在不同进程上的另一个Activity。相关的布局如代码清单4-5所示。布局中实际有三个扮演主要角色的视图。EditText从用户那里获取输入，Button触发对factorial()的调用，而以result为ID的TextView则用来显示从factorial那里得到的结果。

代码清单 4-5 res/layout/main.xml

```
<?xml version="1.0" encoding="utf-8"?>
<LinearLayout xmlns:android="http://schemas.android.com/apk/res/android"
  android:orientation="vertical" android:layout_width="fill_parent"
  android:layout_height="match_parent">
  <TextView android:layout_width="match_parent"
    android:layout_height="wrap_content"
    android:text="Android CookBook RPC Demo"
    android:textSize="22dp" />
  <LinearLayout
  android:orientation="horizontal"
  android:layout_width="match_parent"
  android:layout_height="wrap_content">
  <EditText android:layout_width="wrap_content"
    android:layout_height="wrap_content" android:id="@+id/value1"
    android:hint="0-30"></EditText>
  <Button android:layout_width="wrap_content"
    android:layout_height="wrap_content" android:id="@+id/buttonCalc"
    android:text="GET"></Button>
    </LinearLayout>
  <TextView android:layout_width="wrap_content"
    android:layout_height="wrap_content" android:text="result"
    android:textSize="36dp" android:id="@+id/result"></TextView>
</LinearLayout>
```

**AndroidManifest.xml** 文件如代码清单 4-6 所示。在 `service` 标签里，有一个额外的属性：`android:process=".remoteService"`。该属性要求系统创建一个名为 `remoteService` 的新进程来运行第二个 Activity。

代码清单 4-6 AndroidManifest.xml

```
<?xml version="1.0" encoding="utf-8"?>
<manifest xmlns:android="http://schemas.android.com/apk/res/android"
  package="com.cookbook.advance.rpc"
  android:versionCode="1" android:versionName="1.0">
  <application android:icon="@drawable/icon"
               android:label="@string/app_name" >
    <activity android:name=".rpc" android:label="@string/app_name">
      <intent-filter>
        <action android:name="android.intent.action.MAIN" />
        <category android:name="android.intent.category.LAUNCHER" />
      </intent-filter>
    </activity>

    <service android:name=".rpcService" android:process=".remoteService"/>
  </application>
  <uses-sdk android:minSdkVersion="7" />
</manifest>
```

第二个 Activity 如代码清单 4-7 所示。它需要调用 `bindService()` 读取 `rpcService` 里提供的 `factorial()` 方法。`bindService()` 方法要求一个服务连接实例作为接口以监控应用程序服务的状态。因此，这个 Activity 包含一个实现了服务连接的内部类 `myServiceConnection`。

`myServiceConnection` 和 `IAdditionService` 类都在 `rpc` Activity 中被实例化。`myServiceConnection` 对 `onServiceConnected` 和 `onServiceDisconnected` 回调方

法进行监听。onServiceConnected 回调函数将 IBinder 实例传递给 IAdditionService 实例。onServiceDisconnected 回调函数将 IAdditionService 实例置为空（null）。

rpc Activity 里还定义了另两个方法：initService()和 releaseService()。initService()方法尝试初始化一个新的 myServiceConnection。接下来，它创建一个面向特定包名和类名的新 Intent，并将其与 myServiceConnection 实例以及名为 BIND_AUTO_CREATE 的标志一起传递给 bingService 方法。当服务被绑定之后，onService Connection 回调函数被触发，并将 IBinder 方法传递给 IAdditionService 实例，这样 rpc Activity 就可以开始调用 factorial 方法。输出结果如图 4-1 所示。

图 4-1 AIDL 应用程序的输出结果

**代码清单 4-7 src/com/cookbook/advance/rpc/rpc.java**

```
package com.cookbook.advance.rpc;

import android.app.Activity;
import android.content.ComponentName;
import android.content.Context;
import android.content.Intent;
import android.content.ServiceConnection;
import android.os.Bundle;
import android.os.IBinder;
import android.os.RemoteException;
import android.view.View;
import android.view.View.OnClickListener;
import android.widget.Button;
import android.widget.EditText;
import android.widget.TextView;
import android.widget.Toast;

public class rpc extends Activity {
  IAdditionService service;
  myServiceConnection connection;

  class myServiceConnection implements ServiceConnection {

    public void onServiceConnected(ComponentName name,
                                   IBinder boundService) {
      service = IAdditionService.Stub.asInterface((IBinder) boundService);
      Toast.makeText(rpc.this, "Service connected", Toast.LENGTH_SHORT)
           .show();
    }

    public void onServiceDisconnected(ComponentName name) {
      service = null;
      Toast.makeText(rpc.this, "Service disconnected", Toast.LENGTH_SHORT)
           .show();
    }
  }

  private void initService() {
```

```
        connection = new myServiceConnection();
        Intent i = new Intent();
        i.setClassName("com.cookbook.advance.rpc",
                       com.cookbook.advance.rpc.rpcService.class.getName());
        if(!bindService(i, connection, Context.BIND_AUTO_CREATE)) {
            Toast.makeText(rpc.this, "Bind Service Failed", Toast.LENGTH_LONG)
            .show();
        }
    }

    private void releaseService() {
        unbindService(connection);
        connection = null;
    }

    @Override
    public void onCreate(Bundle savedInstanceState) {
        super.onCreate(savedInstanceState);
        setContentView(R.layout.main);

        initService();

        Button buttonCalc = (Button) findViewById(R.id.buttonCalc);

        buttonCalc.setOnClickListener(new OnClickListener() {
            TextView result = (TextView) findViewById(R.id.result);
            EditText value1 = (EditText) findViewById(R.id.value1);

            public void onClick(View v) {
                int v1, res = -1;
                try {
                    v1 = Integer.parseInt(value1.getText().toString());
                    res = service.factorial(v1);
                } catch (RemoteException e) {
                    e.printStackTrace();
                }
                result.setText(Integer.toString(res));
            }
        });
    }

    @Override
    protected void onDestroy() {
        releaseService();
    }
}
```

AIDL 有助于在不同用户 ID 下（意味着不同的应用程序）的进程间进行完整的远程过程调用。不足之处在于 AIDL 是完全同步的，且比较缓慢。如果一个或多个 Activity 需要与一个服务通信，向其传递查询或命令，并接收返回结果的话，有两种更简单快捷的方法，这将在随后的两个技巧中介绍。

## 技巧 35：使用 Messenger

Messenger 类是远程进程中的 Handler 的一个引用，可用于向该 Handler 发送消息。Handler 允许消息被送入一个队列中，并被逐个处理。通过提供两个 Handler，一个在 Activity 中而另一个在服务中，这两个类就可以通过在 Handler 之间发送消息来实现通信。这可以通过调用 Messenger.send(msg) 来实现，它将消息注入到远程 Handler 之中。Message 类拥有一个名为 Message.replyTo 的特殊域，该域可以持有一个消息引用。然后可以通过该引用将操作结果传回给原来的线程。

下面是创建服务的步骤。

（1）创建一个 Service 类，并定义一些整型常量，这些常量描述了用于注册客户端和移除注册客户端以及发送结果的消息。

（2）创建一个对上述信息做出反应的 Handler。

（3）创建一个引用 Handler 的 messenger 实例。

（4）在服务的 onBind 函数里，通过调用 Messenger.getBinder() 来返回绑定器对象。

代码清单 4-8 中给出了上述过程。

**代码清单 4-8　Messenger 服务**

```
package com.cookbook.messenger_service;

/**
 * MessageControlledService is an abstract service implementation that
 * communicates with clients via Messenger objects. Messages are passed
 * directly into the handler of the server/client.
 */
public class MessageControlledService extends Service {

    public static final int MSG_INVALID             = Integer.MIN_VALUE;
    public static final int MSG_REGISTER_CLIENT     = MSG_INVALID+1;
    public static final int MSG_UNREGISTER_CLIENT   = MSG_INVALID+2;
    public static final int MSG_RESULT              = MSG_INVALID+3;

    /** Make sure your message constants are MSG_FIRST_USER+n**/
    public static final int MSG_FIRST_USER=1;

    private static final String LOG_TAG =
        MessageControlledService.class.getCanonicalName();

    /** keeps track of all current registered clients */
    ArrayList<Messenger> mClients = new ArrayList<Messenger>();

    /**
     * handler of incoming messages from clients
     */
    private class CommandHandler extends Handler {

        @Override
        public void handleMessage(Message msg) {
            switch (msg.what) {
            case MSG_REGISTER_CLIENT:
                mClients.add(msg.replyTo);

                break;
            case MSG_UNREGISTER_CLIENT:
                mClients.remove(msg.replyTo);
```

```
                break;
            default:
                handleNextMessage(msg);
            }
        }
    }

    private final Handler mHandler=new CommandHandler();

    /**
     * target we publish for clients to send messages to IncomingHandler
     */
    private final Messenger mMessenger = new Messenger(mHandler);

    /**
     * Call this to send an arbitrary message to the registered clients
     * @param what
     * @param arg1
     * @param arg2
     * @param object
     */
    protected final void sendMessageToClients(final int what,final int arg1,
        final int arg2,final Object object) {
        for (int i = mClients.size() - 1; i >= 0; i--) {
            try {
                Message msg = Message.obtain(null, what,arg1,arg2, object);
                mClients.get(i).send(msg);
            } catch (RemoteException e) {
                // The client is dead. Remove it from the list;
                // we are going through the list from back to front
                // so this is safe to do inside the loop.
                mClients.remove(i);
            }
        }
    }

    //------service stuff

    @Override
    public IBinder onBind(Intent arg0) {
        return mMessenger.getBinder();
    }

    /**
     * This is your main method
     *
     * @param msg the next message in the queue
     */
    public void handleNextMessage(final Message msg){
        String echo="ECHO: "+(String)msg.obj;
        sendMessageToClients(MSG_RESULT, -1, -1, echo);
    }
}
```

已连接的客户端可通过 ArrayList<Messenger>列表来追踪。在 Handler 中，如果收到 MSG_REGISTER_CLIENT，就将客户端添加到该列表；反之，如果收到 MSG_UNREGISTER_CLIENT，则将客户端从列表中移除。sendMessageToClients(..)方法对列表中的 messenger 进行循环，并调用每个 messenger 的 send 方法。这样就会将结果广播给所有已连接的客户端，并允许同时与多个 Activity 通信。所有其他的消息都作为调用 handleNextMessage(..)的参数，此调用会完成所有实际工作。在本例中，message.obj 域作为字符串读入，并添加 **Echo** 一词，再传回给调用者。

在本例的 Activity 里，当收到 MSG_RESULT 类型的消息时，message.obj 转换为字符串并以 Toast 的形式显示出来。如果布局上的按钮被点击，文本编辑域中的内容会发送给服务。这样就创建了一个回应服务（echo service）。

Activity 中用于处理连接的代码比用于服务本身的代码多得多，因为它需要启动并绑定服务，以读取它的 messenger。创建服务连接的步骤如下。

（1）创建一个 ServiceConnection 的新实例，并实现 onServiceConnected()和 onServiceDisconnected()调用。

（2）实现一个 Handler，并创建一个引用该 Handler 的 messenger 实例。这一 messenger 实例将在稍后交给服务。

（3）在 onCreate()中通过调用 startService(..)来启动服务。

（4）在 onResume()中实现与服务的绑定。

（5）在 onPause 中通过调用 unbindService()来断开与服务的连接。

MessengerActivity 的代码如代码清单 4-9 所示。

**代码清单 4-9　MessengerActivity**

```
public class MessengerActivity extends Activity {
    private static final String LOG_TAG = null;

    EditText editText;

    /** messenger for communicating with service */
    Messenger mServiceMessenger = null;
    /** flag indicating whether we have called bind on the service */
    boolean mIsBound;

    protected Handler mHandler = new Handler() {
        @Override
        public void handleMessage(Message msg) {
            if(msg.what==MessageControlledService.MSG_RESULT){
                Toast.makeText(getApplicationContext(),
                                (String)msg.obj,
                                Toast.LENGTH_SHORT
                                ).show();
            }
        }
    };

    Messenger mLocalMessageReceiver=new Messenger(mHandler);

    /**
     * class for interacting with the main interface of the service
     */
    private final ServiceConnection mServiceConnection = new ServiceConnection() {
        @Override
        public void onServiceConnected(ComponentName className, IBinder service){
```

```
            // This is called when the connection with the service has been
            // established, giving us the service object we can use to
            // interact with the service. We are communicating with our
            // service through an IDL interface, so get a client-side
            // representation of that from the raw service object.
            mServiceMessenger = new Messenger(service);

            // We want to monitor the service for as long as we are
            // connected to it
            try {
                Message msg = Message.obtain(null,
                 MessageControlledService.MSG_REGISTER_CLIENT);
                msg.replyTo = mLocalMessageReceiver;
                mServiceMessenger.send(msg);

            } catch (RemoteException e) {
                // In this case the service has crashed before we could even
                // do anything with it; we can count on soon being
                // disconnected (and then reconnected if it can be restarted)
                // so there is no need to do anything here.
            }

            Log.v(LOG_TAG, "service connected");
        }

        @Override
        public void onServiceDisconnected(ComponentName className) {
            // This is called when the connection with the service has been
            // unexpectedly disconnected--that is, its process crashed
            mServiceMessenger = null;
            Log.v(LOG_TAG, "service disconnected");
        }
    };

    @Override
    protected void onResume() {
        super.onResume();
        bindMessengerService();
    }

    @Override
    protected void onPause() {
        super.onPause();
        unbindAccountService();
    }

    void bindMessengerService() {
        Log.v(LOG_TAG,"binding accountservice");
        // Establish a connection with the service. We use an explicit
        // class name because there is no reason to let other
        // applications replace our component.
        bindService(new Intent(getApplicationContext(),
            MessageControlledService.class),
            mServiceConnection, Context.BIND_AUTO_CREATE);
        mIsBound = true;
    }

    protected void unbindAccountService() {
        if (mIsBound) {
            Log.v(LOG_TAG,"unbinding accountservice");
            // If we have received the service, and hence registered with
            // it, now is the time to unregister
```

```
            if (mServiceMessenger != null) {

                try
                {
                  Message msg=Message.obtain(
                         null,
                         MessageControlledService.MSG_UNREGISTER_CLIENT
                         );

                  msg.replyTo = mServiceMessenger;
                  mServiceMessenger.send(msg);

                } catch (RemoteException e) {

                // There is nothing special we need to do if the service

                // has crashed
              }
            }
            // Detach our existing connection
            unbindService(mServiceConnection);
            mIsBound = false;
        }
    }

    protected boolean sendMessageToService(final int what,
         final int arg1,final int arg2,
         final Object object) {
         try {
              Message msg = Message.obtain(null, what, arg1, arg2, object);
              mServiceMessenger.send(msg);
         } catch (RemoteException e) {
              Log.e(LOG_TAG,"unable to send message to account service",e);
              //Retry binding
              bindMessengerService();
              return false;
         }
         return true;

    }
    @Override
    protected void onCreate(Bundle savedInstanceState) {
         super.onCreate(savedInstanceState);
         startService(new Intent(getApplicationContext(),
         MessageControlledService.class));
         setContentView(R.layout.main);

        editText=(EditText) findViewById(R.id.editText1);

         Button sendButton = (Button) findViewById(R.id.Button01);
         sendButton.setOnClickListener(new View.OnClickListener() {
              public void onClick(View view){
                   String text=editText.getText().toString();
                   sendMessageToService(MessageControlledService.MSG_FIRST_USER,
                        -1,-1,text);
              }
         });
    }
}
```

只要 Activity 被绑定到服务上，`ServiceConnection.onServiceConnected()`方法就被调用。服务返回的 `Ibinder` 是它的 messenger 实例（请参考代码清单 4-8），因此它保存在一个域变量中。代码中包括一个 `sendMessageToService()`函数，它与服务中的 `SendMessageToClients()`

函数类似，但它使用的是刚收到的 messenger。接下来，通过传送一个 MSG_REGISTER_CLIENT 并把 messenger 的 Message.replyTo 域设置为 Activity 的本地 Handler 的引用，可以将 Activity 注册到服务上。

在 Activity 的 onPause 方法中，必须解除绑定服务，并允许它在不再有客户端与之连接的情况下停止工作。在调用解除绑定方法之前，一定要向服务发送一条 MSG_UNREGISTER_CLIENT，否则服务可用的 messenger 列表将不同步。

## 技巧 36：使用 ResultReceiver

ResultReceiver 是一种可打包类，在其内部含有可以直接进行不同进程间调用的 IPC 绑定器。当接收器的实例在另一个进程中创建时，可以调用某个进程的 ResultReceiver.send(int ResultCode, Bundle data)方法。这样另一个进程就能够读取该方法的参数并对其给予响应。由于 ResultReceiver 是可打包的，它可以被作为参数传递给某个 Intent。该 Intent 可以用于开启另外的 Activity 或服务，本例就是这样做的。ResultReceiver 接受一个 bundle 作为参数，这就可以发送实现 parcelable 接口的更复杂的对象。

第 3 章中出现过的 IntentService 在此被用来实现与前一技巧相似的回应服务。IntentService 类堪称 ResultReceiver 的黄金搭档，因为它接受一些 Intent 作为命令，并将它们排成有序队列来执行。在那些 Intent 中传递的 ResultReceiver 可以随后被用于将结果发回给发起调用的 Activity。整个服务代码在代码清单 4-10 中给出。

**代码清单 4-10 ResultReceiverIntentService**

```
public class ResultReceiverIntentService extends IntentService {

     public static final String EXTRA_RESULT_RECEIVER = "EXTRA_RESULT_RECEIVER";

     public ResultReceiverIntentService() {
          super("SimpleIntentService");
     }

     @Override
     protected void onHandleIntent(Intent intent) {
          ResultReceiver rr=intent.getParcelableExtra(EXTRA_RESULT_RECEIVER);
          if(intent.hasExtra("msg")){
               String text= "Echo: "+intent.getStringExtra("msg");
               Bundle resultBundle=new Bundle();
               resultBundle.putString("msg",text);
               rr.send(1, resultBundle);
          }

     }

}
```

服务只需要实现两个函数：一个是构造函数，其中队列通过调用 super("name")来获得一个名字；另一个是 onHandleIntent(..)函数。这里，使用一个定义为 EXTRA_RESULT_RECEIVER 的关键字常量调用 getParcelableExtra()，从而将 ResultReceiver 从 Intent 中提取出来。随后将修改过的文本存入一个 bundle 中，再将 bundle 通过接收器的 send(..)方法发送给初始的 Activity。使用整型常量作为结果代码，可以把多种类型的消息发回给 Activity。这个 Activity 如代码清单 4-11 所示。

**代码清单 4-11 主 Activity**

```
public class SimpleActivity extends Activity {
    EditText editText;

    Handler handler=new Handler();

    ResultReceiver resultReceiver=new ResultReceiver(handler){

        @Override
        protected void onReceiveResult(int resultCode, Bundle resultData) {
            super.onReceiveResult(resultCode, resultData);
            Toast.makeText(
                                SimpleActivity.this,
                                resultData.getString("msg"),
                                Toast.LENGTH_SHORT
                                ).show();
        }
    };

    @Override
    protected void onCreate(Bundle savedInstanceState) {
        super.onCreate(savedInstanceState);
        setContentView(R.layout.main);

        editText=(EditText) findViewById(R.id.editText1);

        Button sendButton = (Button) findViewById(R.id.Button01);
        sendButton.setOnClickListener(new View.OnClickListener() {
            public void onClick(View view){
                Intent intent=new Intent(
                                SimpleActivity.this,
                                ResultReceiverIntentService.class
                                );
                intent.putExtra("msg",editText.getText().toString());

                    intent.putExtra(
                  ResultReceiverIntentService.EXTRA_RESULT_RECEIVER,
                          resultReceiver
                          );
                startService(intent);
            }
        });

    }
}
```

该 Activity 拥有 ResultReceiver 的一个实现，其中重写了 onReceiveResult() 方法。该方法被内部绑定器对象调用，然后另一个进程执行 ResultReceiver 的 send() 函数。必须把一个 Handler 实例传给 ResultReceiver，ResultReceiver 会在这个 Handler 线程上执行 onReceiveResult 方法。本例完成的实现只是读取由服务发回的文本并将其在一个 Toast 中显示出来。

要发送一条命令，须将编辑域中的字符串读出并放入 Intent 中，同时把 ResultReceiver 实例设为 Intent 的附加参数。startService 调用 IntentService，它在 IntentService 尚未运行的情况下将其启动，并将 Intent 和其参数一起传递到队列之中。将 IntentService 与 ResultReceiver 组合使用是在 Android 中进行进程间通信的最简便的方法。

# 第 5 章

# 用户界面布局

Android 用户界面（UI）由屏幕视图、屏幕触摸事件和按键事件组成。用于指定 UI 的框架被设计成能支持各种不同 Android 设备。本章关注该框架在初始化和改变图形布局时的运用，而按键和手势的处理则放到第 6 章中讲述。

## 5.1 资源目录和常规属性

UI 的显示要用到由开发者生成的资源文件，其中有一些在第 2 章介绍 Android 项目目录结构的部分已经有所涉及。为完整起见，在此将全部的资源目录总结如下。

- **res/anim/**：逐帧动画（frame-by-frame animation）或渐变动画（tweened animation）对象。
- **res/animator/**：属性动画（property animation）使用的可扩展标记语言（eXtensible Markup Language，XML）文件。
- **res/color/**：指定颜色状态列表的 XML 文件。
- **res/drawable/**：图像资源，包括位图、可拉伸小图片（nine-patch）、形状、动画、状态列表等。要注意，这些图像在编译期间可能被修改和优化。
- **res/layout/**：指定屏幕布局的 XML 文件。
- **res/menu/**：指定菜单的 XML 文件。
- **res/values/**：含有资源描述符（descriptor）的 XML 文件。如同其他资源目录一样，其中的文件名可以任意命名，但常见的像本书中所使用的文件名有 **arrays.xml**、**colors.xml**、**dimens.xml**、**strings.xml** 和 **styles.xml** 等。
- **res/xml/**：前面没有涵盖到的其他 XML 文件。
- **res/raw/**：其他一些前面没有涵盖到的资源，包括不允许修改和优化的图像。

每个 UI 对象都有三种可定义的属性，用于设定 UI 带给用户的观感，分别是：对象的尺度、对象中的文本以及对象的颜色。表 5-1 总结了这三种常见 UI 属性可能的取值。注意，对于尺度而言，最好使用 **dp** 或 **sp** 以保证设备独立性。

**表 5-1　三种常见 UI 属性的可能取值**

| 属　性 | 可 能 取 值 |
|---|---|
| 尺度 | 任何带有下列尺度单位之一的数值。<br>● px：屏幕上实际的像素值。<br>● dp（或 dip）：独立于设备的、相对一个 160dpi 屏幕的像素值。<br>● sp：独立于设备的像素值，根据用户字体大小的偏好进行调整。<br>● in：基于屏幕物理尺寸的英寸值。<br>● mm：基于屏幕物理尺寸的毫米值。<br>● pt：以 1/72 英寸大小的点来度量的值，同样基于屏幕的物理尺寸。 |
| 字符串 | 任何字符串，只要是撇号或引号被转义的，如 Don\'t worry<br>任何用双引号括起来的字符串，如"Don't worry"<br>任何带格式的字符串，如 Population: %1$d<br>可以包含 HTML 标签，如<b>、<i>或<u><br>可以包含特殊字符，比如由© 表示的符号© |
| 颜色 | 可能的取值有以下几个。<br>● 12 位颜色值#rgb。<br>● 带 alpha 透明通道的 16 位颜色值#argb。<br>● 24 位颜色值#rrggbb。<br>● 带 alpha 透明通道的 32 位颜色值#aarrggbb。<br>还可以使用在 Java 文件的 Color 类中预定义的颜色，如 Color.CYAN |

要将上述应用程序的外观感觉统一起来，可以将每一类属性放到一个全局性的 XML 资源文件中，这也可以让后面重新定义属性变得简便，因为它们是被收集在三个文件中的。

- 控件的尺度和度量方式在 XML 资源文件 **res/values/dimens.xml** 中声明。例如：
  - XML 声明——<dimen name="large">48sp</dimen>；
  - XML 引用——@dimen/large；
  - Java 引用——getResources().getDimension(R.dimen.large)。
- 控件的标签和文本在 XML 资源文件 **res/values/strings.xml** 中声明，例如：
  - XML 声明——<string name="start_pt">I\'m here</string>；
  - XML 引用——@string/start_pt；
  - Java 引用——getBaseContext().getString(R.string.start_pt)。
- 控件的颜色在 XML 资源文件 **res/values/colors.xml** 中声明，例如：
  - XML 声明——<color name="red">#f00</color>；
  - XML 引用——@color/red；
  - Java 引用——getResources().getColor(R.color.red)。

## 技巧 37：指定替代资源

前一节中描述的资源提供了 Android 系统默认使用的通用配置方法。开发者能为由不同的限定符区分的特定的配置指定不同的值。

为支持多种语言，字符串可以被翻译，并在不同语言的 **values** 目录下被使用。例如，美式英语、英式英语、法语、简体中文（中国大陆）、繁体中文（中国台湾）及德语分别使用下面的文件来添加：

**res/values-en-rUS/strings.xml**

**res/values-en-rGB/strings.xml**

**res/values-fr/strings.xml**

**res/values-zh-rCN/strings.xml**

**res/values-zh-rTW/strings.xml**

**res/values-de/strings.xml**

并不是所有字符串都需要在这些文件中重新定义，对于那些选定的语言文件中缺失的字符串，会返回到系统默认的 **res/values/string.xml** 文件中查找，该文件中会包含应用程序用到的字符串的完整集合。如果任何可绘制（drawable）资源含有文本并需要特定的语言形式，它们也要使用类似的目录结构（比如 **res/drawables-zh-hdpi/**）。

要支持多种不同像素密度的屏幕，需要把可绘制项目以及原始资源（根据需要）调整大小，并在不同 dpi 值所对应的目录下使用。例如，一个图像文件可能在下面每个文件夹中都存在：

**res/drawable-ldpi/**

**res/drawable-mdpi/**

**res/drawable-hdpi/**

**res/drawable-nodpi/**

其中低、中、高密度屏幕分别被定义为 120dpi、160 dpi 和 240dpi。并非所有的 dpi 选项都需要填写。在运行时，Android 会决定最接近的可绘制类型，并适当调整其大小。可以给位图图像使用 `nodpi` 选项，以避免其大小被调整。如果同时指定了语言和 dpi 选项，那么目录名会同时包含两种修饰符：**drawable-en-rUS-mdpi/**。

Android 设备可用的不同类型屏幕在第 1 章中已经讨论过。为不同的屏幕类型分别定义 XML 布局文件通常是有益的。最常用的修饰符有以下几个。

- 屏幕方向的横或纵：**-port** 和 **-land**。
- 常规屏（QVGA、HVGA 和 VGA）及宽屏（WQVGA、FWVGA 和 WVGA）：**-notlong** 和 **-long**。
- 小（对角线长不超过 3.0 英寸）、中（对角线长不超过 4.5 英寸）、大（对角线长为 4.5 英寸以上）屏幕尺寸：**-small**、**-normal** 和 **-large**。

如果未定义屏幕方向或纵横比，Android 系统会为屏幕自动调整 UI 尺寸（虽然有时效果并不美观）。然而，如果为不同屏幕定义了布局，应当将一个特殊元素添加进 **AndroidManifest.xml** 文件的 application 元素之中，以确保能被系统正确地支持：

```
<supports-screens
  android:largeScreens="true"
  android:normalScreens="true"
  android:smallScreens="true"
  android:resizable="true"
  android:anyDensity="true" />
```

注意，如果 android:minSdkVersion 或 android:targetSdkVersion 的值为 3（对应 Android 1.5），则默认只有 normalScreens（对应 G1 手机的屏幕）被设为 true。因此，有必要为应用程序显式声明 supports-screens 元素，以使 UI 在更多的新手机上能有合适的尺寸。

还应注意，从 Android 3.2 开始，一些 7 英寸平板电脑的屏幕尺寸类型实际与 5 英寸设备是一致的（原本应当使用 large 布局）。为适应此状况，加入了一些新的选项：最小高度（sm<N>dp）、有效屏幕宽度（w<N>dp）和有效屏幕高度（h<N>dp），其中<N>是以 dp 值描述的支持的尺寸大小。

## 5.2　View 和 ViewGroup

视图是图形化布局的基本组成模块。每个视图由一个 View 对象来描述，该对象负责绘出视图的矩形区域，并处理区域中的事件。View 是与用户交互的对象的基类，这些对象又被称为微件。比如按钮或复选框就属于微件。

ViewGroup 对象是 View 的一种，它们被用作容纳多个 View（或其他 ViewGroup）的容器。例如，ViewGroup 对象可以容纳一组垂直或水平放置的视图或微件，如图 5-1 所示。ViewGroup 是屏幕布局的基类。

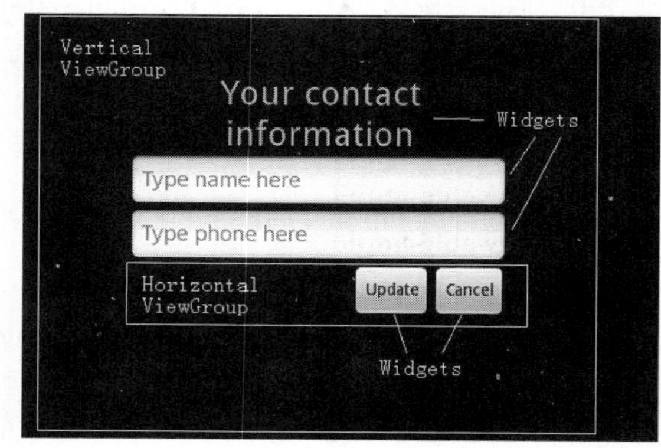


图 5-1　包含 ViewGroup 和微件的 View 示例

布局定义了用户界面元素以及它们的位置和行为，这些可以由 XML 或 Java 来指定。最常见的是用 XML 声明一个初始的基本布局，而用 Java 处理运行时的变化。这样做兼具使用 XML 设定 View 和 ViewGroup 对象总体位置的简便性，以及用 Java 修改应用程序中任意组件的灵活性。

将 XML 编写的布局和 Java 编写的 Activity 分开的另一个好处是，同一个 Android 应用程序可以根据屏幕方向、设备类型（如手机或平板电脑）、本地环境（如英文或中文）等的不同，产生不同的行为。这些定制内容可以被抽象到各个 XML 资源文件中，而不会搞乱下层的 Activity。

## 技巧 38：用 Eclipse 编辑器生成布局

一个快速生成布局的方法是使用 Eclipse 图形化布局编辑器这一便捷的工具。对于一个

Activity，打开它的布局资源 XML 文件（这里我们用 **main.xml** 文件）。单击 Layout 标签，就会展示图形化的布局效果。让我们从擦除组件开始，在黑色屏幕上单击，删除上面的所有东西，然后按以下步骤操作。

（1）从 Layouts Selector 中拖曳一个布局到屏幕区域上。例如，选择 `TableLayout`，它可以容纳成列的 `View` 或 `ViewGroup`。

（2）拖曳出其他布局，将它们嵌套放置在上一步的布局之中。例如，选择 `TableRow`，它可以容纳成行的 `View` 或 `ViewGroup`。本例添加三个 `TableRow`。

（3）在 Outline 视图中分别右击每个 `TableRow`，从 View Selector 中为其添加视图元素。例如，为第一个 `TableRow` 添加一个 `Button` 和一个 `CheckBox`，为第二个 `TableRow` 添加两个 `TextView`，为第三个 `TableRow` 添加一个 `TimePicker`。

（4）在三个 `TableRow` 元素每个底下添加一个 `Spinner` 和一个 `VideoView`。

运行结果会像图 5-2 中所示的那样，可以切换横向或纵向视图来观察布局的变化。单击 **main.xml** 标签可以将 XML 代码显示出来，如代码清单 5-1 所示。以上提供了一种创建具有 Android 观感的 UI 的简单方法。

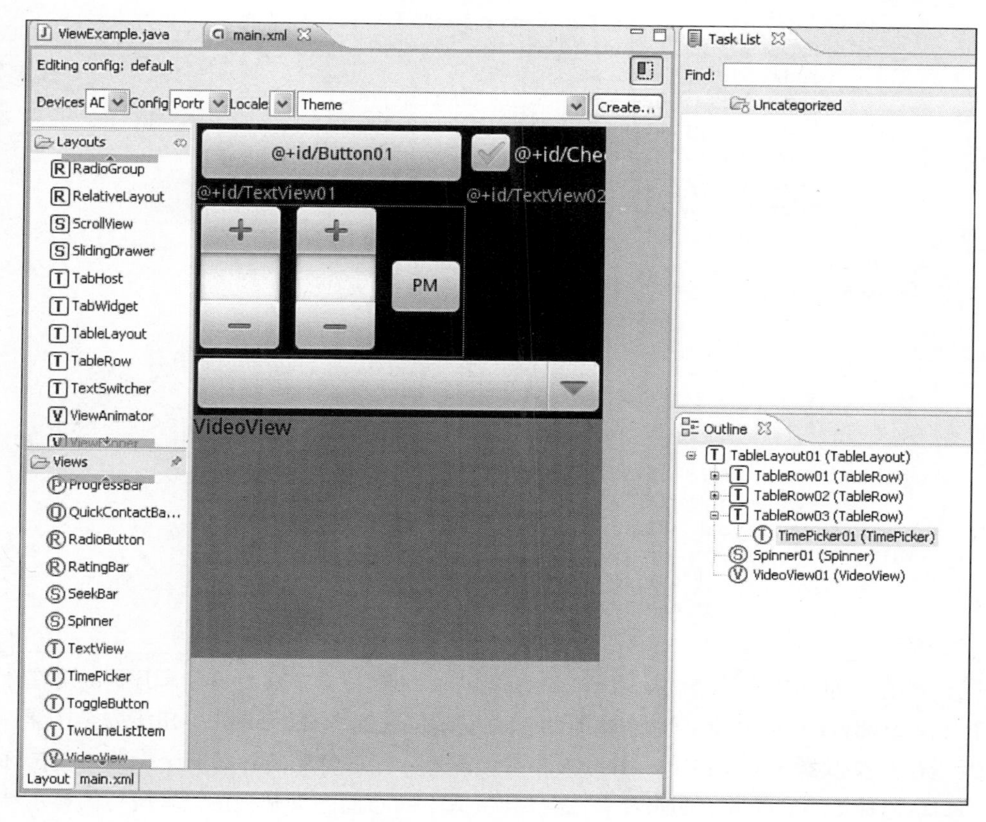


图 5-2　Eclipse 中的 Android 布局生成器范例

**代码清单 5-1 main.xml**

```
<?xml version="1.0" encoding="utf-8"?>
<TableLayout android:id="@+id/TableLayout01"
    android:layout_width="match_parent"
    android:layout_height="match_parent"
    xmlns:android="http://schemas.android.com/apk/res/android">
    <TableRow android:id="@+id/TableRow01"
        android:layout_width="wrap_content"
        android:layout_height="wrap_content">
        <Button android:text="@+id/Button01"
            android:id="@+id/Button01"
            android:layout_width="wrap_content"
            android:layout_height="wrap_content" />
        <CheckBox android:text="@+id/CheckBox01"
            android:id="@+id/CheckBox01"
            android:layout_width="wrap_content"
            android:layout_height="wrap_content" />
    </TableRow>
    <TableRow android:id="@+id/TableRow02"
        android:layout_width="wrap_content"
        android:layout_height="wrap_content">
        <TextView android:text="@+id/TextView01"
            android:id="@+id/TextView01"
            android:layout_width="wrap_content"
            android:layout_height="wrap_content" />
        <TextView android:text="@+id/TextView02"
            android:id="@+id/TextView02"
            android:layout_width="wrap_content"
            android:layout_height="wrap_content" />
    </TableRow>
    <TableRow android:id="@+id/TableRow03"
        android:layout_width="wrap_content"
        android:layout_height="wrap_content">
        <TimePicker android:id="@+id/TimePicker01"
            android:layout_width="wrap_content"
            android:layout_height="wrap_content" />
    </TableRow>
    <Spinner android:id="@+id/Spinner01"
        android:layout_width="wrap_content"
        android:layout_height="wrap_content" />
    <VideoView android:id="@+id/VideoView01"
        android:layout_width="wrap_content"
        android:layout_height="wrap_content" />
</TableLayout>
```

另一种查看布局的方法是利用 Hierarchy Viewer。在模拟器中运行应用程序时，可在命令行中运行 **hierarchyviewer** 命令，该命令位于 SDK 安装位置的 **tools/**目录下。出于安全因素，该命令只能运行在模拟器虚拟设备下，因为在真实设备上运行它可能会暴露安全设置。单击你感兴趣的窗口，选择 **Load View Hierarchy**。这将会生成不同布局的关系视图。本技巧的显示结果如图 5-3 所示。

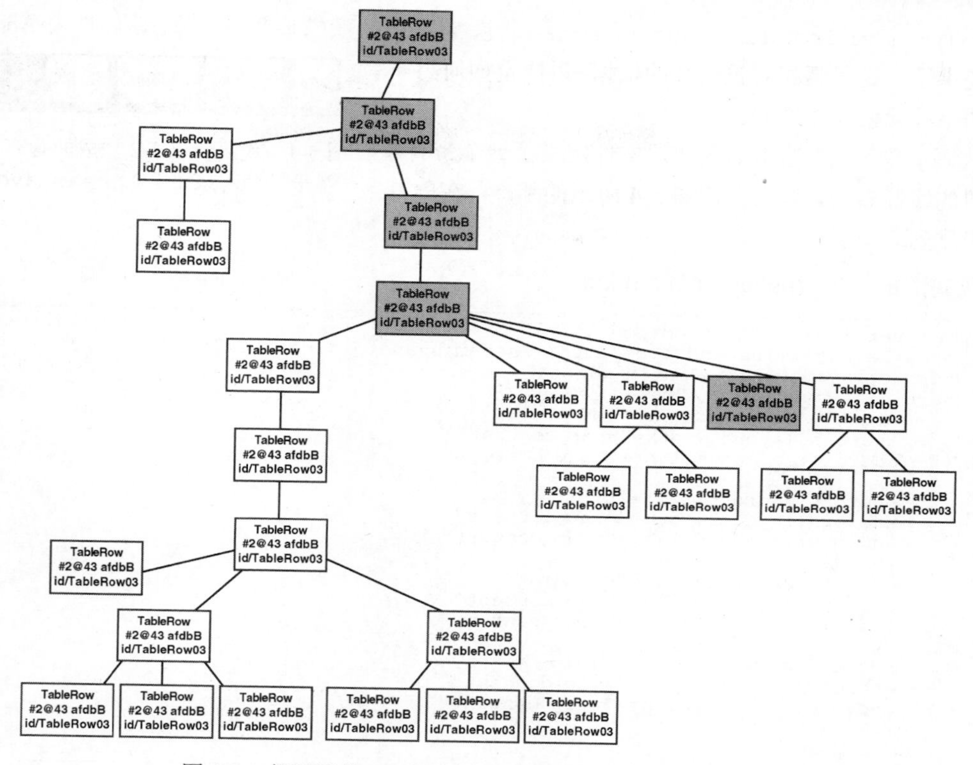


图 5-3 代码清单 5-1 程序对应的 Android Hierarchy Viewer

## 技巧 39：控制 UI 元素的宽度和高度

这个技巧显示了为 UI 元素指定宽度和高度如何对布局整体产生影响。每个 View 对象都需要为其指定总高度（android:layout_width）和总宽度（android:layout_height），可以用下列三种方法之一实现。

- 精确尺度（exact dimension）：提供准确控制，但对不同屏幕类型的可调节性不好。
- wrap_content：大小恰好能够装下其元素的内容加上 padding 值。
- match_parent：令尺寸最大化，加上 padding 值后能填满其父元素。match_parent 代替了从 Android 2.2 版以后被弃用的 fill_parent。

padding 是元素周围的空白区域，如未特别指定，默认值为 0。它是 UI 元素的一部分，需要被指定为准确的值，可以用两类属性来指定其大小。

- padding：将元素四边的 padding 设为同一个值。
- paddingLeft、paddingRight、paddingTop 和 paddingBottom：分别设定上下左右四边的 padding 值。

一些开发者将 padding 与 margin 混淆。margin 是元素周围的空白区域，但并不被计入 UI 元素的大小之中。

还有一个 android:layout_weight 属性，它可以被赋成某个整型值。该属性为 Android 系统提供了界定元素相对重要性的依据，以决定如何为布局中的不同元素安排空间。

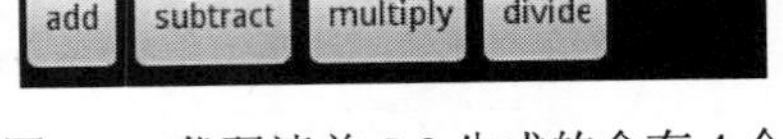


图 5-4 代码清单 5-2 生成的含有 4 个水平排列的按钮的 LinearLayout

代码清单 5-2 中给出的主布局文件定义了一个带有 4 个按钮的线性布局，按钮会像图 5-4 所示的那样，在屏幕上水平排列。

**代码清单 5-2 res/layout/main.xml**

```
<?xml version="1.0" encoding="utf-8"?>
<LinearLayout xmlns:android="http://schemas.android.com/apk/res/android"
 android:layout_width="match_parent"
 android:layout_height="match_parent">
       <Button android:text="add"
       android:layout_width="wrap_content"
       android:layout_height="wrap_content"
       />
       <Button android:text="subtract"
       android:layout_width="wrap_content"
       android:layout_height="wrap_content"
       />
       <Button android:text="multiply"
       android:layout_width="wrap_content"
       android:layout_height="wrap_content"
       />
       <Button android:text="divide"
       android:layout_width="wrap_content"
       android:layout_height="wrap_content"
       />
</LinearLayout>
```

如果将“add”按钮的高度改为 match_parent，该按钮就会在垂直方向上填满其父组件的空间，同时保持文字对齐。如果将任一按钮的宽度改为 match_parent，那么水平布局中所有位于它后面的按钮都会被挤出去。图 5-5 为我们展示了上述情况。

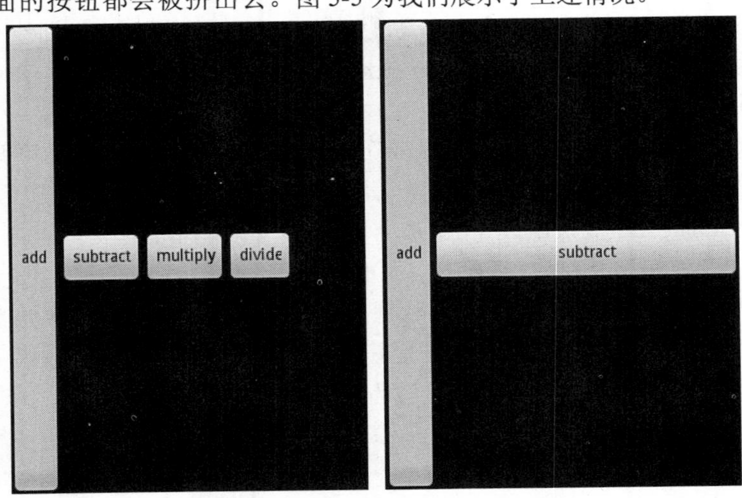


图 5-5 将高度设为 match_parent 还能继续保持水平方向对齐，若将宽度设为 match_parent 将会把剩下的按钮全部挤走

还要注意一点，图 5-4 中的“multiply”和“divide”按钮的最后一个字母有一部分没有显示全。可以通过在文本的最后添加一个空格来修正这个问题，比如“multiply”和“divide ”。然而更常用的方法是通过布局来解决。请看图 5-6 所示的几种按钮格式。

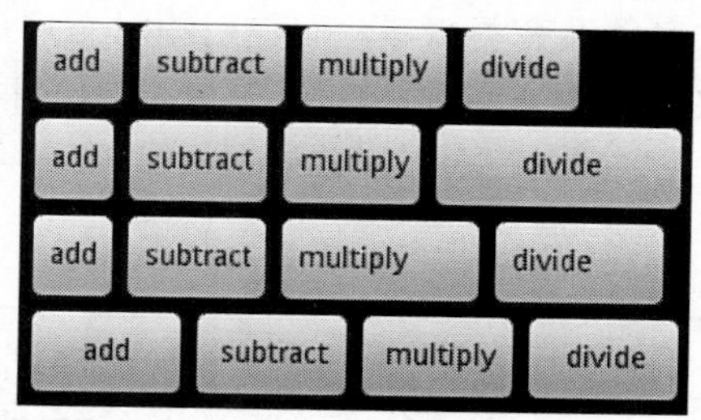


图 5-6 几种带有 4 个按钮的布局的调整方法

图 5-6 中的 4 行按钮分别如下。

- 第一行与代码清单 5-2 中创建的相同，但在每个单词末尾都添加了空格。
- 第二行中，最后一个按钮的布局宽度被改为 match_parent，为按钮提供了所需的空间，但该办法不适用于前面那些按钮，因为就会出现图 5-5 右边一幅的状况，某些按钮被挤掉。

```
<Button android:text="divide"
        android:layout_width="match_parent"
        android:layout_height="wrap_content"
/>
```

- 第三行为“multiply”按钮添加了 padding，使按钮变大了。但它并没有为文字本身增加空间，因为它被声明成了 wrap_conent 类型：

```
<Button android:text="multiply"
        android:layout_width="wrap_content"
        android:layout_height="wrap_content"
        android:paddingRight="20sp"
/>
```

- 第四行中，所有按钮都被设为 match_parent，但同时添加了 layout_weight 属性，并被赋予相同的属性值。这样产生的布局结果最令人满意：

```
<Button android:text="add"
        android:layout_width="match_parent"
        android:layout_height="wrap_content"
        android:layout_weight="1"
/>
<Button android:text="subtract"
  android:layout_width="match_parent"
  android:layout_height="wrap_content"
  android:layout_weight="1"
/>
<Button android:text="multiply"
  android:layout_width="match_parent"
  android:layout_height="wrap_content"
  android:layout_weight="1"
/>
<Button android:text="divide"
  android:layout_width="match_parent"
  android:layout_height="wrap_content"
  android:layout_weight="1"
/>
```

## 技巧 40：设置相对布局和布局 ID

有时将布局位置设定为与起始对象或父对象的某种相对关系，比使用绝对的规则要更加方便。另外，如果 UI 启用了嵌套 LinearLayout，它可能会变得低效，而使用相对布局可能更为简单。实现方法是使用 RelativeLayout 视图，如代码清单 5-3 所示，而布局效果则如图 5-7 所示。

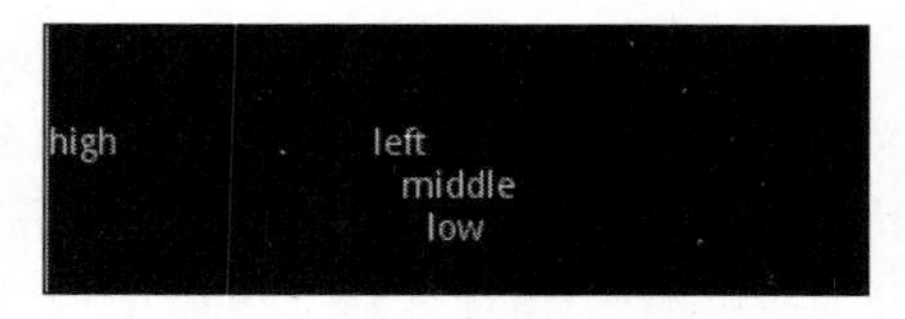


图 5-7　带 4 个文本视图的 RelativeLayout 示例

**代码清单 5-3　RelativeLayout**

```
<?xml version="1.0" encoding="utf-8"?>
<RelativeLayout xmlns:android="http://schemas.android.com/apk/res/android"
    android:layout_width="match_parent"
    android:layout_height="match_parent">
    <TextView android:id="@+id/mid" android:text="middle"
        android:layout_width="wrap_content"
        android:layout_height="wrap_content"
        android:layout_centerInParent="true"/>
    <TextView android:id="@+id/high" android:text="high"
        android:layout_width="wrap_content"
        android:layout_height="wrap_content"
        android:layout_above="@id/mid"/>
    <TextView android:id="@+id/low" android:text="low"
        android:layout_width="wrap_content"
        android:layout_height="wrap_content"
        android:layout_centerHorizontal="true"
        android:layout_below="@id/mid"/>
    <TextView android:id="@+id/left" android:text="left"
        android:layout_width="wrap_content"
        android:layout_height="wrap_content"
        android:layout_alignBottom="@id/high"
        android:layout_toLeftOf="@id/low"/>
</RelativeLayout>
```

对于清单中属性的解释，以及可用于相对布局的不同规则被收集在表 5-2 中。因为每个布局都可能有一部分在 XML 文件中声明，另一部分用 Java 代码声明，表中给出了两种不同的引用布局的方法。表的前三行给出了需要被指向某个视图 ID 的属性，而最后两行则是布尔型的属性。

**表 5-2　相对布局子控件可使用的规则**

| XML 属性（全部以 android:标签开头） | Java 常量 | 相对布局规则 |
|---|---|---|
| layout_above<br>layout_below<br>layout_toRightOf<br>layout_toLeftOf | ABOVE<br>BELOW<br>RIGHT_OF<br>LEFT_OF | 将本视图的边缘与给定的视图的边缘对齐 |
| layout_alignTop<br>layout_alignBottom<br>layout_alignRight<br>layout_alignLeft | ALIGN_TOP<br>ALIGN_BOTTOM<br>ALIGN_RIGHT<br>ALIGN_LEFT | 将本视图的边缘与给定的视图的边缘对齐 |

续表

| XML 属性（全部以 android:标签开头） | Java 常量 | 相对布局规则 |
| --- | --- | --- |
| layout_alignBaseline | ALIGN_BASELINE | 将本视图文本基线（baseline）与给定视图文本基线对齐 |
| layout_alignParentTop<br>layout_alignParentBottom<br>layout_alignParentRight<br>layout_alignParentLeft | ALIGN_PARENT_TOP<br>ALIGN_PARENT_BOTTOM<br>ALIGN_PARENT_RIGHT<br>ALIGN_PARENT_LEFT | 将本视图边缘与父视图边缘对齐 |
| layout_centerInParent<br>layout_centerHorizontal<br>layout_centerVertical | CENTER_IN_PARENT<br>CENTER_HORIZONTAL<br>CENTER_VERTICAL | 将本视图置于父视图的中央 |

## 技巧 41：通过编程声明布局

在 Android 中，XML 布局框架是我们首选的方法，因为它既可以适应设备变化，又能简化开发过程。然而，有时需要通过手工编程（比如使用 Java）来对布局的某些方面进行修改。事实上，整个布局都可用 Java 来声明。为说明这一点，我们把上个技巧中的布局的一部分用 Java 代码来实现，如代码清单 5-4 所示。

需要强调的是，用 Java 来编码布局不仅繁琐，而且会令人沮丧。该方法并未利用资源目录的模块化方式，而该方式不修改 Java 代码就能改变布局，这在技巧 37 中介绍过。

**代码清单 5-4 src/com/cookbook/programmaticlayout/ProgrammaticLayout.java**

```
package com.cookbook.programmatic_layout;

import android.app.Activity;
import android.os.Bundle;
import android.view.ViewGroup;
import android.view.ViewGroup.LayoutParams;
import android.widget.RelativeLayout;
import android.widget.TextView;

public class ProgrammaticLayout extends Activity {
    private int TEXTVIEW1_ID = 100011;
    @Override
    public void onCreate(Bundle savedInstanceState) {
        super.onCreate(savedInstanceState);

        //Here is an alternative to: setContentView(R.layout.main);
        final RelativeLayout relLayout = new RelativeLayout( this );
        relLayout.setLayoutParams( new RelativeLayout.LayoutParams(
                                    LayoutParams.MATCH_PARENT,
                                    LayoutParams.MATCH_PARENT ) );
        TextView textView1 = new TextView( this );
        textView1.setText("middle");
        textView1.setTag(TEXTVIEW1_ID);
```

```
        RelativeLayout.LayoutParams text1layout = new
            RelativeLayout.LayoutParams( LayoutParams.WRAP_CONTENT,
                                         LayoutParams.WRAP_CONTENT );
        text1layout.addRule( RelativeLayout.CENTER_IN_PARENT );
        relLayout.addView(textView1, text1layout);
        TextView textView2 = new TextView( this );
        textView2.setText("high");

        RelativeLayout.LayoutParams text2Layout = new
            RelativeLayout.LayoutParams( LayoutParams.WRAP_CONTENT,
                                         LayoutParams.WRAP_CONTENT );
        text2Layout.addRule(RelativeLayout.ABOVE, TEXTVIEW1_ID );
        relLayout.addView( textView2, text2Layout );

        setContentView( relLayout );
    }
}
```

## 技巧 42：通过独立线程更新布局

我们在第 3 章中已经讨论过，当一个耗时的 Activity 在运行的时候，必须注意确保 UI 线程能够保持响应。这可以通过为耗时任务创建独立的线程并让 UI 线程在高优先级上继续运行来实现。如果随后这个独立线程需要更新 UI，可以用一个 Handler 向 UI 线程发送更新信息。

本技巧使用一个按钮来触发一项分为两部分的耗时计算工作，每部分完成之后会更新屏幕。代码清单 5-5 所示的 XML 文件给出了布局定义，其中包含名为 `computation_status` 的状态文本，以及名为 `action` 的触发按钮。程序使用的字符串定义在 **strings.xml** 中，如代码清单 5-6 所示。

代码清单 5-5 res/layout/main.xml

```
<?xml version="1.0" encoding="utf-8"?>
<LinearLayout
    xmlns:android="http://schemas.android.com/apk/res/android"
    android:orientation="vertical"
    android:layout_width="match_parent"
    android:layout_height="match_parent">
    <TextView android:id="@+id/computation_status"
        android:layout_width="match_parent"
        android:layout_height="wrap_content"
        android:text="@string/hello" android:textSize="36sp"
        android:textColor="#000" />
    <Button android:text="@string/action"
        android:id="@+id/action"
        android:layout_width="wrap_content"
        android:layout_height="wrap_content" />
</LinearLayout>
```

代码清单 5-6 res/layout/strings.xml

```
<?xml version="1.0" encoding="utf-8"?>
<resources>
    <string name="hello">Hello World, HandlerUpdateUi!</string>
    <string name="app_name">HandlerUpdateUi</string>
```

```
    <string name="action">Press to Start</string>
    <string name="start">Starting...</string>
    <string name="first">First Done</string>
    <string name="second">Second Done</string>
</resources>
```

通过后台线程更新 UI 的步骤如下。

（1）初始化一个指向被后台线程所更新的 UI 对象（本例中为 av）的 Handler。

（2）定义一个可以根据需要更新 UI 的 Runnable 函数（本例中为 mUpdateResults）。

（3）声明一个用于处理线程间信息的 Handler（本例中为 mHandler）。

（4）在后台线程中，设置适当的标志来传达状态的变化（本例中要被改变的是 textString 和 backgroundColor）。

（5）在后台线程中，让 Handler 将 UI 更新函数传送给主线程。

经由上述步骤产生的 Activity 如代码清单 5-7 所示。

**代码清单 5-7　src/com/cookbook/handler_ui/HandlerUpdateUi.java**

```java
package com.cookbook.handler_ui;

import android.app.Activity;
import android.graphics.Color;
import android.os.Bundle;
import android.os.Handler;
import android.view.View;
import android.widget.Button;
import android.widget.TextView;

public class HandlerUpdateUi extends Activity {
    TextView av; //UI reference
    int textString = R.string.start;
    int backgroundColor = Color.DKGRAY;

    final Handler mHandler = new Handler();
    // Create runnable for posting results to the UI thread
    final Runnable mUpdateResults = new Runnable() {
        public void run() {
            av.setText(textString);
            av.setBackgroundColor(backgroundColor);
        }
    };

    @Override
    public void onCreate(Bundle savedInstanceState) {
        super.onCreate(savedInstanceState);
        setContentView(R.layout.main);
        av = (TextView) findViewById(R.id.computation_status);

        Button actionButton = (Button) findViewById(R.id.action);
        actionButton.setOnClickListener(new View.OnClickListener() {
            public void onClick(View view) {
                doWork();
            }
        });
    }

    //example of a computationally intensive action with UI updates
```

```
    private void doWork() {
        Thread thread = new Thread(new Runnable() {
            public void run() {
                textString=R.string.start;
                backgroundColor = Color.DKGRAY;
                mHandler.post(mUpdateResults);

                computation(1);
                textString=R.string.first;
                backgroundColor = Color.BLUE;
                mHandler.post(mUpdateResults);

                computation(2);
                textString=R.string.second;
                backgroundColor = Color.GREEN;
                mHandler.post(mUpdateResults);
            }
        });
        thread.start();
    }

    final static int SIZE=1000; //large enough to take some time
    double tmp;
    private void computation(int val) {
        for(int ii=0; ii<SIZE; ii++)
            for(int jj=0; jj<SIZE; jj++)
                tmp=val*Math.log(ii+1)/Math.log1p(jj+1);
    }
}
```

## 5.3　文本操作

在诸如 TextView、EditText 和 Button 一类的包含文本的视图中，文本是由 XML 布局文件中的 android:text 元素来表示的。如本章开头讨论的那样，最好通过在 **string.xml** 中定义字符串来初始化文本，让所有的字符串集中到同一个地方。因此，向 TextView 一类的 UI 元素中添加文本的方法可能像下面这样：

```
<TextView android:text="@string/myTextString"
        android:id="@+id/my_text_label"
        android:layout_width="wrap_content"
        android:layout_height="wrap_content" />
```

默认字体取决于 Android 设备和用户偏好。要指定确切的字体，可使用表 5-3 列出的那些元素。在第四栏中用粗体标出的是元素的默认值。

表 5-3　TextView 的常用属性

| 属　性 | XML 元素 | Java 方法 | 可　取　值 |
|---|---|---|---|
| 显示字符串 | android:text | setText(CharSequence) | 任意字符串 |
| 字体大小 | android:textSize | setTextSize(float) | 任意尺寸 |
| 字体颜色 | android:textColor | setTextColor(int) | 任意颜色 |
| 背景颜色 | N/A | setBackgroundColor(int) | 任意颜色 |

续表

| 属 性 | XML 元素 | Java 方法 | 可 取 值 |
|---|---|---|---|
| 字体样式 | android:textStyle | setTypeface(Typeface) | Bold、Italic、Bold Italic |
| 字体类型 | android:typeface | setTypeface(Typeface) | **Normal**、Sans serif、Monospace |
| 文本在显示区域中的位置 | android:gravity | setGravity(int) | Top、Bottom、**Left**、Right（还有……） |

## 技巧 43：设置和改变文本属性

这个技巧实现的功能是，当按钮被点击时改变显示文本的颜色。很容易将功能扩展为改变字体大小或字体风格，这在本技巧的结尾处会加以探讨。

主布局只是带有一个 TextView 和一个 Button 的垂直排列的 LinearLayout，如代码清单 5-8 所示，文本控件 ID 被设为 mod_text，所显示的文本是在 **strings.xml** 中定义的名为 changed_text 的字符串，这些在代码清单 5-9 中给出。按钮的 ID 被设为 change，按钮上的文本显示是 **strings.xml** 中的 button_text 字符串。

**代码清单 5-8 res/layout/main.xml**

```
<?xml version="1.0" encoding="utf-8"?>
<LinearLayout xmlns:android="http://schemas.android.com/apk/res/android"
    android:orientation="vertical"
        android:layout_width="match_parent"
        android:layout_height="match_parent">
        <TextView android:text="@string/changed_text"
                  android:textSize="48sp"
                  android:id="@+id/mod_text"
                  android:layout_width="wrap_content"
                  android:layout_height="wrap_content" />
        <Button android:text="@string/button_text"
                  android:textSize="48sp"
                  android:id="@+id/change"
                  android:layout_width="wrap_content"
                  android:layout_height="wrap_content" />
</LinearLayout>
```

**代码清单 5-9 res/values/strings.xml**

```
<?xml version="1.0" encoding="utf-8"?>
<resources>
    <string name="app_name">ChangeFont</string>
    <string name="changed_text">Rainbow Connection</string>
    <string name="button_text">Press to change the font color</string>
</resources>
```

代码清单 5-10 给出的 Activity 使用了 **main.xml** 布局，并用 mod_text 这一 ID 来标识 TextView Handler。接着又重写了按钮的 OnClickListener 方法，以便用表 5-3 中给出的办法来设定文本颜色。

代码清单 5-10　src/com/cookbook/change_font/ChangeFont.java

```
package com.cookbook.change_font;

import android.app.Activity;
import android.os.Bundle;
import android.view.View;
import android.widget.Button;
import android.widget.TextView;

public class ChangeFont extends Activity {
    TextView tv;
    private int colorVals[]={R.color.start, R.color.mid, R.color.last};
    int idx=0;
    /** called when the activity is first created */
    @Override
    public void onCreate(Bundle savedInstanceState) {
        super.onCreate(savedInstanceState);
        setContentView(R.layout.main);
        tv = (TextView) findViewById(R.id.mod_text);

        Button changeFont = (Button) findViewById(R.id.change);
        changeFont.setOnClickListener(new View.OnClickListener() {
            public void onClick(View view) {
                tv.setTextColor(getResources().getColor(colorVals[idx]));
                idx = (idx+1)%3;
            }
        });
    }
}
```

可用的颜色资源被定义在全局文件 **colors.xml** 中，如代码清单 5-11 所示。定义的颜色有红、绿和蓝，但它们被根据功能命名为 `start`、`mid` 和 `last`。这就提供一种简便的方法，在稍后无需改变 Handler 的名称，就能改变颜色。

代码清单 5-11　res/values/colors.xml

```
<?xml version="1.0" encoding="utf-8"?>
<resources>
        <color name="start">#f00</color>
        <color name="mid">#0f0</color>
        <color name="last">#00f</color>
</resources>
```

对本技巧给出的程序略做修改，就可以实现对文本大小（或文本风格）的改变。例如，可将 `colorVals[]`变为 `sizeVals[]`，并将其指向 `R.dimen` 资源：

```
private int sizeVals[]={R.dimen.small, R.dimen.medium, R.dimen.large};
tv.setTextSize(getResources().getDimension(sizeVals[idx]));
```

同时也要把 **color.xml** 替换为 **dimens.xml**，如代码清单 5-12 所示。

代码清单 5-12　dimens.xml 文件的用法

```
<?xml version="1.0" encoding="utf-8"?>
<resources>
        <dimen name="start">12sp</dimen>
        <dimen name="mid">24sp</dimen>
        <dimen name="last">48sp</dimen>
</resources>
```

如果要改变的是文本字符串，则将 colorVals[]改为 textVals[]，并像如下这样指向 R.string 包含的资源：

```
private int textVals[]={R.string.first_text,
                        R.string.second_text, R.string.third_text};
tv.setText(getBaseContext().getString(textVals[idx]));
```

在此要用到 **string.xml**，如代码清单 5-13 所示。

**代码清单 5-13 strings.xml 文件的用法**

```
<?xml version="1.0" encoding="utf-8"?>
<resources>
    <string name="app_name">ChangeFont</string>
    <string name="changed_text">Rainbow Connection</string>
    <string name="buttoN_text">Press to Change the Font Color</string>
    <string name="first_text">First</string>
    <string name="second_text">Second</string>
    <string name="third_text">Third</string>
</resources>
```

## 技巧 44：提供文本输入

EditText 类提供了一种用于文本输入的简单视图。它可以像 TextView 那样被声明，其最常用的属性如表 5-4 所示。尽管每种属性都有对应的 Java 方法，但把它们都展示在这里启发性会比较差。属性的默认值仍然在最后一列中用粗体标出。

**表 5-4 EditText 的常用属性**

| 属 性 | XML 元素 | 可 取 值 |
|---|---|---|
| 最小显示行数 | android:minLines | 任意整数 |
| 最大显示行数 | android:maxLines | 任意整数 |
| 值为空时的提示文本 | android:hint | 任意字符串 |
| 输入类型 | android:inputType | **text**<br>textCapSentences<br>textAutoCorrect<br>textAutoComplete<br>textEmailAddress<br>textNoSuggestions<br>textPassword<br>number<br>phone<br>date<br>time<br>（还有……） |

例如，在布局文件中使用下面的 XML 代码可以显示一个文本输入窗口，其中包含灰色的 **“Type text here”** 作为提示语。在不带键盘或者键盘被隐藏的设备上，选择编辑窗口后会弹出软键盘以进行文本输入，如图 5-8 所示。

```
<EditText android:id="@+id/text_result"
  android:inputType="text"
  android:textSize="30sp"
  android:hint="Type text here"
  android:layout_width="match_parent"
  android:layout_height="wrap_content" />
```

通过使用 `android:inputType="phone"`或=`"textEmailAddress"`，可以在用户选择输入窗口时，显示用于输入电话号码的软键盘或是用于输入电子邮件地址的软键盘。这两种情况见图 5-9，其中对提示文本做了相应修改。

图 5-8　使用软键盘输入文本

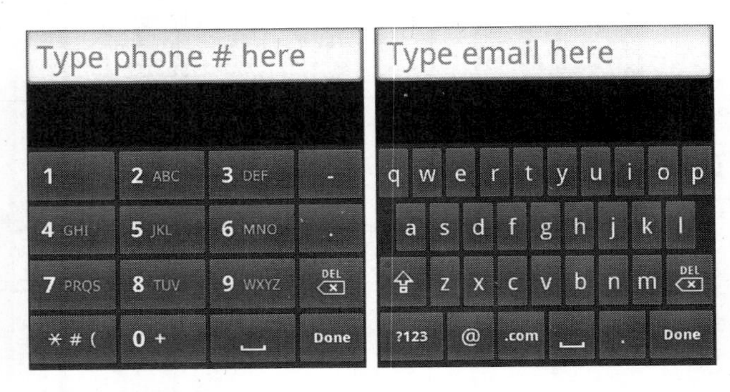


图 5-9　当 `inputType` 被设为`"phone"`或`"textEmailAddress"`时，使用软键盘

还有一点要注意，可以利用表 5-4 中给出的属性设定，在文本输入时使每句的首字母自动大写，或自动更正拼写错误的单词，或在输入时关闭单词提示。可以按照文本输入时的不同需求来控制上述这些选项。

## 技巧 45：创建表单

表单是一种图形布局，它带有一些具有文本输入或选择功能的区域。要想输入文本，可以使用 `EditText` 对象。在 `EditText` 被声明之后，一些 Java 代码需要在运行时捕获文本输入。代码清单 5-14 给出了这一功能的实现。注意，本例中输入的文本 `textResult` 不能被修改。要想修改，需要建立它的副本，在副本上进行修改。

**代码清单 5-14　从 EditText 对象中捕获文本**

```
CharSequence phoneNumber;
EditText textResult = (EditText) findViewById(R.id.text_result);
textResult.setOnKeyListener(new OnKeyListener() {
    public boolean onKey(View v, int keyCode, KeyEvent event) {
```

```
        // Register the text when "enter" is pressed
        if ((event.getAction() == KeyEvent.ACTION_UP) &&
            (keyCode == KeyEvent.KEYCODE_ENTER)) {
          // Grab the text for use in the activity
          phoneNumber = textResult.getText();
          return true;
        }
        return false;
      }
});
```

如果 onKey 方法返回 true，则意味着向其父函数表明按键按下事件已经完成，无需进一步处理。

为了提供表单中通常使用的不同的用户选择形式，实现了几种标准的微件，比如复选框、单选按钮以及下拉选择菜单。这些标准微件是使用了下一节中介绍的微件实现的。

## 5.4 其他微件：从按钮到拖动条

Android 系统提供了一些标准图形微件，开发者用它们可以创建贯穿整个应用程序的具有结合力的用户体验。最常见的微件有以下几个。

- Button：一个矩形图形，当其边界范围内的屏幕区域被触碰时予以响应。其中可以包含用户定义的文本或图像。
- CheckBox：一类带有一个选中标记图形及描述文字的按钮，当被触碰时可以切换选中或取消状态。还有一种 ToggleButton 与之类似，这里我们也会讨论。
- RadioButton：一类带有一个圆点图形的按钮，触碰时会被选中，但无法对其取消选中。多个 RadioButton 可以被组合为一个 RadioGroup（作为一个 LinearLayout），某个时刻一组中只有一个单选按钮可以被选中。
- Spinner：一种可显示当前选中项目，并带有一个箭头可引出一个下拉菜单的按钮。当下拉菜单被触碰时，会显示可选项列表。重新选择后，新选项会在下拉菜单上显示出来。
- ProgressBar：一个加亮的横条，可以直观地呈现某个操作的进度（也可以显示二级进度）。进度条本身不具有交互性，如果不能确定进度的量化值，可以把 ProgressBar 设为不确定模式，这样它会呈现出循环的动作效果。
- SeekBar：一种交互式的进度条，其进度可以通过拖动来修改，这在媒体播放等情境中很有用。它可以显示媒体已播放的时长，而用户可以通过拖动来定位到文件更靠前或更靠后的位置。

下面的几个技巧将给出上述控件的实际应用案例。

## 技巧 46：在表格布局中使用图像按钮

我们在第 2 章中就介绍过按钮。与其他视图一样，可以通过 android:background 属性在按钮上添加背景图片。然而，使用 ImageButton 这一专门微件可以带来额外的布局灵活性。ImageButton 通过 android:src 属性来指定显示的图像，如下所示：

```
<ImageButton android:id="@+id/imagebutton0"
            android:src="@drawable/android_cupcake"
            android:layout_width="wrap_content"
            android:layout_height="wrap_content" />
```

使用上述方式，图像会显示在 Button 微件的顶部。ImageButton 微件继承了 ImageView 微件放置图片的方式，即采用 android:scaleType。该属性可取的值及其对给定图片的改动效果如图 5-10 所示。

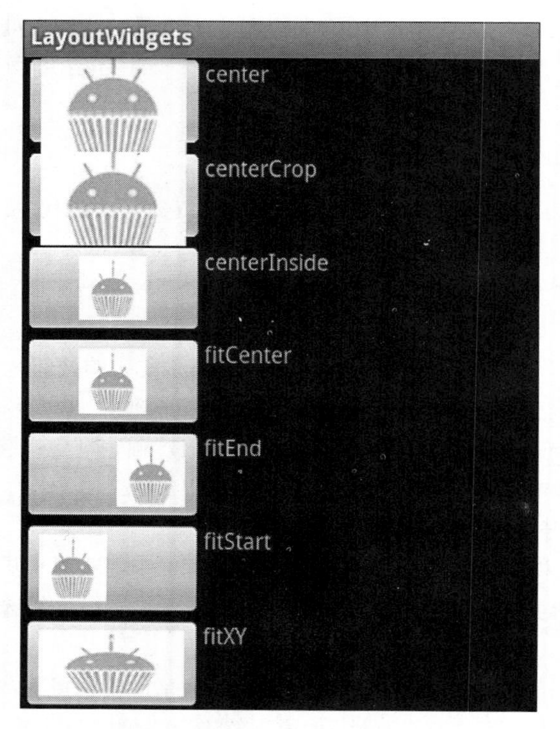


图 5-10　图像视图的 android:scaleType 属性效果示例

下面给出可对图像按钮进行的其他一些操作。

- 使用 android:padding 来避免按钮重叠，或增加按钮的间距。
- 将 android:background 设为 null（在 XML 布局文件中为@null）可以隐藏按钮，只显示图像。

按钮被隐藏后，默认情况下当图像按钮被按下后不会有视觉效果反馈。可以创建一个仅包含一个选择器（selector）元素的可绘制 XML 文件来弥补这一不足：

```
<?xml version="1.0" encoding="utf-8"?>
<selector xmlns:android="http://schemas.android.com/apk/res/android">
    <item android:drawable="@drawable/myimage_pressed"
```

```
        android:state_pressed="true" />
    <item android:drawable="@drawable/myimage_focused"
        android:state_focused="true" />
    <item android:drawable="@drawable/myimage_normal" />
</selector>
```

以上代码为按钮的按下状态、获得焦点状态及正常状态分别指定了三种不同的图像。此例中的三幅图像应当放在可绘制资源目录（如 **res/drawable-mdpi**）中。这样就可以将选择器中的文件指定为 ImageButton 的 android:src 属性。

当一个布局中同时含有多个图像按钮时，通常使用表格布局比较有益，本技巧也会对此加以介绍。TableLayout 视图组与垂直方向的 LinearLayout 相似。可以通过在每行上使用 TableRow 视图组来指定多行。代码清单 5-15 给出了布局的示例，其中为每行指定了一个 ImageButton 和一个 TextView，产生的屏幕布局效果如图 5-11 所示。

**代码清单 5-15　res/layout/ibutton.xml**

```
<?xml version="1.0" encoding="utf-8"?>
<TableLayout
    xmlns:android="http://schemas.android.com/apk/res/android"
    android:layout_width="match_parent"
    android:layout_height="match_parent">
    <TableRow>
        <ImageButton android:id="@+id/imagebutton0"
            android:src="@drawable/android_cupcake"
            android:scaleType="fitXY"
            android:background="@null"
            android:padding="5dip"
            android:layout_width="wrap_content"
            android:layout_height="90dip" />
        <TextView android:text="Cupcake"
            android:layout_width="wrap_content"
            android:layout_height="wrap_content" />
    </TableRow>
    <TableRow>
        <ImageButton android:id="@+id/imagebutton1"
            android:src="@drawable/android_donut"
            android:scaleType="fitXY"
            android:background="@null"
            android:padding="5dip"
            android:layout_width="wrap_content"
            android:layout_height="90dip" />
        <TextView android:text="Donut"
            android:layout_width="wrap_content"
            android:layout_height="wrap_content" />
    </TableRow>
    <TableRow>
        <ImageButton android:id="@+id/imagebutton2"
            android:src="@drawable/android_eclair"
            android:scaleType="fitXY"
            android:background="@null"
            android:padding="5dip"
            android:layout_width="wrap_content"
            android:layout_height="90dip" />
        <TextView android:text="Eclair"
            android:layout_width="wrap_content"
            android:layout_height="wrap_content" />
    </TableRow>
```

```
        <TableRow>
            <ImageButton android:id="@+id/imagebutton3"
                android:src="@drawable/android_froyo"
                android:scaleType="fitXY"
                android:background="@null"
                android:padding="5dip"
                android:layout_width="wrap_content"
                android:layout_height="90dip" />
            <TextView android:text="FroYo"
                android:layout_width="wrap_content"
                android:layout_height="wrap_content" />
        </TableRow>
        <TableRow>
            <ImageButton android:id="@+id/imagebutton4"
                android:src="@drawable/android_gingerbread"
                android:scaleType="fitXY"
                android:background="@null"
                android:padding="5dip"
                android:layout_width="wrap_content"
                android:layout_height="90dip" />
            <TextView android:text="Gingerbread"
                android:layout_width="wrap_content"
                android:layout_height="wrap_content" />
        </TableRow>
    </TableLayout>
```

图 5-11　由 ImageButton 和 TextView 组成的 TableLayout

## 技巧 47：使用复选框和开关按钮

复选框预定义了对勾图形、选中或未选中时的颜色以及按下时的颜色。这样就为 Android 应用程序带来了统一的观感。如果需要自定义选中或未选中时的图形，可以通过 setButtonDrawable() 来实现。

要实现本技巧中的复选框范例，需要在布局文件中声明 CheckBox，如代码清单 5-16 所示。android:text 属性定义了复选框后的文字标签。为方便说明，我们还在布局上添加了一些文本视图。

**代码清单 5-16 res/layout/ckbox.xml**

```
<?xml version="1.0" encoding="utf-8"?>
<LinearLayout
        xmlns:android="http://schemas.android.com/apk/res/android"
        android:orientation="vertical"
        android:layout_width="match_parent"
        android:layout_height="match_parent">
        <CheckBox android:id="@+id/checkbox0"
                android:text="Lettuce"
                android:layout_width="wrap_content"
                android:layout_height="wrap_content" />
        <CheckBox android:id="@+id/checkbox1"
                android:text="Tomato"
                android:layout_width="wrap_content"
                android:layout_height="wrap_content" />
        <CheckBox android:id="@+id/checkbox2"
                android:text="Cheese"
                android:layout_width="wrap_content"
                android:layout_height="wrap_content" />
        <TextView android:text="Lettuce, Tomato, Cheese choices:"
                android:layout_width="wrap_content"
                android:layout_height="wrap_content" />
        <TextView android:id="@+id/status"
                android:layout_width="wrap_content"
                android:layout_height="wrap_content" />
</LinearLayout>
```

布局文件中的视图可以与 Java 文件中的视图实例相关联，如代码清单 5-17 所示。这里使用了一个私有的内部类来登记三明治中的馅料。三个复选框都拥有 onClickListener 方法，以跟踪馅料的变化，从而更新文本视图中的显示。图 5-12 给出了在某种选择组合下的最终输出。

**代码清单 5-17 src/com/cookbook/layout_widgets/CheckBoxExample.java**

```
package com.cookbook.layout_widgets;

import android.app.Activity;
import android.os.Bundle;
import android.view.View;
import android.view.View.OnClickListener;
import android.widget.CheckBox;
import android.widget.TextView;

public class CheckBoxExample extends Activity {
    private TextView tv;

    @Override
    public void onCreate(Bundle savedInstanceState) {
        super.onCreate(savedInstanceState);
        setContentView(R.layout.ckbox);
        tv = (TextView) findViewById(R.id.status);

        class Toppings {private boolean LETTUCE, TOMATO, CHEESE;}

        final Toppings sandwichToppings = new Toppings();
        final CheckBox checkbox[] = {
```

```
                (CheckBox) findViewById(R.id.checkbox0),
                (CheckBox) findViewById(R.id.checkbox1),
                (CheckBox) findViewById(R.id.checkbox2)};

        checkbox[0].setOnClickListener(new OnClickListener() {
            @Override
            public void onClick(View v) {
                if (((CheckBox) v).isChecked()) {
                    sandwichToppings.LETTUCE = true;
                } else {
                    sandwichToppings.LETTUCE = false;
                }
                tv.setText(""+sandwichToppings.LETTUCE + " "
                        +sandwichToppings.TOMATO + " "
                        +sandwichToppings.CHEESE + " ");
            }
        });
        checkbox[1].setOnClickListener(new OnClickListener() {
            @Override
            public void onClick(View v) {
                if (((CheckBox) v).isChecked()) {
                    sandwichToppings.TOMATO = true;
                } else {
                    sandwichToppings.TOMATO = false;
                }
                tv.setText(""+sandwichToppings.LETTUCE + " "
                        +sandwichToppings.TOMATO + " "
                        +sandwichToppings.CHEESE + " ");
            }
        });
        checkbox[2].setOnClickListener(new OnClickListener() {
            @Override
            public void onClick(View v) {
                if (((CheckBox) v).isChecked()) {
                    sandwichToppings.CHEESE = true;
                } else {
                    sandwichToppings.CHEESE = false;
                }
                tv.setText(""+sandwichToppings.LETTUCE + " "
                        +sandwichToppings.TOMATO + " "
                        +sandwichToppings.CHEESE + " ");
            }
        });
    }
}
```

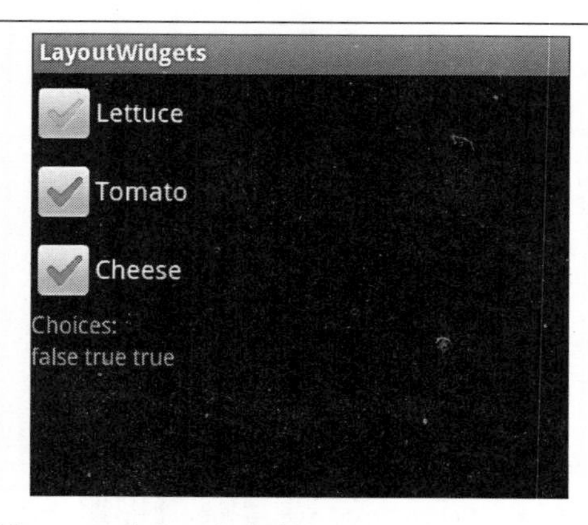


图 5-12　选中及未选中状态的 CheckBox 示例

开关按钮与复选框类似，但使用不同的图形，而且文本与按钮是集成在一起的，而不是放在一边的。可以修改代码清单 5-16 及 5-17，将其中的 CheckBox 替换成 ToggleButton：

```
<ToggleButton android:id="@+id/ToggleButton0"
              android:textOff="No Lettuce"
              android:textOn="Lettuce"
              android:layout_width="wrap_content"
              android:layout_height="wrap_content" />
```

注意，android:text 元素被 android:textOff（如果未指定，其默认值为“OFF”）及 android:textOn（如果未指定，其默认值为“ON”）元素所取代，从而根据选择状态来改变开关按钮上显示的内容。图 5-13 给出了一种输出示例。

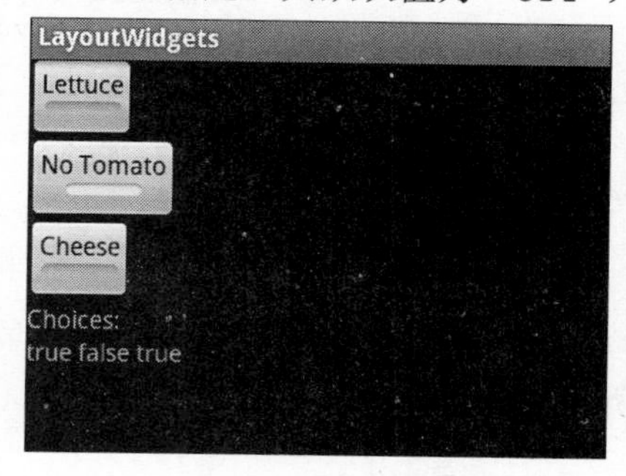

图 5-13 一组选中或未选中状态的 ToggleButton 示例

## 技巧 48：使用单选按钮

单选按钮就像是无法被取消勾选的复选框。选择某个单选按钮后会自动取消先前选中的其他单选按钮的选中状态。通常把一组单选按钮放到一个 RadioGroup 中，以确保同一时刻其中只有一个被选中。这在代码清单 5-18 中的布局文件中有所体现。

**代码清单 5-18 res/layout/rbutton.xml**

```
<?xml version="1.0" encoding="utf-8"?>
      <RadioGroup android:id="@+id/RadioGroup01"
            android:layout_width="wrap_content"
            android:layout_height="wrap_content">
            <RadioButton android:text="Republican"
                  android:id="@+id/RadioButton02"
                  android:layout_width="wrap_content"
                  android:layout_height="wrap_content" />
            <RadioButton android:text="Democrat"
                  android:id="@+id/RadioButton03"
                  android:layout_width="wrap_content"
                  android:layout_height="wrap_content" />
            <RadioButton android:text="Independent"
                  android:id="@+id/RadioButton01"
                  android:layout_width="wrap_content"
                  android:layout_height="wrap_content" />
      </RadioGroup>
```

这个例子中的 Activity 与代码清单 5-17 中的相似，只是用 RadioButton 代替了 CheckBox。代码清单 5-18 的布局显示效果如图 5-14 所示。

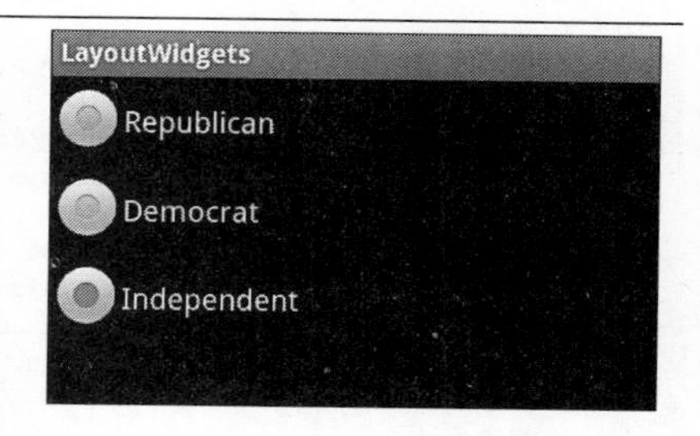

图 5-14 含有三个单选按钮的 RadioGroup 示例

## 技巧 49：创建下拉菜单

下拉菜单被称作 spinner，它是一种定义在像代码清单 5-19 中给出的那种通常的屏幕布局中的微件。

**代码清单 5-19　res/layout/spinner.xml**

```
<?xml version="1.0" encoding="utf-8"?>
<LinearLayout
        xmlns:android="http://schemas.android.com/apk/res/android"
        android:layout_width="wrap_content"
        android:layout_height="wrap_content">
        <Spinner android:id="@+id/spinner"
                android:prompt="@string/oceaN_prompt"
                android:layout_width="wrap_content"
                android:layout_height="wrap_content" />
</LinearLayout>
```

下拉菜单的标题可以用 `android:prompt` 属性指定，标题对应的字符串需要在 **strings.xml** 文件中定义，比如：

```
<string name="ocean_prompt">Choose your favorite ocean</string>
```

还需要为下拉菜单定义一个单独的布局来指定其外观，如代码清单 5-20 中的 **spinner_entry.xml** 所示。

**代码清单 5-20　res/layout/spinner_entry.xml**

```
<?xml version="1.0" encoding="utf-8"?>
<TextView
  xmlns:android="http://schemas.android.com/apk/res/android"
  android:gravity="center"
  android:textColor="#000"
  android:textSize="40sp"
  android:layout_width="match_parent"
  android:layout_height="wrap_content">
</TextView>
```

注意，下拉菜单条目布局的内容并不限于文本，还可以包含图像或任何其他在布局中支持的对象。

调用下拉菜单的 Activity 需要声明一个 `Adapter` 构造器，用布局文件中定义的视图填充下拉菜单。代码清单 5-21 给出了这种 Activity 的一个例子。

**代码清单 5-21　src/com/cookbook/layout_widgets/SpinnerExample.java**

```
package com.cookbook.layout_widgets;

import android.app.Activity;
import android.os.Bundle;
import android.widget.ArrayAdapter;
import android.widget.Spinner;

public class SpinnerExample extends Activity {
    private static final String[] oceans = {
        "Pacific", "Atlantic", "Indian",
        "Arctic", "Southern" };

    @Override
    protected void onCreate(Bundle savedInstanceState) {
        super.onCreate(savedInstanceState);
```

```
        setContentView(R.layout.spinner);

        Spinner favoriteOcean = (Spinner) findViewById(R.id.spinner);

        ArrayAdapter<String> mAdapter = new
            ArrayAdapter<String>(this, R.layout.spinner_entry, oceans);
        mAdapter.setDropDownViewResource(R.layout.spinner_entry);
        favoriteOcean.setAdapter(mAdapter);
    }
}
```

在前一个例子中，下拉菜单的条目由字符串数组 oceans[]来定义，该数组被传给 ArrayAdapter 构造器。这一实现假定下拉菜单条目在运行时不会变化。要指定更为一般的情况，即允许增加或修改条目，mAdapter 需要使用其 add()方法来构建。代码清单 5-21 中 onCreate()方法的粗体部分可以被改为如下内容：

```
Spinner favoriteOcean = (Spinner) findViewById(R.id.spinner);
ArrayAdapter<String> mAdapter = new
    ArrayAdapter<String>(this, R.layout.spinner_entry);
mAdapter.setDropDownViewResource(R.layout.spinner_entry);
for(int idx=0; idx<oceans.length; idx++)
    mAdapter.add(oceans[idx]);
favoriteOcean.setAdapter(mAdapter);
```

这一 ArrayAdapter 构造器允许 add()、remove()和 clear()方法在运行时修改选择列表，同时允许 getView()通过为每个下拉菜单选项重用布局视图来提升运行速度。

## 技巧 50：使用进度条

本技巧演示了进度条的使用，前面在代码清单 5-7 中使用文本来显示计算进度，而此处我们将用图形化的显示取而代之。我们通过向布局中添加一个进度条对象来实现，比如：

```
<ProgressBar android:id="@+id/ex_progress_bar"
        style="@android:attr/progressBarStyleHorizontal"
        android:layout_width="270dp"
        android:layout_height="50dp"
        android:progress="0"
        android:secondaryProgress="0" />
```

随着进度的变化，android:progress 属性会随之变化，呈现为一个横跨屏幕的彩色进度条。设定可选属性 android:secondaryProgress 会显示浅色的进度条，可用于表明进度中的里程碑等情形。

负责更新进度条的 Activity 如代码清单 5-22 所示。该 Activity 与代码清单 5-7 中的类似，只是用 ProgressBar 代替了对背景颜色的改变。在本例中，结果更新函数通过 Java 代码来更新进度属性。

**代码清单 5-22　src/com/cookbook/handler_ui/HandlerUpdateUi.java**

```
package com.cookbook.handler_ui;

import android.app.Activity;
```

```
import android.os.Bundle;
import android.os.Handler;
import android.view.View;
import android.widget.Button;
import android.widget.ProgressBar;

public class HandlerUpdateUi extends Activity {
    private static ProgressBar m_progressBar; //UI reference
    int percentDone = 0;

    final Handler mHandler = new Handler();
    // Create runnable for posting results to the UI thread
    final Runnable mUpdateResults = new Runnable() {
        public void run() {
            m_ progressBar.setProgress(percentDone);
        }
    };

    @Override
    public void onCreate(Bundle savedInstanceState) {
        super.onCreate(savedInstanceState);
        setContentView(R.layout.main);
        m_progressBar = (ProgressBar) findViewById(R.id.ex_progress_bar);

        Button actionButton = (Button) findViewById(R.id.action);
        actionButton.setOnClickListener(new View.OnClickListener() {
            public void onClick(View view) {
                doWork();
            }
        });
    }

    //example of a computationally intensive action with UI updates
    private void doWork() {
        Thread thread = new Thread(new Runnable() {
            public void run() {
                percentDone = 0;
                mHandler.post(mUpdateResults);

                computation(1);
                percentDone = 50;
                mHandler.post(mUpdateResults);

                computation(2);
                percentDone = 100;
                mHandler.post(mUpdateResults);
            }
        });
        thread.start();
    }

    final static int SIZE=1000; //large enough to take some time
    double tmp;
    private void computation(int val) {
        for(int ii=0; ii<SIZE; ii++)
            for(int jj=0; jj<SIZE; jj++)
                tmp=val*Math.log(ii+1)/Math.log1p(jj+1);
    }
}
```

如果需要更频繁地显示更新状况，可用 Handler 的 `postDelayed` 方法来代替 `post` 方法，并在 Runnable 的结果更新函数的结尾添加 postDelayed 方法（这与第 3 章的技巧 20 中所用的类似）。

## 技巧 51：使用拖动条

拖动条与进度条相似，但可以根据用户输入来改变进度值。当前进度通过一个被称为 thumb 的小滑块来指示。用户可以点击并拖动小滑块，确定新的进度位置，从而改变进度值。代码清单 5-23 给出了本例的主 Activity。

**代码清单 5-23 src/com/cookbook/seekbar/SeekBarEx.java**

```
package com.cookbook.seekbar;

import android.app.Activity;
import android.os.Bundle;
import android.widget.SeekBar;

public class SeekBarEx extends Activity {
    private SeekBar m_seekBar;
    boolean advancing = true;

    @Override
    public void onCreate(Bundle savedInstanceState) {
        super.onCreate(savedInstanceState);
        setContentView(R.layout.main);

        m_seekBar = (SeekBar) findViewById(R.id.SeekBar01);
        m_seekBar.setOnSeekBarChangeListener(new
                    SeekBar.OnSeekBarChangeListener() {
            public void onProgressChanged(SeekBar seekBar,
                    int progress, boolean fromUser) {
                if(fromUser) count = progress;
            }

            public void onStartTrackingTouch(SeekBar seekBar) {}
            public void onStopTrackingTouch(SeekBar seekBar) {}
        });

        Thread initThread = new Thread(new Runnable() {
            public void run() {
                show_time();
            }
        });
        initThread.start();
    }

    int count;
    private void show_time() {
        for(count=0; count<100; count++) {
            m_seekBar.setProgress(count);

            try {
                Thread.sleep(100);
            } catch (InterruptedException e) {
                e.printStackTrace();
            }
```

```
            }
        }
    }
```

在布局 XML 文件中对微件的定义如代码清单 5-24 所示。注意，我们用一幅杯型蛋糕的图像取代了默认的 thumb 按钮，效果如图 5-15 所示。

**代码清单 5-24　res/layout/main.xml**

```
<?xml version="1.0" encoding="utf-8"?>
<RelativeLayout
    xmlns:android="http://schemas.android.com/apk/res/android"
    android:layout_width="match_parent"
    android:layout_height="match_parent">
    <TextView android:layout_width="match_parent"
        android:layout_height="wrap_content"
        android:textSize="24sp" android:text="Drag the cupcake"
        android:layout_alignParentTop="true" />
    <SeekBar android:id="@+id/SeekBar01"
        android:layout_centerInParent="true"
        android:layout_width="match_parent"
        android:layout_height="wrap_content"
        android:thumb="@drawable/pink_cupcake_no_bg" />
</RelativeLayout>
```

图 5-15　SeekBar 示例，其中采用自定义的杯型蛋糕图片作为小滑块

# 第 6 章

# 用户界面事件

用户界面包括屏幕布局和事件处理两个方面。我们在第 5 章已经讨论了如何用文本或按钮等 View 对象组成视图。本章则介绍如何处理用户事件，比如物理按键、触摸事件或菜单导航等。本章还会介绍如何使用某些高级用户界面库，诸如手势和 3D 图形等。

## 6.1 事件处理器和事件监听器

大多数与 Android 设备进行的用户交互会被系统捕获并发送给相应的回调方法。例如，如果物理键盘上的“BACK”（后退）键被按下，将调用 onBackPressed() 方法。这类事件可以通过扩展相关类并重写类方法来实现，这些方法被称作事件处理器（event handler）。

与 View 及 ViewGroup 对象的交互还支持事件监听器（event listener），该类方法会等待已注册的事件发生，并触发系统向相关回调方法发送事件信息。例如，onClickListener 事件监听器可通过 setOnclickListener() 注册给某个按钮，用于监听其按下事件。

应当尽可能地使用事件监听器，因为这样可以避免类的过度继承。此外，如果 Activity 实现了事件监听器，它获得它所包含的所有布局对象的回调，有助于编写更简洁的代码。本章将通过处理物理按键事件及屏幕触摸事件来对事件监听器和事件处理器的使用进行演示。

## 技巧 52：截取物理按键事件

标准 Android 设备拥有多个可以触发事件的物理按键，表 6-1 列出了这些按键。

表 6-1 Android 设备可能带有的物理按键

| 物理按键 | 按键事件 | 描述 |
|---|---|---|
| 电源键 | KEYCODE_POWER | 启动设备或将设备从睡眠状态唤醒；显示锁屏时的用户界面 |

续表

| 物理按键 | 按键事件 | 描　述 |
| --- | --- | --- |
| 后退键 | KEYCODE_BACK | 导航到上一屏 |
| 菜单键 | KEYCODE_MENU | 显示当前活动应用的菜单 |
| Home 键 | KEYCODE_HOME | 转到主屏幕 |
| 搜索键 | KEYCODE_SEARCH | 为当前活动应用启动搜索 |
| 拍照键 | KEYCODE_CAMERA | 启动相机 |
| 音量键 | KEYCODE_VOLUME_UP<br>KEYCODE_ VOLUME_DOWN | 根据上下文情境调节相关媒体音量（在通话时调整语音音量，在媒体播放时调整音乐音量，也可调整铃声音量） |
| 方向键（DPAD） | KEYCODE_DPAD_CENTER<br>KEYCODE_DPAD_UP<br>KEYCODE_DPAD_DOWN<br>KEYCODE_DPAD_LEFT<br>KEYCODE_DPAD_RIGHT | 某些设备上的方向键 |
| 轨迹球 | KEYCODE_DPAD_CENTER<br>KEYCODE_DPAD_UP<br>KEYCODE_DPAD_DOWN<br>KEYCODE_DPAD_LEFT<br>KEYCODE_DPAD_RIGHT | 某些设备带有的摇杆 |
| 键盘 | KEYCODE_0，…，KEYCODE_9<br>KEYCODE_A，…，KEYCODE_Z | 某些设备带有的滑出式键盘 |
| 媒体键 | KEYCODE_HEADSETHOOK | 耳机上的播放/暂停按钮 |

系统首先会将 KeyEvent 发送给当前获得焦点的 Activity 或视图中适宜的回调方法。下面给出这些回调方法。

- onKeyUp()、onKeyDown()、onKeyLongPress()：物理按键按下时的回调方法。
- onTrackballEvent()、onTouchEvent()：轨迹球及触屏操作的回调方法。
- onFocusChanged()：视图获得或失去焦点时调用的回调方法。

以上方法可以被应用程序重写以自定义不同的动作。例如，要关闭拍照键（以防被意外碰到），只要在 Activity 的 onKeyDown() 方法中把事件“消耗”（consume）掉即可。具体操作为截获 KeyEvent.KEYCODE_CAMERA 事件的方法并返回 true：

```
public boolean onKeyDown(int keyCode, KeyEvent event) {
     if (keyCode == KeyEvent.KEYCODE_CAMERA) {
          return true; // Consume event, hence do nothing on camera button
     }
     return super.onKeyDown(keyCode, event);
}
```

因为事件被消耗掉以后，就不会再被传递给其他 Android 组件，但有几个例外。

- 电源键和 HOME 键是由系统截获的，不会触及应用程序的自定义。
- 对后退键、菜单键、HOME 键和搜索键不应拦截其 KeyDown 事件，而应该拦截 KeyUp。这点与 Android 2.0 的建议一致，因为在某些平台上这些按钮不见得是真实的物理按键。

代码清单 6-1 显示了几个截获不同的物理按键事件的例子，包括以下几种：

- 在 `onKeyDown()` 方法中截获拍照和 DPAD 向左按键的相应事件，从而在屏幕上显示一条消息，然后将事件消耗掉（通过返回 `true` 实现）。
- 截获音量增加按键事件，并在屏幕上显示一条消息，但事件并不被消耗（返回值为 `false`），但音量仍然会增加。
- 在 `onKeyDown()` 方法中截获搜索键的相应事件，然后使用 `startTracking()` 方法来追踪直到按键被松开，此时再将一条消息发送到屏幕。
- 在 `onBackPressed()` 方法中对后退键的相应事件进行截获。

对最后一项要说明一点：Android 的一条可用性指导方针是，一般不应当对后退键自定义处理。然而，如果在 Activity 或对话框中由于某种原因确实需要对后退键进行自定义，在 API Level 5（代号 Eclair）或更高版本中，有一个独立的方法 `onBackPressed()` 可用来截获后退键事件。

**代码清单 6-1 src/com/cookbook/PhysicalKeyPress.java**

```
package com.cookbook.physkey;
import android.app.Activity;
import android.os.Bundle;
import android.view.KeyEvent;
import android.widget.Toast;

public class PhysicalKeyPress extends Activity {
    @Override
    public void onCreate(Bundle savedInstanceState) {
        super.onCreate(savedInstanceState);
        setContentView(R.layout.main);
    }
    @Override
    public boolean onKeyDown(int keyCode, KeyEvent event) {
        switch (keyCode) {
            case KeyEvent.KEYCODE_CAMERA:
            Toast.makeText(this, "Pressed Camera Button",
                    Toast.LENGTH_LONG).show();
            return true;
        case KeyEvent.KEYCODE_DPAD_LEFT:
            Toast.makeText(this, "Pressed DPAD Left Button",
                    Toast.LENGTH_LONG).show();
            return true;
        case KeyEvent.KEYCODE_VOLUME_UP:
            Toast.makeText(this, "Pressed Volume Up Button",
                    Toast.LENGTH_LONG).show();
            return false;
        case KeyEvent.KEYCODE_SEARCH:
            //example of tracking through to the KeyUp
            if(event.getRepeatCount() == 0)
                event.startTracking();
            return true;
```

```
        case KeyEvent.KEYCODE_BACK:
        // Make new onBackPressed compatible with earlier SDKs
            if (android.os.Build.VERSION.SDK_INT
                    < android.os.Build.VERSION_CODES.ECLAIR
                    && event.getRepeatCount() == 0) {
                onBackPressed();
            }
        }
        return super.onKeyDown(keyCode, event);
    }
    @Override
    public void onBackPressed() {
        Toast.makeText(this, "Pressed BACK Key",
                         Toast.LENGTH_LONG).show();
    }
    @Override
    public boolean onKeyUp(int keyCode, KeyEvent event) {
        if (keyCode == KeyEvent.KEYCODE_SEARCH && event.isTracking()
                && !event.isCanceled()) {
            Toast.makeText(this, "Pressed SEARCH Key",
                    Toast.LENGTH_LONG).show();
            return true;
        }
        return super.onKeyUp(keyCode, event);
    }
}
```

为了向后兼容早期版本的 SDK，可以截获 `KeyEvent.KEYCODE_BACK`；而且在早期 SDK 中，可以显式地调用 `onBackPressed()` 方法，如代码清单 6-1 所示。（注意：由于显式地使用了 Eclair，这段代码只能在 Android 2.0 或更高版本下编译，但在运行时对所有设备都是向后兼容的）。要在视图中截获后退键（本例中并未涉及），需要使用 `startTracking()` 方法，这与代码清单 6-1 中对搜索键的处理相似。

## 技巧 53：构建菜单

在 Android 中，开发者可以实现三种类型的菜单，本技巧中将逐一给出示例。

- 选项菜单（options menu）：Activity 的主菜单，当按下菜单键时就会显示出来。对于 Android API Level 10 或更低版本，它包含一个图标菜单（icon menu），还可能包含一个会在“更多”菜单项被选中时弹出的扩展菜单（expanded menu）。较新的 Android 版本只有原始的选项菜单。
- 上下文菜单（context menu）：一个浮动的菜单项列表，当长按某个视图时出现。
- 子菜单（submenu）：一个浮动菜单项列表，当某个菜单项被选中时会出现。

选项菜单是在某个 Activity 里按下菜单键的时候首次构建的。此时将会启动 `onCreateOptionsMenu()` 方法，此方法中通常包含对 menu 类方法的调用，比如：

```
menu.add(GROUP_DEFAULT, MENU_ADD, 0, "Add")
     .setIcon(R.drawable.ic_launcher);
```

`add()` 方法的第一个参数为菜单项组设定名字标签。同一组中的菜单项可以被一起处理。第二个参数是代表菜单项的整型 ID，它被传递给回调函数，以确定是哪个菜单项被选中。第三个参数是菜单中项目的顺序。如果不给定该参数，则按项目被添加到 Menu 对象上的次序来决定项目顺序。最后一个参数是菜单项显示的文本，该参数可以是一个 String 型变量或是一个像 R.string.myLabel 这样的字符串资源。选项菜单是唯一一种支持为菜单项添加图标的菜单，可以通过 setIcon()方法来实现。

onCreateOptionsMenu()方法只被调用一次，在 Activity 的其他部分中不需要被再次构建。然而，如果在运行时需要改变任意菜单项，可以随时调用 onPrepareOptionsMenu()方法。

如果选项菜单中的某一项被点击，onOptionsItemSelected()方法会被调用。调用时会传递选中的菜单项的 ID，可以使用 switch 语句来确定被选中的选项是哪个。

在本技巧中的菜单项为“添加笔记”、“删除笔记”和“发送笔记”。本例将这些功能实现为简单的模拟函数，具体操作为增加计数器(itemNum)的值、减少计数器的值，以及通过 Toast 在屏幕上显示计数器的当前值。为给出在运行时改变菜单项的例子，“删除笔记”选项被设定为只有在已有笔记被添加的情况下才可用。实现方法是，将“删除笔记”选项放到一个独立的组中，并在 itemNum 值为 0 时将该组隐藏。Activity 的代码如代码清单 6-2 所示。

**代码清单 6-2 src/com/cookbook/building_menus/BuildingMenus.java**

```
package com.cookbook.building_menus;

import android.app.Activity;
import android.os.Bundle;
import android.view.ContextMenu;
import android.view.Menu;
import android.view.MenuItem;
import android.view.SubMenu;
import android.view.View;
import android.view.ContextMenu.ContextMenuInfo;
import android.widget.TextView;
import android.widget.Toast;

public class BuildingMenus extends Activity {
    private final int MENU_ADD=1, MENU_SEND=2, MENU_DEL=3;
    private final int GROUP_DEFAULT=0, GROUP_DEL=1;
    private final int ID_DEFAULT=0;
    private final int ID_TEXT1=1, ID_TEXT2=2, ID_TEXT3=3;
    private String[] choices = {"Press Me", "Try Again", "Change Me"};

    private static int itemNum=0;
    private static TextView bv;

    @Override
    public void onCreate(Bundle savedInstanceState) {
        super.onCreate(savedInstanceState);
        setContentView(R.layout.main);
        bv = (TextView) findViewById(R.id.focus_text);

        registerForContextMenu((View) findViewById(R.id.focus_text));
    }

    @Override
    public boolean onCreateOptionsMenu(Menu menu) {
        menu.add(GROUP_DEFAULT, MENU_ADD, 0, "Add")
            .setIcon(R.drawable.ic_launcher); //example of adding icon
```

```
        menu.add(GROUP_DEFAULT, MENU_SEND, 0, "Send");
        menu.add(GROUP_DEL, MENU_DEL, 0, "Delete");

        return super.onCreateOptionsMenu(menu);
    }

    @Override
    public boolean onPrepareOptionsMenu(Menu menu) {
        if(itemNum>0) {
            menu.setGroupVisible(GROUP_DEL, true);
        } else {
            menu.setGroupVisible(GROUP_DEL, false);
        }
        return super.onPrepareOptionsMenu(menu);
    }

    @Override
    public boolean onOptionsItemSelected(MenuItem item) {
        switch(item.getItemId()) {
        case MENU_ADD:
            create_note();
            return true;
        case MENU_SEND:
            send_note();
            return true;
        case MENU_DEL:
            delete_note();
            return true;
        }
        return super.onOptionsItemSelected(item);
    }

    @Override
    public void onCreateContextMenu(ContextMenu menu, View v,
            ContextMenuInfo menuInfo) {
        super.onCreateContextMenu(menu, v, menuInfo);
        if(v.getId() == R.id.focus_text) {
            SubMenu textMenu = menu.addSubMenu("Change Text");
            textMenu.add(0, ID_TEXT1, 0, choices[0]);
            textMenu.add(0, ID_TEXT2, 0, choices[1]);
            textMenu.add(0, ID_TEXT3, 0, choices[2]);
            menu.add(0, ID_DEFAULT, 0, "Original Text");
        }
    }

    @Override
    public boolean onContextItemSelected(MenuItem item) {
        switch(item.getItemId()) {
        case ID_DEFAULT:
            bv.setText(R.string.hello);
            return true;
        case ID_TEXT1:
        case ID_TEXT2:
        case ID_TEXT3:
            bv.setText(choices[item.getItemId()-1]);
            return true;
        }
        return super.onContextItemSelected(item);
    }

    void create_note() { // mock code to create note
        itemNum++;
```

```
    }
    void send_note() { // mock code to send note
        Toast.makeText(this, "Item: "+itemNum,
                Toast.LENGTH_SHORT).show();
    }
    void delete_note() { // mock code to delete note
        itemNum--;
    }
}
```

代码清单 6-2 中的 Activity 也给出了上下文菜单和子菜单的例子。名为 focus_text 的文本视图被添加到布局中，如代码清单 6-3 所示，并在 Activity 的 onCreate()方法中使用 registerForContextMenu()将其注册为一个上下文菜单。

**代码清单 6-3 res/layout/main.xml**

```
<?xml version="1.0" encoding="utf-8"?>
<LinearLayout xmlns:android="http://schemas.android.com/apk/res/android"
    android:orientation="vertical"
    android:layout_width="match_parent"
    android:layout_height="match_parent"
    >
<TextView android:id="@+id/focus_text"
    android:layout_width="match_parent"
    android:layout_height="wrap_content"
    android:textSize="40sp"
    android:text="@string/hello"
    />
</LinearLayout>
```

当该视图被长按时，会调用 onCreateContextMenu 来建立上下文菜单。此处通过 addSubMenu()方法来为 Menu 实例实现 SubMenu。子菜单项随主菜单项一同被指定，无论哪个菜单中的选项被选中，都会调用 onContextItemSelected()方法。本技巧在此展示了如何根据菜单项来改变文本。

图 6-1 和图 6-2 显示了不同情形下菜单的不同样子。

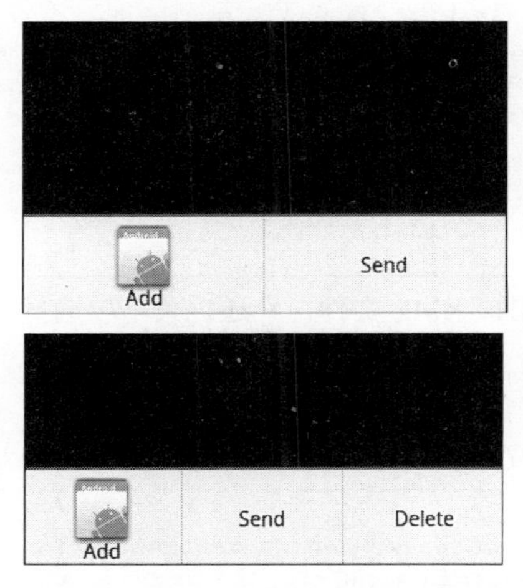


图 6-1 选项菜单（上图）和在运行时添加的选项（下图）

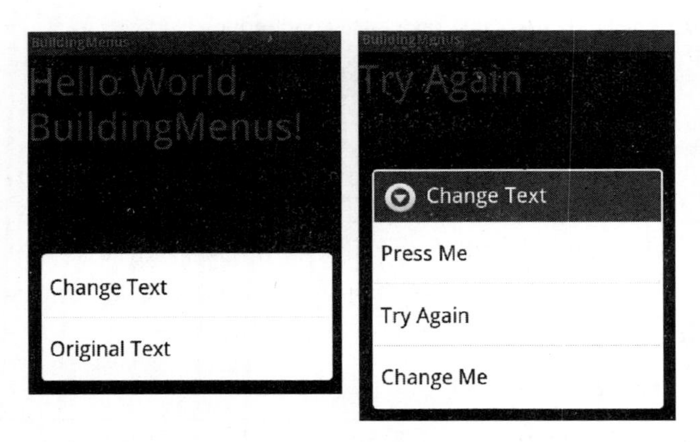


图 6-2　当文本被长按时显示的上下文菜单（左图）以及“Change Text”选项下属的子菜单，为文本视图提供三个不同的字符串（右图）

## 技巧 54：在 XML 文件中定义菜单

菜单也可以在 XML 文件中构建，并用前面技巧中提到的合适的回调方法来使用它。对于较大的菜单这种方法比较好用，并且仍然可以用 Java 来处理动态选项。

菜单文件通常被保存在 **res/menu** 资源目录下。例如，要实现上一节中的上下文菜单，只要创建一个带有嵌套菜单定义的 XML 文件，如代码清单 6-4 所示。

**代码清单 6-4　res/menu/context_menu.xml**

```
<?xml version="1.0" encoding="utf-8"?>
<menu xmlns:android="http://schemas.android.com/apk/res/android">
  <item android:id="@+id/submenu" android:title="Change Text">
  <menu xmlns:android="http://schemas.android.com/apk/res/android">
   <item android:id="@+id/text1" android:title="Press Me" />
   <item android:id="@+id/text2" android:title="Try Again" />
   <item android:id="@+id/text3" android:title="Change Me" />
  </menu>
  </item>
  <item android:id="@+id/orig" android:title="Original Text" />
</menu>
```

然后，当创建菜单时使用该 XML 文件，并通过菜单项选择方法引用菜单项的 ID。为实现这一点，代码清单 6-2 中有两个方法需要被修改，如代码清单 6-5 所示。

**代码清单 6-5　主 Activity 中被改变的方法**

```
@Override
    public void onCreateContextMenu(ContextMenu menu, View v,
            ContextMenuInfo menuInfo) {
        super.onCreateContextMenu(menu, v, menuInfo);
```

```
        MenuInflater inflater = getMenuInflater();
        inflater.inflate(R.menu.context_menu, menu);
    }
@Override
    public boolean onContextItemSelected(MenuItem item) {
        switch(item.getItemId()) {
        case R.id.orig:
            bv.setText(R.string.hello);
            return true;
        case R.id.text1:
            bv.setText(choices[0]);
            return true;
        case R.id.text2:
            bv.setText(choices[1]);
            return true;
        case R.id.text3:
            bv.setText(choices[2]);
            return true;
        }
        return super.onContextItemSelected(item);
    }
```

## 技巧 55：创建操作栏

操作栏（action bar）是在 Android 的 3.0 发布版本（代号 Honeycomb）中被引入的一种窗口特性。Android 从 3.0 版本开始不再需要使用专门的菜单按钮，取而代之的是 ActionBar。操作栏可以用来显示用户动作以及全局的菜单选项。它还可以被用来增强标识（这将会显示一个图标或者应用 logo），或者在 Fragment 之间切换、提供下拉导航栏、显示诸如搜索和分享一类的用户动作。

要使用操作栏，须使用 Holo 主题，还要将 android:targetSdkVersion 设为 11 或者更高。对于 11 以前的版本，则可以使用 ActionBarSherlock，这将在本章后面涉及。下面的代码段是一个，可以在 **AndroidManifest.xml** 中使用的范例：

```
<uses-sdk android:minSdkVersion="11" android:targetSdkVersion="16" />
```

本技巧展示了如何创建一个操作栏。代码清单 6-6 中给出了菜单的定义代码。

**代码清单 6-6 res/menu/activity_action_bar.xml**

```
<menu xmlns:android="http://schemas.android.com/apk/res/android">

<item android:id="@+id/menu_share"
  android:title="Share"
  android:icon="@drawable/ic_launcher"
  android:orderInCategory="0"
  android:showAsAction="ifRoom|withText" />
<item
  android:id="@+id/menu_settings"
  android:orderInCategory="100"
```

```
    android:showAsAction="never"
    android:title="@string/menu_settings"/>

</menu>
```

操作栏的工作方式与菜单类似，且也在 XML 中定义。该 XML 会在 ActionBarActivity 中调用，如代码清单 6-7 所示。

**代码清单 6-7 src/com/cookbook/actionbar/ActionBarActivity.java**

```
package com.cookbook.actionbar;

import android.app.Activity;
import android.os.Bundle;
import android.view.Menu;
import android.view.MenuInflater;
import android.view.MenuItem;
import android.widget.Toast;

public class ActionBarActivity extends Activity {

  @Override
  protected void onCreate(Bundle savedInstanceState) {
    super.onCreate(savedInstanceState);
    setContentView(R.layout.activity_action_bar);
  }

  @Override
  public boolean onCreateOptionsMenu(Menu menu) {
    MenuInflater inflater = getMenuInflater();
    inflater.inflate(R.menu.activity_action_bar, menu);
    return true;
  }

  @Override
  public boolean onOptionsItemSelected(MenuItem item) {
    switch (item.getItemId()) {
    case R.id.menu_share:
      Toast.makeText(this, "Implement share options here", Toast.LENGTH_SHORT).show();
    default:
        return super.onOptionsItemSelected(item);
    }
  }
}
```

代码清单 6-7 用粗体标记的部分是被添加到默认的 Activity 中、用于处理与 Share 菜单项的交互的代码段。对于本例，当 Share 按钮被按下时，会显示一个 Toast。但可以将其修改为与提供器进行协作，以实现与其他应用程序及服务的集成。

图 6-3 和图 6-4 分别给出了本例在 Ice Cream Sandwich 版系统的平板电脑和 Jelly Bean[①]版的手机上的运行效果。

① Ice Cream Sandwich 和 Jelly Bean 分别对应 Android 版本号 4.0 和 4.1。——译者注

Implment share options here

图 6-3 在 Ice Cream Sandwich 版系统的平板电脑上显示的含有一条 Toast 信息操作栏效果

图 6-4 在 Jelly Bean 版系统的手机上显示的含有一条 Toast 信息操作栏效果

## 技巧 56：使用 ActionBarSherlock

上一节讲述了如何向在 Android 3.0 及以上版本的设备上运行的应用程序中添加操作栏。那么使用老式设备的用户又该怎么办呢？这就需要 ActionBarSherlock 出马了。ActionBarSherlock 填补了不同 API 级别之间的鸿沟，使得在 11 以下 API 级别的系统中使用操作栏成为可能。

首先要到 http://actionbarsherlock.com/去下载 ActionBarSherlock。将下载的文件解压缩，并通过"Android Project from Existing Code"将解压出的库文件夹添加到项目中。当项目被导入某个工作区后，还需要在项目属性页的 Android 这一部分将其添加为一个库。也可以使用 Ant 或 Maven 将 ActionBarSherlock 构建到项目中，具体方法请参见 http://actionbarsherlock.com/usage.html。

在项目中设置好使用 ActionBarSherlock 库之后，就可以对代码清单 6-7 中的项目做一些修改。注意，Android 支持库也需要被添加进项目，因为 ActionBarSherlock 要用到它才能运行在 Honeycomb（API Level 11）以前的 Android 版本上。如果用 Eclipse 作为集成开发环境，要实现这一点，只需在项目文件夹上右击，并选择 **Android Tools→Add Support Library**，然后将启动下载过程，过程结束后，支持库就添加到项目中了。

接下来只要对代码清单 6-7 中的代码进行几处小改动，就能把 ActionBarSherlock 添加到项目中。代码清单 6-8 显示了需对 `ActionBarActivity` 进行的改动。

**代码清单 6-8　src/com/cookbook/actionbar/ActionBarActivity.java**

```
package com.cookbook.actionbar;

import com.actionbarsherlock.app.SherlockActivity;
import android.os.Bundle;
import com.actionbarsherlock.view.Menu;
import com.actionbarsherlock.view.MenuInflater;
import com.actionbarsherlock.view.MenuItem;
import android.widget.Toast;

public class ActionBarActivity extends SherlockActivity {

  @Override
  protected void onCreate(Bundle savedInstanceState) {
    super.onCreate(savedInstanceState);
    setContentView(R.layout.activity_action_bar);
  }

  @Override
  public boolean onCreateOptionsMenu(Menu menu) {
    MenuInflater inflater = getSupportMenuInflater();
    inflater.inflate(R.menu.activity_action_bar, menu);
    return true;
  }

  @Override
  public boolean onOptionsItemSelected(MenuItem item) {
    switch (item.getItemId()) {
    case R.id.menu_share:
      Toast.makeText(this, "Implement share options here", Toast.LENGTH_SHORT).show();
    default:
        return super.onOptionsItemSelected(item);
    }
```

```
    }
}
```

代码中的粗体部分显示了对代码清单 6-7 的改动之处。这些改动包括需要添加的 `import` 语句，以及将 `extends Activity` 改为 `extends SherlockActivity`。对菜单的 inflater 也需要进行修改，如今需要使用 `getSupportMenuInflater` 来代替 `getMenuInflater`。

要想让操作栏运行在 Android 的早期版本上，必须修改 **AndroidManifest.xml** 文件，下面一行代码是修改的例子：

```
<uses-sdk android:minSdkVersion="7" android:targetSdkVersion="16" />
```

除 **AndroidManifest.xml** 要被修改之外，还要修改应用程序的主题，以允许 ActionBarSherlock 的显示和正常工作。下面是应用程序设置的例子：

```
<application
  android:allowBackup="true"
  android:icon="@drawable/ic_launcher"
  android:label="@string/app_name"
  android:theme="@style/Theme.Sherlock" >
```

注意，用粗体标出的主题设置语句被修改成使用 Sherlock 主题。关于伴随 AndroidBarSherlock 使用的主题的更多信息，可以在 http://actionbarsherlock.com/theming.html 上找到。

图 6-5 和图 6-6 分别给出了运行 Gingerbread 版系统的设备（即 Android 2.3，API Level 9）在横向模式和纵向模式下的运行效果截图。注意，在纵向模式下 SHARE 一词被移除，以节约空间。

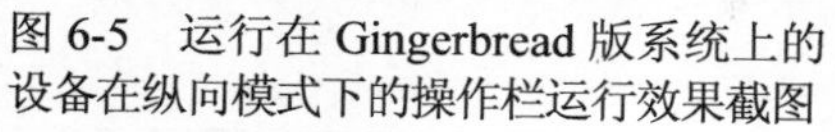
图 6-5　运行在 Gingerbread 版系统上的设备在纵向模式下的操作栏运行效果截图

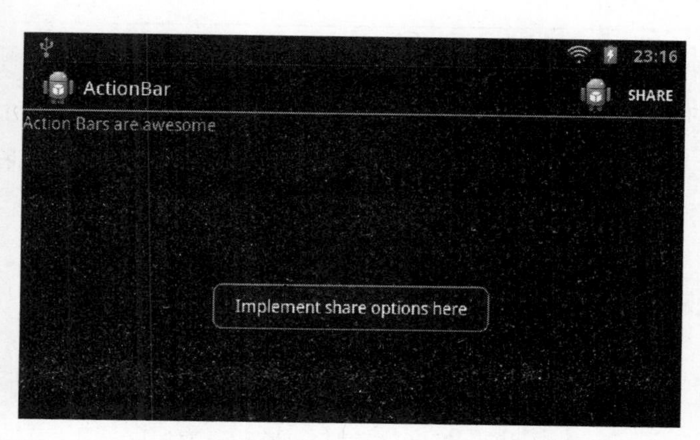


图 6-6　运行在 Gingerbread 版系统上的设备在横向模式下的操作栏运行效果截图

## 技巧57：使用搜索键

从Android 4.1（Jelly Bean）起，所有设备的搜索键已被Google硬编码为调用Google Now服务的快捷键。这意味着开发者不能再在他们的应用程序中对该键的功能进行重定义。然而，那些为Jelly Bean以前的版本构建应用程序的人可以对搜索键进行功能映射，来触发自定义的行为。如果某个获得了焦点的应用程序的Activity被定义为可搜索的，搜索键就可将其激活。应当始终用一个菜单项或者与之等价的其他办法来作为调用可搜索Activity的备用方案，以兼容那些不带有搜索键的设备。该菜单项需要调用`onSearchRequested()`方法。

可搜索的Activity在理想状况下应该被声明为`singleTop`启动模式，该模式在第2章中讨论过。这样就能确保当多个搜索任务进行时，栈不会被上述多个可搜索Activity的实例所阻塞。manifest文件应当包含下面几行：

```
<activity android:name=".SearchDialogExample"
          android:launchMode="singleTop" >
  <intent-filter>
      <action android:name="android.intent.action.SEARCH" />
  </intent-filter>
  <meta-data android:name="android.app.searchable"
             android:resource="@xml/my_search"/>
</activity>
```

包含了搜索配置细节的XML文件在代码清单6-9中给出。在应用程序中定义搜索时必须将该文件包含进来。

**代码清单6-9　res/xml/my_search.xml**

```
<?xml version="1.0" encoding="utf-8"?>
<searchable xmlns:android="http://schemas.android.com/apk/res/android"
   android:label="@string/app_name" android:hint="Search MyExample Here">
</searchable>
```

本技巧提供了一个搜索界面。代码清单6-10给出了一个最简单的主Activity，它使用默认的**main.xml**文件。

**代码清单6-10　src/com/cookbook/search_diag/MainActivity.java**

```
package com.cookbook.search_diag;

import android.app.Activity;
import android.os.Bundle;

public class MainActivity extends Activity {
     @Override
     protected void onCreate(Bundle savedInstanceState) {
          super.onCreate(savedInstanceState);
          setContentView(R.layout.main);
     }
}
```

接下来，如果搜索键被选中，可搜索 Activity 就会被激活。onCreate()方法会检查 Intent 是否是 ACTION_SEARCH 类型，如果是，就会在其上执行。代码清单 6-11 显示了主 Activity，其功能只是将查询内容显示到屏幕上。

**代码清单 6-11 src/com/cookbook/search_diag/SearchDialogExample.java**

```
package com.cookbook.search_diag;

import android.app.Activity;
import android.app.SearchManager;
import android.content.Intent;
import android.os.Bundle;
import android.widget.Toast;

public class SearchDialogExample extends Activity {
    /** called when the activity is first created */
    @Override
    public void onCreate(Bundle savedInstanceState) {
        super.onCreate(savedInstanceState);
        setContentView(R.layout.main);
        Intent intent = getIntent();

        if (Intent.ACTION_SEARCH.equals(intent.getAction())) {
          String query = intent.getStringExtra(SearchManager.QUERY);
          Toast.makeText(this, "The QUERY: " + query,
                                       Toast.LENGTH_LONG).show();
        }
    }
}
```

## 技巧 58：响应触摸事件

任何与显示屏的交互，无论是触摸还是用轨迹球进行的导航选择，都是与处于该位置的相应视图元素的交互。如第 5 章所述，屏幕布局由分层级的一系列视图组成，系统从分层结构（或称为树）的顶部开始向下传递事件，直到其被某个视图处理为止。有的事件，如果没有被消耗掉，则会在被处理后继续向树的下层传递。

代码清单 6-12 给出了名为 ex_button 的按钮，它设置了两个事件监听器，会处理点击或长按事件。当事件发生时，相应的回调方法被调用，从而在屏幕上显示一个 Toast 来表示方法已被触发。

**代码清单 6-12 src/com/cookbook/touch_examples/TouchExamples.java**

```
package com.cookbook.touch_examples;

import android.app.Activity;
import android.os.Bundle;
import android.view.View;
import android.view.View.OnClickListener;
import android.view.View.OnLongClickListener;
import android.widget.Button;
import android.widget.Toast;

public class TouchExamples extends Activity {
```

```
    @Override
    public void onCreate(Bundle savedInstanceState) {
        super.onCreate(savedInstanceState);
        setContentView(R.layout.main);
        Button ex = (Button) findViewById(R.id.ex_button);

        ex.setOnClickListener(new OnClickListener() {
            public void onClick(View v) {
                Toast.makeText(TouchExamples.this, "Click",
                        Toast.LENGTH_SHORT).show();
            }
        });
        ex.setOnLongClickListener(new OnLongClickListener() {
            public boolean onLongClick(View v) {
                Toast.makeText(TouchExamples.this, "LONG Click",
                        Toast.LENGTH_SHORT).show();
                return true;
            }
        });
    }
}
```

包含按钮的布局则由代码清单 6-13 给出。

**代码清单 6-13 res/layout/main.xml**

```
<?xml version="1.0" encoding="utf-8"?>
<LinearLayout
    xmlns:android="http://schemas.android.com/apk/res/android"
    android:orientation="vertical"
    android:layout_width="match_parent"
    android:layout_height="match_parent">
    <Button android:id="@+id/ex_button"
        android:text="Press Me"
        android:layout_width="wrap_content"
        android:layout_height="wrap_content" />
</LinearLayout>
```

为了代码结构紧凑，将回调方法以内部方法的形式定义在代码清单 6-12 中，但若考虑可读性和重用性，也可以将其显式地定义：

```
View.OnClickListener myTouchMethod = new View.OnClickListener() {
    public void onClick(View v) {
        //Insert relevant action here
    }
};
ex.setOnClickListener(myTouchMethod);
```

还有一种方法是让 Activity 实现 `OnClickListener` 接口，这样一来回调方法就属于 Activity 层级，避免了对额外类的调用：

```
public class TouchExamples extends Activity implements OnClickListener {
    @Override
    public void onCreate(Bundle savedInstanceState) {
        super.onCreate(savedInstanceState);
        setContentView(R.layout.main);
        Button ex = (Button) findViewById(R.id.ex_button);
        ex.setOnClickListener(this);
    }
```

```
        public void onClick(View v) {
            if(v.getId() == R.id.directory_button) {
                // Insert relevant action here
            }
        }
    }
```

onClick()方法在 Activity 层级的实现有助于展示父视图如何处理多个子控件的触摸事件。

## 技巧 59：监听滑动手势

我们在本章开头说过，每个视图都有与之关联的 onTouchEvent()方法。在本技巧中，将会用一个手势检测器（gesture detector）来重写该方法，以设置一个手势监听器。OnGestureListener 接口可以响应的手势有以下几个。

- onDown()：当按下事件发生时发出通知。
- onFling()：当“按下并滑动一段距离再释放”这一事件发生后发出通知。
- onLongPress()：长按事件发生时发出通知。
- onScroll()：发生滚动时发出通知。
- onShowPress()：当按下事件在任何动作或释放之前发生时发出通知。
- onSingleTapUp()：当“按下后放开”事件发生时发出通知。

如果只需要上述手势的一部分，可以继承 SimpleOnGestureListener 类。如果前面提到的任一方法没有显式地实现，它就会返回 false。

滑动包括两个事件：按下（第一个 MotionEvent）和释放（第二个 MotionEvent）。每个动作事件在屏幕上都有特定的位置，由一个$(x, y)$坐标对给出，其中 $x$ 为横轴坐标，$y$ 为纵轴坐标。事件的$(x, y)$速度也一并给出。

代码清单 6-14 显示了一个实现了 onFling()方法的 Activity。当移动幅度足够大（此处以 60 像素为限），事件就会被消耗掉，并将对该事件的描述语句添加到屏幕上。

**代码清单 6-14 src/com/cookbook/fling_ex/FlingExample.java**

```
package com.cookbook.fling_ex;

import android.app.Activity;
import android.os.Bundle;
import android.view.GestureDetector;
import android.view.MotionEvent;
import android.view.GestureDetector.SimpleOnGestureListener;
import android.widget.TextView;

public class FlingExample extends Activity {
    private static final int LARGE_MOVE = 60;
    private GestureDetector gestureDetector;
    TextView tv;

    @Override
    public void onCreate(Bundle savedInstanceState) {
        super.onCreate(savedInstanceState);
```

```
        setContentView(R.layout.main);
        tv = (TextView) findViewById(R.id.text_result);

        gestureDetector = new GestureDetector(this,
                new SimpleOnGestureListener() {
            @Override
            public boolean onFling(MotionEvent e1, MotionEvent e2,
                    float velocityX, float velocityY) {

                if (e1.getY() - e2.getY() > LARGE_MOVE) {
                    tv.append("\nFling Up with velocity " + velocityY);
                    return true;

                } else if (e2.getY() - e1.getY() > LARGE_MOVE) {
                    tv.append("\nFling Down with velocity " + velocityY);
                    return true;

                } else if (e1.getX() - e2.getX() > LARGE_MOVE) {
                    tv.append("\nFling Left with velocity " + velocityX);
                    return true;

                } else if (e2.getX() - e1.getX() > LARGE_MOVE) {
                    tv.append("\nFling Right with velocity " + velocityX);
                    return true;
                }

                return false;
        } });
    }

    @Override
    public boolean onTouchEvent(MotionEvent event) {
        return gestureDetector.onTouchEvent(event);
    }
}
```

前面Activity中包含描述性文字的TextView在主XML布局中进行定义，如代码清单6-15所示。

**代码清单 6-15 res/layout/main.xml**

```
<?xml version="1.0" encoding="utf-8"?>
<LinearLayout
    xmlns:android="http://schemas.android.com/apk/res/android"
    android:orientation="vertical"
    android:layout_width="match_parent"
    android:layout_height="match_parent">
    <TextView android:id="@+id/text_result"
    android:layout_width="match_parent"
    android:layout_height="match_parent"
    android:textSize="16sp"
    android:text="Fling right, left, up, or down\n" />
</LinearLayout>
```

## 技巧60：使用多点触控

多点触控事件是指当有不止一个触摸点（比如手指）同时触碰屏幕时发生的事件。多点触控事件通过触摸监听器 OnTouchListener 来识别，该监听器能捕捉多种类型的动作事件。

- ACTION_DOWN：主要（第一个）触摸点（手指）按下，标志着手势的开始。
- ACTION_POINTER_DOWN：辅助（第二个）触摸点（手指）按下。
- ACTION_MOVE：在按压手势做出期间发生按压位置的变化。
- ACTION_POINTER_UP：辅助触摸点被放开。
- ACTION_UP：主要触摸点被放开，触摸手势结束。

这个技巧将在屏幕上显示一幅图像，并允许用户通过多点触控来缩放图像。它还将监听单点触控事件，从而在屏幕上拖动图片。代码清单 6-16 给出了 Activity 的实现。首先，Activity 在 onCreate()方法中设置了 onTouchListener 接口。当触摸事件发生时，onTouch()方法会检查动作事件并根据其做出如下动作。

- 如果第一个触摸点触及屏幕，则触摸状态（touch state）被声明为拖动动作，同时保存按压的位置及 Matrix。
- 如果当第一个触摸点仍处于按下状态，又有第二个触摸点触及屏幕，则将计算两个触摸点之间的距离。如果距离大于某一阈值（本例设为 50 像素），则将触摸状态变为缩放动作，并将两事件间的距离、二者的中点位置以及 Matrix 保存起来。
- 如果发生移动，单点触摸模式下将会移动图片，而多点事件状态下则会缩放图片。
- 如果某个触摸点抬起，触摸状态将被设为无动作。

**代码清单 6-16 src/com/cookbook/multitouch/MultiTouch.java**

```
package com.cookbook.multitouch;

import android.app.Activity;
import android.graphics.Matrix;
import android.os.Bundle;
import android.util.FloatMath;

import android.view.MotionEvent;
import android.view.View;
import android.view.View.OnTouchListener;
import android.widget.ImageView;

public class MultiTouch extends Activity implements OnTouchListener {
    // Matrix instances to move and zoom image
    Matrix matrix = new Matrix();
    Matrix eventMatrix = new Matrix();

    // possible touch states
    final static int NONE = 0;
    final static int DRAG = 1;
    final static int ZOOM = 2;
    int touchState = NONE;

    @Override
    public void onCreate(Bundle savedInstanceState) {
        super.onCreate(savedInstanceState);
        setContentView(R.layout.main);
        ImageView view = (ImageView) findViewById(R.id.imageView);
        view.setOnTouchListener(this);
```

```
}

final static float MIN_DIST = 50;
static float eventDistance = 0;
static float centerX =0, centerY = 0;
@Override
public boolean onTouch(View v, MotionEvent event) {
    ImageView view = (ImageView) v;

    switch (event.getAction() & MotionEvent.ACTION_MASK) {
    case MotionEvent.ACTION_DOWN:
        //Primary touch event starts: remember touch-down location
        touchState = DRAG;
        centerX = event.getX(0);
        centerY = event.getY(0);
        eventMatrix.set(matrix);
        break;

    case MotionEvent.ACTION_POINTER_DOWN:
        //Secondary touch event starts: remember distance and center
        eventDistance = calcDistance(event);
        calcMidpoint(centerX, centerY, event);
        if (eventDistance > MIN_DIST) {
            eventMatrix.set(matrix);

            touchState = ZOOM;
        }
        break;

    case MotionEvent.ACTION_MOVE:
        if (touchState == DRAG) {
            //single finger drag, translate accordingly
            matrix.set(eventMatrix);
            matrix.setTranslate(event.getX(0) - centerX,
                                event.getY(0) - centerY);
        } else if (touchState == ZOOM) {
            //multifinger zoom, scale accordingly around center
            float dist = calcDistance(event);

            if (dist > MIN_DIST) {
                matrix.set(eventMatrix);
                float scale = dist / eventDistance;

                matrix.postScale(scale, scale, centerX, centerY);
            }
        }

        // Perform the transformation
        view.setImageMatrix(matrix);
        break;

    case MotionEvent.ACTION_UP:
    case MotionEvent.ACTION_POINTER_UP:
        touchState = NONE;
        break;
    }

    return true;
}

private float calcDistance(MotionEvent event) {
    float x = event.getX(0) - event.getX(1);
    float y = event.getY(0) - event.getY(1);
```

```
            return FloatMath.sqrt(x * x + y * y);
        }

        private void calcMidpoint(float centerX, float centerY,
                                  MotionEvent event) {
            centerX = (event.getX(0) + event.getX(1))/2;
            centerY = (event.getY(0) + event.getY(1))/2;
        }
    }
```

代码清单 6-17 则给出了指定可缩放图片的布局代码。在本技巧中，该图片被设为 **icon.png**，它是在 Eclipse 中自动创建的；你可以用任意其他图片取而代之。

**代码清单 6-17 res/layout/main.xml**

```
<?xml version="1.0" encoding="utf-8"?>
<FrameLayout
       xmlns:android="http://schemas.android.com/apk/res/android"
       android:layout_width="match_parent"
       android:layout_height="match_parent" >
    <ImageView android:id="@+id/imageView"
           android:layout_width="match_parent"
           android:layout_height="match_parent"
           android:src="@drawable/ic_launcher"
           android:scaleType="matrix" >
    </ImageView>
</FrameLayout>
```

# 6.2 高级用户界面库

有些用户界面特性需要复杂的算术运算。对这些运算进行优化有时是富有挑战且耗费精力的。因此开发者对利用已有的 UI 库饶有兴趣。下面两个技巧将给出一些这方面说明性的例子，权作抛砖引玉。

## 技巧 61：使用手势

手势是在触摸屏上手绘的形状。android.gesture 包提供了用简单方法识别和处理手势的库。首先，每个 SDK 都有一个可以用来构建手势集的示例程序，位于 **platforms/android-2.0/samples/GestureBuilder/** 目录下。手势构造器（Gesture Builder）项目可以导入并运行在 Android 设备上。该项目会产生一个名为 **/sdcard/gestures** 的文件，可以将其从设备上复制出来，作为本技巧的原始资源使用。

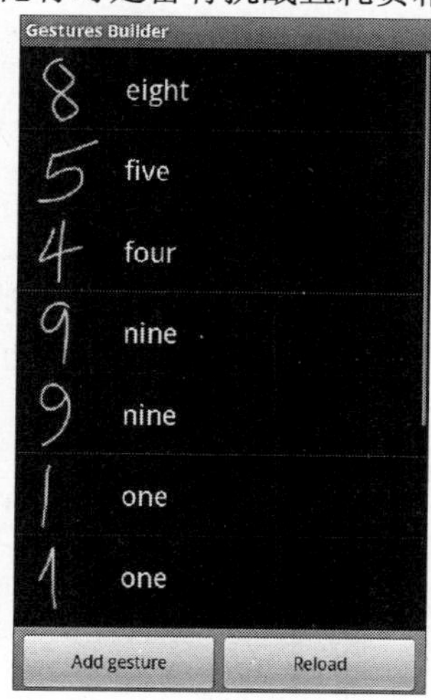


图 6-7 Gesture Builder 应用程序，由 Android SDK 附带，可用来创建手势库

作为例子，该程序可以生成一个由手写数字构成的文件，如图 6-7 所示。多个手势可以共用一个名字，因此提供同一手势的不同样本有助于改善模式识别的效果。

当这个包含了 0～9 的各个数字的不同变体的文件被创建出来后，可以将其复制到诸如 **res/raw/number** 这样的目录下。代码清单 6-18 给出了布局，而代码清单 6-19 则给出了主 Activity。在 Activity 中，利用上述原始资源来创建 GestureLibrary。

本技巧在屏幕顶端添加一个 GestureOverlayView，并实现一个 OnGesturePerformedListener。当手势被绘制出来，它会被传递给 onGesturePerformed() 方法，在方法中将其与库中所有的手势相比较，并返回按照相似度顺序由高到低排列的预测结果列表。每个预测结果包含其在库中定义的名字及其与输入手势的相关度分数。只要列表第一项的分数大于 1，通常就是代表匹配成功。

**代码清单 6-18　res/layout/main.xml**

```
<?xml version="1.0" encoding="utf-8"?>
<LinearLayout xmlns:android="http://schemas.android.com/apk/res/android"
    android:orientation="vertical"
    android:layout_width="match_parent"
    android:layout_height="match_parent"
    >

<TextView
    android:layout_width="match_parent"
    android:layout_height="wrap_content"
    android:gravity="center_horizontal" android:textSize="20sp"
    android:text="Draw a number"
    android:layout_margin="10dip"/>

<android.gesture.GestureOverlayView
    android:id="@+id/gestures"
    android:layout_width="match_parent"
    android:layout_height="0dip"
    android:layout_weight="1.0" />

<TextView android:id="@+id/prediction"
    android:layout_width="match_parent"
    android:layout_height="wrap_content"
    android:gravity="center_horizontal" android:textSize="20sp"
    android:text=""
    android:layout_margin="10dip"/>
</LinearLayout>
```

为清楚起见，本技巧将所有预测结果都转换为字符串并将其在屏幕上显示。图 6-8 给出了一种可能的输出结果。该例子说明即使一个视觉匹配并不完整，一个只写了部分的数字仍然可以与库进行良好地匹配。

**代码清单 6-19　src/com/cookbook/gestures/Gestures.java**

```
package com.cookbook.gestures;

import java.text.DecimalFormat;
import java.text.NumberFormat;
import java.util.ArrayList;

import android.app.Activity;
import android.gesture.Gesture;
import android.gesture.GestureLibraries;
import android.gesture.GestureLibrary;
import android.gesture.GestureOverlayView;
```

```
import android.gesture.Prediction;
import android.gesture.GestureOverlayView.OnGesturePerformedListener;
import android.os.Bundle;
import android.widget.TextView;

public class Gestures extends Activity
                    implements OnGesturePerformedListener {
    private GestureLibrary mLibrary;
    private TextView tv;

    @Override
    public void onCreate(Bundle savedInstanceState) {
        super.onCreate(savedInstanceState);
        setContentView(R.layout.main);
        tv = (TextView) findViewById(R.id.prediction);

        mLibrary = GestureLibraries.fromRawResource(this, R.raw.numbers);
        if (!mLibrary.load()) finish();

        GestureOverlayView gestures =
                        (GestureOverlayView) findViewById(R.id.gestures);
        gestures.addOnGesturePerformedListener(this);
    }

    public void onGesturePerformed(GestureOverlayView overlay,
                                   Gesture gesture) {
        ArrayList<Prediction> predictions = mLibrary.recognize(gesture);
        String predList = "";
        NumberFormat formatter = new DecimalFormat("#0.00");
        for(int i=0; i<predictions.size(); i++) {
            Prediction prediction = predictions.get(i);
            predList = predList + prediction.name + " "
                        + formatter.format(prediction.score) + "\n";
        }
        tv.setText(predList);
    }
}
```

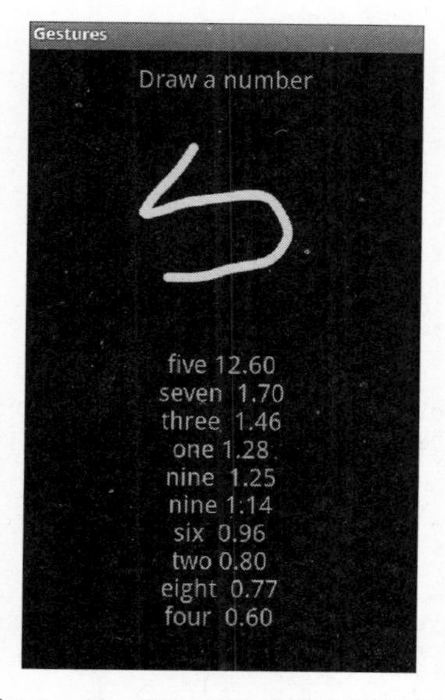


图 6-8 显示预测分数的手势识别示例

## 技巧 62：绘制 3D 图像

Android 支持针对嵌入式系统的开放图形库（即 OpenGL ES）。本技巧基于一个 Android API 演示程序，将展示如何利用 OpenGL ES 来创建一个三维金字塔形状，让其在屏幕中跳来跳去，碰到屏幕边缘就转向，同时还进行旋转运动。主 Activity 需要两个各自独立的支持类：一个定义形状（在代码清单 6-20 中给出）、一个渲染形状（在代码清单 6-21 中给出）。

**代码清单 6-20　src/com/cookbook/open_gl/Pyramid.java**

```
package com.cookbook.open_gl;

import java.nio.ByteBuffer;
import java.nio.ByteOrder;
import java.nio.IntBuffer;

import javax.microedition.khronos.opengles.GL10;

class Pyramid {
    public Pyramid() {
        int one = 0x10000;
        /* square base and point top to make a pyramid */
        int vertices[] = {
                -one, -one, -one,
                -one,  one, -one,
                 one,  one, -one,
                 one, -one, -one,
                 0, 0, one
        };

        /* purple fading to white at the top */
        int colors[] = {
                one, 0, one, one,
                one, 0, one, one,
                one, 0, one, one,
                one, 0, one, one,
                one, one, one, one
        };

        /* triangles of the vertices above to build the shape */
        byte indices[] = {
                0, 1, 2, 0, 2, 3, //square base
                0, 3, 4, // side 1
                0, 4, 1, // side 2
                1, 4, 2, // side 3
                2, 4, 3  // side 4
        };

        // buffers to be passed to gl*Pointer() functions
        ByteBuffer vbb = ByteBuffer.allocateDirect(vertices.length*4);
        vbb.order(ByteOrder.nativeOrder());
        mVertexBuffer = vbb.asIntBuffer();
        mVertexBuffer.put(vertices);
        mVertexBuffer.position(0);

        ByteBuffer cbb = ByteBuffer.allocateDirect(colors.length*4);
        cbb.order(ByteOrder.nativeOrder());
        mColorBuffer = cbb.asIntBuffer();
        mColorBuffer.put(colors);
        mColorBuffer.position(0);
```

```
        mIndexBuffer = ByteBuffer.allocateDirect(indices.length);
        mIndexBuffer.put(indices);
        mIndexBuffer.position(0);
    }

    public void draw(GL10 gl) {
        gl.glFrontFace(GL10.GL_CW);
        gl.glVertexPointer(3, GL10.GL_FIXED, 0, mVertexBuffer);
        gl.glColorPointer(4, GL10.GL_FIXED, 0, mColorBuffer);
        gl.glDrawElements(GL10.GL_TRIANGLES, 18, GL10.GL_UNSIGNED_BYTE,
                          mIndexBuffer);
    }

    private IntBuffer mVertexBuffer;
    private IntBuffer mColorBuffer;
    private ByteBuffer mIndexBuffer;
}
```

注意，金字塔有 5 个顶点，正方形底面上有 4 个，尖顶上有 1 个。很重要的一点是，顶点应当是有序的，可以被一条贯穿图形的线遍历（而不仅仅是随机列出）。形状的中心位于原点(0, 0, 0)处。

这 5 个顶点分别被设为几种 RGBA 格式的颜色，底面上的顶点被设为紫色，而尖顶的顶点设为白色。图形库以颜色渐变的方式填充整个图形。不同的颜色或阴影有助于体现三维视觉效果。

核心方法 draw()用于绘制三角形元素。正方形底面可由两个三角形拼成，而每个向尖端延伸的侧面各自都是一个三角形，共有 6 个三角形，18 个索引点。图 6-9 从两个不同的视角给出了金字塔在弹跳运动过程中的状态。

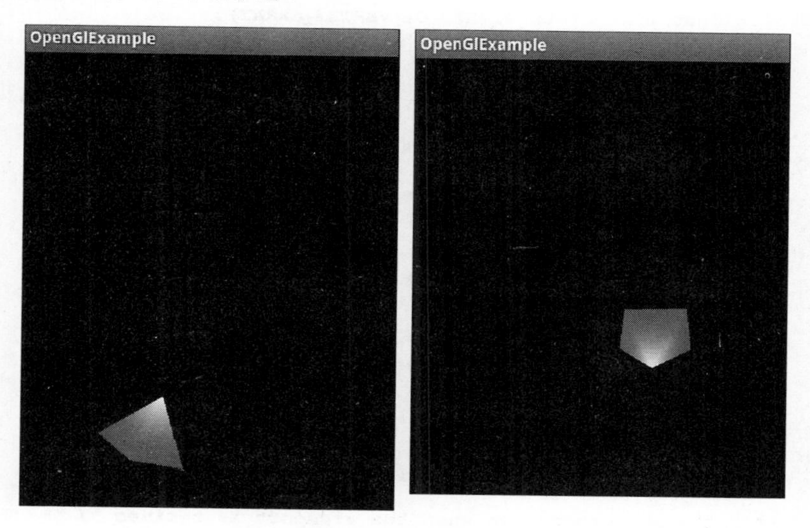


图 6-9 用 OpenGL ES 创建的会旋转和弹跳的金字塔

接下来要创建一个独立的、扩展了 GLSurfaceView.Renderer 的类，以使用 OpenGL ES 库来渲染金字塔，如代码清单 6-21 所示。需要实现三个方法。

- onSurfaceCreated()：一次性地初始化 OpenGL 框架。
- onSurfaceChanged()：在启动或当视口（viewport）大小发生改变时，设置投影。

- onDrawFrame()：逐帧绘制图形。

**代码清单 6-21 src/com/cookbook/open_gl/PyramidRenderer.java**

```
package com.cookbook.open_gl;

import javax.microedition.khronos.egl.EGLConfig;
import javax.microedition.khronos.opengles.GL10;

import android.opengl.GLSurfaceView;

/**
 * Render a tumbling pyramid
 */
class PyramidRenderer implements GLSurfaceView.Renderer {
    public PyramidRenderer(boolean useTranslucentBackground) {
        mTranslucentBackground = useTranslucentBackground;
        mPyramid = new Pyramid();
    }

    public void onDrawFrame(GL10 gl) {
        /* Clear the screen */
        gl.glClear(GL10.GL_COLOR_BUFFER_BIT | GL10.GL_DEPTH_BUFFER_BIT);

        /* Draw a pyramid rotating */
        gl.glMatrixMode(GL10.GL_MODELVIEW);
        gl.glLoadIdentity();
        gl.glTranslatef(mCenter[0], mCenter[1], mCenter[2]);
        gl.glRotatef(mAngle,       0, 1, 0);
        gl.glRotatef(mAngle*0.25f, 1, 0, 0);

        gl.glEnableClientState(GL10.GL_VERTEX_ARRAY);
        gl.glEnableClientState(GL10.GL_COLOR_ARRAY);
        mPyramid.draw(gl);

        mAngle += mAngleDelta;

        /* Draw it bouncing off the walls */
        mCenter[0] += mVel[0];
        mCenter[1] += mVel[1];

        if(Math.abs(mCenter[0])>4.0f) {
            mVel[0] = -mVel[0];
            mAngleDelta=(float) (5*(0.5-Math.random()));
        }
        if(Math.abs(mCenter[1])>6.0f) {
            mVel[1] = -mVel[1];
            mAngleDelta=(float) (5*(0.5-Math.random()));
        }
    }

    public void onSurfaceChanged(GL10 gl, int width, int height) {
        gl.glViewport(0, 0, width, height);

        /* Set a new projection when the viewport is resized */
        float ratio = (float) width / height;
        gl.glMatrixMode(GL10.GL_PROJECTION);
        gl.glLoadIdentity();
        gl.glFrustumf(-ratio, ratio, -1, 1, 1, 20);
    }

    public void onSurfaceCreated(GL10 gl, EGLConfig config) {
        gl.glDisable(GL10.GL_DITHER);

        /* one-time OpenGL initialization */
```

```
            gl.glHint(GL10.GL_PERSPECTIVE_CORRECTION_HINT,
                    GL10.GL_FASTEST);

            if (mTranslucentBackground) {
                gl.glClearColor(0,0,0,0);
            } else {
                gl.glClearColor(1,1,1,1);
            }
            gl.glEnable(GL10.GL_CULL_FACE);
            gl.glShadeModel(GL10.GL_SMOOTH);
            gl.glEnable(GL10.GL_DEPTH_TEST);
        }
        private boolean mTranslucentBackground;
        private Pyramid mPyramid;
        private float mAngle, mAngleDelta=0;
        private float mCenter[]={0,0,-10};
        private float mVel[]={0.025f, 0.03535227f, 0f};
    }
```

金字塔跳跃的动态在 onDrawFrame()方法中捕获。为绘制新图像首先要清屏，然后在 mCenter[]中设置金字塔中心。屏幕中心被定义为原点，所以起始点(0,0,-10)是将图形从右上方开始建立。每次更新时，图形将旋转 mAngleDelta 所代表的角度，并平移 mVel[]所代表的量。mVel 的 $x$ 和 $y$ 方向分量被设置得有足够的差异，使得围绕四壁弹跳的运动富有随机性。当形状到达屏幕边缘时，速度变为原来的相反数，使之弹回。

最终，主 Activity 必须将其内容视图（content view）设为 OpenGL ES 对象，如代码清单 6-22 所示。

**代码清单 6-22　src/com/cookbook/open_gl/OpenGlExample.java**

```
package com.cookbook.open_gl;

import android.app.Activity;
import android.opengl.GLSurfaceView;
import android.os.Bundle;
/* wrapper activity demonstrating the use of GLSurfaceView, a view
 * that uses OpenGL drawing into a dedicated surface */
public class OpenGlExample extends Activity {
    @Override
    protected void onCreate(Bundle savedInstanceState) {
        super.onCreate(savedInstanceState);

        // Set our Preview view as the activity content
        mGLSurfaceView = new GLSurfaceView(this);
        mGLSurfaceView.setRenderer(new PyramidRenderer(true));
        setContentView(mGLSurfaceView);
    }

    @Override
    protected void onResume() {
        super.onResume();
        mGLSurfaceView.onResume();
    }

    @Override
    protected void onPause() {
        super.onPause();
        mGLSurfaceView.onPause();
    }

    private GLSurfaceView mGLSurfaceView;
}
```

# 第 7 章

# 高级用户界面技术

Android 设备的横向模式近年来发生了剧烈的变化。手机一度在 Android 设备中占支配地位，但高分辨率小型屏幕、手表、平板电脑以至电视屏幕（通过使用 Google TV 得以实现）等的加入，将应用程序及其布局推向了新的高度。本章将讲述自定义视图的创建、动画、访问可存取特性、使用手势、绘制 3D 图像等。随后，本章将讨论平板电脑，以及如何同时显示多个 Fragment，还兼顾 Activity wrapper 及对话框 Fragment 的使用。

## 7.1 Android 自定义视图

如我们在第 5 章讨论过的，Android 有两种类型的视图：View 对象和 ViewGroup 对象。自定义视图既可以从头创建，也可以通过继承现有的视图结构来创建。在 Android 框架中，一些标准的微件被定义为 View 或 ViewGroup 的子类，如若可能，自定义工作可以由以下类中的某一个作为起点。

- View 类的子类：Button、EditText、TextView、ImageView 等。
- ViewGroup 的子类：LinearLayout、ListView、RelativeLayout、RadioGroup 等。

### 技巧 63：自定义按钮

本技巧要使用名为 MyButton 的类来自定义按钮。该类继承了 Button 微件，因此实现的组件继承了 Button 的大部分特性。对于微件的自定义，最重要的方法是 onMeasure()和 onDraw()。

onMeasure()方法决定了微件所占空间的大小。它接受两个参数：宽度和高度的度量规格。自定义的微件应当基于微件内部内容来计算宽度和高度，然后通过这些值来调用 setMeasuredDimension()。如果不这样做，measure()就会抛出一个 illegalState Exception 异常。

onDraw()方法则允许在微件上进行自定义绘图。绘图通过自顶至底遍历视图树来逐个渲染视图。当所有父视图绘制完成后，子视图才被绘制。如果为视图设定了可绘制的背景，则视图会在回调 onDraw()方法之前先绘制它。

MyButton 类内实现了 8 个成员函数和 2 个构造函数。8 个成员函数具体如下。

- setText()：设置绘制在按钮上的文本。
- setTextSize()：设置文本大小。
- setTextColor()：设置文本颜色。
- measureWidth()：度量 Button 微件的宽度。
- measureHeight()：度量 Button 微件的高度。
- drawArcs()：绘制圆弧。
- onDraw()：在 Button 微件上绘制图形。
- onMeasure()：度量和设置 Button 微件的边界。

setText()、setTextSize()和 setTextColor()方法用于修改文本的属性。每当文本发生变化，需要调用 invalidate()方法以强制视图对按钮微件进行重绘，从而将变化即时反映出来。requestLayout()方法会在 setText()和 setTextSize()方法中调用，但不会在 setTextColor()中调用。这是因为只有当微件的边界发生变化时才会用到布局，而在文本颜色变化时并不需要布局。

onMeasure()方法要把 setMeasuredDimension()方法与 measureWidth()和 measureHeight()这两个方法一起调用，这对于自定义视图是关键一步。

measureWidth()和 measureHeight()两个方法调用时会带上父视图尺寸这一参数，并需要返回基于所需度量模式的、自定义视图的适当宽度值和高度值。如果指定了 EXACTLY 模式，则上述两个方法需要返回由父 View 对象给出的值。如果指定了 AT_MOST 模式，这两个方法会返回内容大小和父视图大小二者中的较小值，以确保内容被适当地调整了大小。否则，两个方法会基于微件中的内容来计算宽度值和高度值。本技巧中，内容大小是基于文本大小的。

drawArcs()方法在按钮上绘制圆弧。当绘制文本时，onDraw()方法会调用 drawArcs()方法。圆弧的动画效果也由该方法产生。每当圆弧被绘制时，其长度会略有增加，而其梯度会旋转变化，这会产生一个不错的动画。

自定义按钮类如代码清单 7-1 所示。我们还需要一个构造方法，而本例中含有两个 MyButton()方法，它们所接受的参数不同，各自用自定义的属性值来初始化标签视图。android.graphics.*库在格式上与 Java 的图形操作库（如 Matrix 和 Paint）相似。

**代码清单 7-1 src/com/cookbook/advance/MyButton.java**

```
package com.cookbook.advance.customcomponent;

import android.content.Context;
import android.graphics.Canvas;
import android.graphics.Color;
import android.graphics.Matrix;
import android.graphics.Paint;
import android.graphics.RectF;
import android.graphics.Shader;
import android.graphics.SweepGradient;
```

```
import android.util.AttributeSet;
import android.util.Log;
import android.widget.Button;

public class MyButton extends Button {
    private Paint mTextPaint, mPaint;
    private String mText;
    private int mAscent;
    private Shader mShader;
    private Matrix mMatrix = new Matrix();
    private float mStart;
    private float mSweep;
    private float mRotate;
    private static final float SWEEP_INC = 2;
    private static final float START_INC = 15;

    public MyButton(Context context) {
        super(context);
        initLabelView();
    }

    public MyButton(Context context, AttributeSet attrs) {
        super(context, attrs);
        initLabelView();
    }

    private final void initLabelView() {
        mTextPaint = new Paint();
        mTextPaint.setAntiAlias(true);
        mTextPaint.setTextSize(16);
        mTextPaint.setColor(0xFF000000);
        setPadding(15, 15, 15, 15);
        mPaint = new Paint();
        mPaint.setAntiAlias(true);
        mPaint.setStrokeWidth(4);
        mPaint.setAntiAlias(true);
        mPaint.setStyle(Paint.Style.STROKE);
        mShader = new SweepGradient(this.getMeasuredWidth()/2,
                                    this.getMeasuredHeight()/2,
                              new int[] { Color.GREEN,
                                          Color.RED,
                                          Color.CYAN,Color.DKGRAY },
                              null);
        mPaint.setShader(mShader);
    }

    public void setText(String text) {
        mText = text;
        requestLayout();
        invalidate();
    }

    public void setTextSize(int size) {
        mTextPaint.setTextSize(size);
        requestLayout();
        invalidate();
    }

    public void setTextColor(int color) {
        mTextPaint.setColor(color);
        invalidate();
    }
```

```
@Override
protected void onMeasure(int widthMeasureSpec, int heightMeasureSpec){
    setMeasuredDimension(measureWidth(widthMeasureSpec),
            measureHeight(heightMeasureSpec));
}

private int measureWidth(int measureSpec) {
    int result = 0;
    int specMode = MeasureSpec.getMode(measureSpec);
    int specSize = MeasureSpec.getSize(measureSpec);

    if (specMode == MeasureSpec.EXACTLY) {
        // We were told how big to be
        result = specSize;
    } else {
        // Measure the text
        result = (int) mTextPaint.measureText(mText)
                + getPaddingLeft()
                + getPaddingRight();
        if (specMode == MeasureSpec.AT_MOST) {
            result = Math.min(result, specSize);
        }
    }

    return result;
}

private int measureHeight(int measureSpec) {
    int result = 0;
    int specMode = MeasureSpec.getMode(measureSpec);
    int specSize = MeasureSpec.getSize(measureSpec);

    mAscent = (int) mTextPaint.ascent();
    if (specMode == MeasureSpec.EXACTLY) {
        // We were told how big to be
        result = specSize;
    } else {
        // Measure the text (beware: ascent is a negative number)
        result = (int) (-mAscent + mTextPaint.descent())
                    + getPaddingTop() + getPaddingBottom();
        if (specMode == MeasureSpec.AT_MOST) {
            Log.v("Measure Height", "At most Height:"+specSize);
            result = Math.min(result, specSize);
        }
    }
    return result;
}

private void drawArcs(Canvas canvas, RectF oval, boolean useCenter,
        Paint paint) {
    canvas.drawArc(oval, mStart, mSweep, useCenter, paint);
}

@Override protected void onDraw(Canvas canvas) {
    mMatrix.setRotate(mRotate, this.getMeasuredWidth()/2,
                    this.getMeasuredHeight()/2);
    mShader.setLocalMatrix(mMatrix);
    mRotate += 3;
    if (mRotate >= 360) {
        mRotate = 0;
    }
```

```
        RectF drawRect = new RectF();
        drawRect.set(this.getWidth()-mTextPaint.measureText(mText),
                  (this.getHeight()-mTextPaint.getTextSize())/2,
                mTextPaint.measureText(mText),
        this.getHeight()-(this.getHeight()-mTextPaint.getTextSize())/2);
        drawArcs(canvas, drawRect, false, mPaint);
        mSweep += SWEEP_INC;
        if (mSweep > 360) {
            mSweep -= 360;
            mStart += START_INC;
            if (mStart >= 360) {
                mStart -= 360;
            }
        }
        if(mSweep >180){
            canvas.drawText(mText, getPaddingLeft(),
                            getPaddingTop() -mAscent, mTextPaint);
        }
        invalidate();
    }
}
```

接下来，可以在代码清单 7-2 给出的布局中使用自定义的 Button 微件。

**代码清单 7-2 res/layout/main.xml**

```
<?xml version="1.0" encoding="utf-8"?>
<LinearLayout xmlns:android="http://schemas.android.com/apk/res/android"
    android:orientation="vertical"
    android:layout_width="match_parent"
    android:layout_height="match_parent"
    android:gravity="center_vertical"
    >
<com.cookbook.advance.customComponent.MyButton
    android:layout_width="wrap_content"
    android:layout_height="wrap_content"
    android:id="@+id/mybutton1"
    />
</LinearLayout>
```

XML 布局中只含有一个 ViewGroup（LinearLayout）和一个 View。通过其定义位置 com.cookbook.advance.customComponent.myButton 可以对其进行调用。该布局被用在 Activity 中，如代码清单 7-3 所示。

**代码清单 7-3 src/com/cookbook/advance/ShowMyButton.java**

```
package com.cookbook.advance.customComponent;

import android.app.Activity;
import android.os.Bundle;

public class ShowMyButton extends Activity{

    @Override
    protected void onCreate(Bundle savedInstanceState) {
        super.onCreate(savedInstanceState);

        setContentView(R.layout.main);
        MyButton myb = (MyButton)findViewById(R.id.mybutton1);
```

```
            myb.setText("Hello Students");
            myb.setTextSize(40);
        }
    }
```

这显示了自定义按钮的使用方法与普通的 Button 微件如出一辙。自定义按钮的运行结果见图 7-1。

图 7-1　自定义按钮的例子

## 7.2 Android 动画

Android 提供了两种动画：逐帧动画和补间动画。逐帧动画的原理是按顺序显示一系列的图片，它使开发者可以定义要显示的图片，并像播放幻灯片那样播放它们。

逐帧动画首先需要在布局文件中有一个 animation-list 元素，该元素包含一个 item 元素的列表，用来按顺序指定要显示的图片。oneshot 属性用来指定动画只播放一次还是会反复播放。定义动画列表的 XML 文件如代码清单 7-4 所示。

**代码清单 7-4　res/anim/animated.xml**

```
<?xml version="1.0" encoding="utf-8"?>
<animation-list xmlns:android="http://schemas.android.com/apk/res/android"
      android:oneshot="false">
    <item android:drawable="@drawable/anddev1" android:duration="200" />
    <item android:drawable="@drawable/anddev2" android:duration="200" />
    <item android:drawable="@drawable/anddev3" android:duration="200" />
</animation-list>
```

要显示逐帧动画，需要将动画设置到视图的背景中：

```
ImageView im = (ImageView) this.findViewById(R.id.myanimated);
im.setBackgroundResource(R.anim.animated);
AnimationDrawable ad = (AnimationDrawable)im.getBackground();
ad.start();
```

设定好视图背景后，就可以通过调用 getBackground()方法来检索到一个 drawable 对象，进而将其投入到 AnimationDrawable 之中。然后，调用 start()方法开始播放动画。

补间动画的实现途径则有所不同，它会通过对单个图像进行一系列的变换来创建动画效果。Android 提供了对下列类的访问机制，作为所有补间动画的实现基础。

- AlphaAnimation：控制透明度变化。
- RotateAnimation：控制旋转。
- ScaleAnimation：控制放大或缩小。
- TranslateAnimation：控制位置平移。

上述 4 个动画类可用来实现不同 Activity、布局、视图等之间的转换。它们在布局 XML 文件中分别被定义为<alpha>、<rotate>、<scale>和<translate>标记。它们都要被<set>这一 AnimationSet（动画集）标记所包含。

- <alpha>属性：

  ```
  android:fromAlpha、android:toAlpha
  ```

  alpha 的值控制不透明度在 0.0～1.0（从透明到不透明）之间变化。

- <rotate>属性：

  ```
  android:fromDegrees、android:toDegrees
  android:pivotX、android:pivotY
  ```

  rotate 指定动画要旋转的角度，以及旋转时围绕的轴（pivot）。

- <scale>属性：

  ```
  android:fromXScale、android:toXScale、
  android:fromYScale、android:toYScale、
  android:pivotX、android:pivotY
  ```

  scale 指定如何在 $x$ 轴或 $y$ 轴方向改变视图的大小。还可以指定在变换时保持固定的中心位置。

- <translate>属性：

  ```
  android:fromXDelta、android:toXDelta、
  android:fromYDelta、android:toYDelta
  ```

  translate 指定对视图进行的平移量大小。

## 技巧 64：创建动画

本技巧创建一个“新邮件”动画，当收到新邮件时播放。主布局文件如代码清单 7-5 所示，而图 7-2 给出了“新邮件”动画的效果示意。

**代码清单 7-5　res/layout/main.xml**

```
<?xml version="1.0" encoding="utf-8"?>
<LinearLayout xmlns:android="http://schemas.android.com/apk/res/android"
    android:orientation="vertical"
    android:layout_width="match_parent"
```

```
    android:layout_height="match_parent"
    android:gravity="center"
    >

    <ImageView
    android:id="@+id/myanimated"
    android:layout_width="wrap_content"
    android:layout_height="wrap_content"
       android:src="@drawable/mail"
    />
    <Button
    android:id="@+id/startanimated"
    android:layout_width="wrap_content"
    android:layout_height="wrap_content"
    android:text="you've got mail"
    />
</LinearLayout>
```

图 7-2　动画所属的基本布局

要让视图动起来，需要定义一个动画集。在 Eclipse 中，右键单击 **res/**文件夹并选择 **New→Android XML File**。接着填写文件名（filename）为 **animated.xml** 并选择文件类型（file type）为 **Animation**。然后就可以编辑该文件来创建如代码清单 7-6 所示的内容。

**代码清单 7-6　res/anim/animated.xml**

```
<?xml version="1.0" encoding="utf-8"?>

<set xmlns:android="http://schemas.android.com/apk/res/android"
android:interpolator="@android:anim/accelerate_interpolator">

    <translate android:fromXDelta="100%p" android:toXDelta="0"
android:duration="5000" />
    <alpha android:fromAlpha="0.0" android:toAlpha="1.0" android:duration="3000" />
        <rotate
        android:fromDegrees="0"
        android:toDegrees="-45"
```

```
              android:toYScale="0.0"
              android:pivotX="50%"
              android:pivotY="50%"
              android:startOffset="700"
              android:duration="3000" />
            <scale
              android:fromXScale="0.0"
              android:toXScale="1.4"
              android:fromYScale="0.0"
              android:toYScale="1.0"
              android:pivotX="50%"
              android:pivotY="50%"
              android:startOffset="700"
              android:duration="3000"
              android:fillBefore="false" />
</set>
```

主 Activity 如代码清单 7-7 所示，它并不复杂，其中使用 AnimationUtils 来装载在 animation 文件中定义的 animationSet，从而创建了一个 Animation 对象。随后，每次用户点击按钮，该按钮会使用 ImageView 对象，后者通过调用 startAnimation()方法来运行动画。在调用 startAnimation()方法时，则会用上已经装载好的 Animation 对象。

**代码清单 7-7　src/com/cookbook/advance/MyAnimation.java**

```
package com.cookbook.advance;

import android.app.Activity;
import android.os.Bundle;
import android.view.View;
import android.view.View.OnClickListener;
import android.view.animation.Animation;
import android.view.animation.AnimationUtils;
import android.widget.Button;
import android.widget.ImageView;

public class MyAnimation extends Activity {
    /** called when the activity is first created */
    @Override
    public void onCreate(Bundle savedInstanceState) {
        super.onCreate(savedInstanceState);
        setContentView(R.layout.main);

        final ImageView im
                = (ImageView) this.findViewById(R.id.myanimated);
        final Animation an
                = AnimationUtils.loadAnimation(this, R.anim.animated);

        im.setVisibility(View.INVISIBLE);
        Button bt = (Button) this.findViewById(R.id.startanimated);
        bt.setOnClickListener(new OnClickListener(){
            public void onClick(View view){
                    im.setVisibility(View.VISIBLE);
                    im.startAnimation(an);
            }
        });
    }
}
```

## 技巧 65：使用属性动画

从 Honeycomb（API Level 11）开始，像按钮这样的对象也可以通过改变属性来实现动画效果。本技巧将创建三个按钮，它们在按下后会基于属性值变化而表现出不同的动画效果。

属性值的改变既可以在 Activity 代码中直接实现，也可以从独立的 XML 文件中装载。当在这两种动画样式中进行选择时，要记得在 Activity 中对动画进行编码能够对动态数据有更多的控制，而 XML 文件则使得复杂的动画变得更容易实现。

关于动画改变的一个重要注意事项是，使用 ValueAnimator、ObjectAnimator 和 AnimatorSet 标签的 XML 文件应当把资源文件放到 **res/animator** 文件夹中，以与遗留的 XML 文件相区别，后者的资源存储在 **res/anim** 文件夹中。

代码清单 7-8 给出了同时使用内联（inline）动画代码和 XML 文件引用两种动画方式的 Activity。其中使用 OnClickListener 来将各个按钮分别与一个为其实现动画效果的函数相绑定。

**代码清单 7-8 src/com/cookbook/propertyanimation/MainActivity.java**

```
package com.cookbook.propertyanimation;

import android.animation.ArgbEvaluator;
import android.animation.ObjectAnimator;
import android.animation.ValueAnimator;
import android.app.Activity;
import android.graphics.Color;
import android.os.Bundle;
import android.view.View;
import android.view.animation.AnimationUtils;
import android.widget.Button;

public class MainActivity extends Activity {

  Button btnShift;
  Button btnRotate;
  Button btnSling;

  @Override
  protected void onCreate(Bundle savedInstanceState) {
    super.onCreate(savedInstanceState);
    setContentView(R.layout.activity_main);

    btnShift = (Button)this.findViewById(R.id.button);
    btnRotate = (Button)this.findViewById(R.id.button1);
    btnSling = (Button)this.findViewById(R.id.button2);

    btnShift.setOnClickListener(new Button.OnClickListener() {
      public void onClick(View v) {
        int start = Color.rgb(0xcc, 0xcc, 0xcc);
          int end = Color.rgb(0x00, 0xff, 0x00);
          ValueAnimator va =
              ObjectAnimator.ofInt(btnShift, "backgroundColor", start, end);
          va.setDuration(750);
          va.setRepeatCount(1);
          va.setRepeatMode(ValueAnimator.REVERSE);
          va.setEvaluator(new ArgbEvaluator());
```

```
            va.start();
        }
    });

    btnRotate.setOnClickListener(new Button.OnClickListener() {
        public void onClick(View v) {
            // Use ValueAnimator
            /*
            ValueAnimator va = ObjectAnimator.ofFloat(btnRotate, "rotation", 0f, 360f);
            va.setDuration(750);
              va.start();
              */
              // Or use an XML-defined animation
        btnRotate.startAnimation(AnimationUtils.loadAnimation(MainActivity.this,
                R.animator.rotation));
        }
    });

    btnSling.setOnClickListener(new Button.OnClickListener() {
        public void onClick(View v) {
            btnSling.startAnimation(AnimationUtils.loadAnimation(MainActivity.this,
                R.animator.sling));
            }
        });
    }
}
```

btnShift 将 ValueAnimator 和 ObjectAnimator 协同使用，以设置当按钮被点击时，应用于其上的值。

btnRotate 含有一段被注释掉的代码，其功能是使用内联代码显示一段旋转动画。取而代之的是 **res/animator** 目录下的 XML 文件，如代码清单 7-9 所示。btnRotate 显示了如何用它来实现相同的动画效果。

### 代码清单 7-9　res/animator/rotation.xml

```
<?xml version="1.0" encoding="utf-8"?>
<rotate
    xmlns:android="http://schemas.android.com/apk/res/android"
    android:fromDegrees="0"
    android:toDegrees="360"
    android:pivotX="50%"
    android:pivotY="50%"
    android:duration="500" android:fillAfter="true">
</rotate>
```

btnSling 也使用 XML 来实现动画，如代码清单 7-10 所示。但它实现了两个而不是一个动画。多个动画可以被包含在同一个 XML 文件中，并且会被同时执行。android:startOffset 属性是用来延迟动画开始的时间，从而让动画逐个“排队”运行。

### 代码清单 7-10　res/animator/sling.xml

```
<?xml version="1.0" encoding="utf-8"?>
<set xmlns:android="http://schemas.android.com/apk/res/android">
  <scale
    android:interpolator="@android:anim/accelerate_decelerate_interpolator"
    android:fromXScale="0.0"
```

```
      android:toXScale="1.8"
      android:fromYScale="0.0"
      android:toYScale="1.4"
      android:pivotX="50%"
      android:pivotY="50%"
      android:fillAfter="false"
      android:duration="1000" />
    <scale
      android:fromXScale="1.8"
      android:toXScale="0.0"
      android:fromYScale="1.4"
      android:toYScale="0.0"
      android:pivotX="50%"
      android:pivotY="50%"
      android:startOffset="1000"
      android:duration="300"
      android:fillBefore="false" />
</set>
```

## 7.3 辅助功能

Android 带有若干辅助功能（accessibility）特征，它们被融入到平台之中。TalkBack 是一个多数 Android 设备都安装了的服务，它通过语音合成来阅读显示在屏幕上的内容。如果设备未安装 TalkBack，可以从 Google Play 上下载。

可以进入设备设置当中的“辅助功能”项来启用 TalkBack。只要安装了 TalkBack，就会有一个相应的开关。开关打开时，用户与 Android 交互的方式就会改变。整个屏幕变成了一个输入层，可以通过手势和划动在不同应用程序和屏幕之间徜徉。要打开一个应用程序，首先要选中它，再在屏幕上双击。

返回键和 Home 键一类的软按键也需要被选中，然后才双击。

### 技巧 66：使用辅助功能特性

Google 建议在创建辅助应用程序前进行下列检查。

- 确保组件被描述。这可以通过在布局 XML 文件中使用 `android:contentDescription` 来完成。
- 确保用到的组件可以获得焦点。
- 如果使用自定义的控制方式，则要利用好辅助功能界面。
- 不要仅仅通过音频来交互。应当始终为应用程序加入可视的提示甚至触觉上的反馈。
- 使用 TalkBack，在不看屏幕的情况下测试应用程序。

当为界面组件添加描述时，`editText` 域会使用 `android:hint` 来取代 `android:contentDescription`。当该域为空时，`android:hint` 的值就会被大声读出来。如果该域包含用户输入的文本，它的值就会取代 `android:hint` 的值而被大声读出。代码清单 7-11 给出的布局文件中，前述的各种域都已被填入了相应的内容。

代码清单 7-11　res/layout/activity_main.xml

```
<RelativeLayout xmlns:android="http://schemas.android.com/apk/res/android"
    xmlns:tools="http://schemas.android.com/tools"
    android:layout_width="match_parent"
    android:layout_height="match_parent"
    tools:context=".MainActivity" >

    <EditText
        android:id="@+id/edittext1"
        android:layout_width="wrap_content"
        android:layout_height="wrap_content"
        android:layout_alignParentTop="true"
        android:layout_centerHorizontal="true"
        android:ems="10"
        android:hint="Enter some text here"
        android:nextFocusDown="@+id/button1" >

        <requestFocus />
    </EditText>

    <TextView
        android:id="@+id/textview1"
        android:layout_width="wrap_content"
        android:layout_height="wrap_content"
        android:layout_below="@+id/editText1"
        android:layout_centerHorizontal="true"
        android:layout_marginTop="35dp"
        android:contentDescription="Text messages will appear here"
        android:text="This is a TextView" />

    <RadioGroup
        android:id="@+id/radiogroup1"
        android:layout_width="wrap_content"
        android:layout_height="wrap_content"
        android:layout_below="@+id/textView1"
        android:layout_centerHorizontal="true"
        android:layout_marginTop="21dp" >

        <RadioButton
            android:id="@+id/radio0"
            android:layout_width="wrap_content"
            android:layout_height="wrap_content"
            android:checked="true"
            android:contentDescription="Select for a banana"
            android:text="Banana" />

        <RadioButton
            android:id="@+id/radio1"
            android:layout_width="wrap_content"
            android:layout_height="wrap_content"
            android:contentDescription="Select for a coconut"
            android:text="Coconut" />

        <RadioButton
            android:id="@+id/radio2"
            android:layout_width="wrap_content"
            android:layout_height="wrap_content"
            android:contentDescription="Select for a grape"
            android:text="Grape" />
    </RadioGroup>

</RelativeLayout>
```

# 7.4 Fragment

在大屏幕设备上，必须考虑应用程序如何能用一种令人愉快并且简单易用的方式来显示信息。既然平板电脑拥有比大多数手机更大的物理视野区域，开发者就可以盘算一下让他们的应用程序如何以不同于小屏幕设备的方式来显示数据。

手机和平板电脑设备当中已经构建有这方面的一些出色范例，比如通讯录或联系人应用程序。在手机上打开某个该类应用程序，会呈现给用户一个列表视图，用户可以在其中的联系人之间上下滚动，点击某条记录可以打开其中的信息。而在平板电脑上，列表视图比手机上的要小一些，且会显示在屏幕左侧，而在屏幕右侧显示联系人的扩展信息。

日历应用程序的功能也有类似之处。在手机上，会显示量少而精的数据；而在平板电脑上则会在屏幕下部显示扩展的信息。

显示扩展数据的功能是通过 Fragment 实现的。Fragment 是 Activity 的模块化组成部分，可以用来改变应用程序的呈现效果。Fragment 的运作方式与 Activity 相像，但生命周期和处理逻辑方式有所不同。本节中的技巧着眼于使用 Fragment 针对不同的屏幕尺寸设备优化内容显示效果。

## 技巧 67：同时显示多个 Fragment

这个技巧包含两个 Fragment。用一个 `ListFragment` 显示列表，用另一个 Fragment 显示 `TextView`。在小屏幕设备上，最佳选择是用两个不同的窗体分别显示列表和文本。而在大屏幕设备上，列表和文本则可以显示在同一个窗体中。

首先，定义了多个 XML 布局文件，其中两个在 **res/layout** 文件夹中，是针对小屏幕设备的；还有一个布局文件在 **res/layout-large** 文件夹中，是针对大屏幕设备的。

代码清单 7-12 显示了 **activity_main.xml** 文件。布局文件非常小，因为 `ListFragment` 会被装载到 `FrameLayout` 之中，并提供所需布局的大部分内容。

**代码清单 7-12　res/layout/activity_main.xml**

```
<?xml version="1.0" encoding="utf-8"?>

<FrameLayout xmlns:android="http://schemas.android.com/apk/res/android"
    android:id="@+id/fragment_container"
    android:layout_width="match_parent"
    android:layout_height="match_parent" />
```

面向小屏幕设备的布局文件还有一个，用来显示文本的值。代码清单 7-13 给出了这个 **text_view.xml**。

**代码清单 7-13　res/layout/text_view.xml**

```
<?xml version="1.0" encoding="utf-8"?>

<TextView xmlns:android="http://schemas.android.com/apk/res/android"
    android:id="@+id/text"
```

```
    android:layout_width="match_parent"
    android:layout_height="match_parent"
    android:padding="16dp"
    android:textSize="18sp" />
```

对于大屏幕设备，需要在 **res** 目录下创建一个名为 **layout-large** 的文件夹。创建好之后，就要建立另一个布局 XML 文件。本例中添加了另一个 **activity_main.xml** 文件。代码清单 7-14 显示了这个用于大屏幕布局的文件的内容。

**代码清单 7-14 res/layout/activity_main.xml**

```
<LinearLayout xmlns:android="http://schemas.android.com/apk/res/android"
    android:orientation="horizontal"
    android:layout_width="match_parent"
    android:layout_height="match_parent">

    <fragment android:name="com.cookbook.fragments.ItemFragment"
              android:id="@+id/item_fragment"
              android:layout_weight="1"
              android:layout_width="0dp"
              android:layout_height="match_parent" />

    <fragment android:name="com.cookbook.fragments.TextFragment"
              android:id="@+id/text_fragment"
              android:layout_weight="2"
              android:layout_width="0dp"
              android:layout_height="match_parent" />

</LinearLayout>
```

代码清单 7-15 中的 Fragment 元素引用了一个类文件，该文件包含用于每个 Fragment 的逻辑。代码清单 7-15 显示了 Fragment 元素的内容，所用到的文件是 **src/com/cookbook/fragments/ItemFragment.java**。**ItemFragment.java** 文件使用 `getFragmentManager` 来决定设备是否使用大布局。还要注意，`ItemFragment` 引用了 `String.Items`，后者是一个在 `Strings.java` 中创建的数组，在 **TextFragment.java** 中被导入。

**代码清单 7-15 src/com/cookbook/fragments/ItemFragment.java**

```
package com.cookbook.fragments;

import android.app.Activity;
import android.os.Build;
import android.os.Bundle;
import android.support.v4.app.ListFragment;
import android.view.View;
import android.widget.ArrayAdapter;
import android.widget.ListView;

public class ItemFragment extends ListFragment {
  OnItemSelectedListener mCallback;

    public interface OnItemSelectedListener {
        public void onItemSelected(int position);
    }

    @Override
    public void onCreate(Bundle savedInstanceState) {
        super.onCreate(savedInstanceState);
```

```
        // Older than Honeycomb requires a different layout
        int layout = Build.VERSION.SDK_INT >= Build.VERSION_CODES.HONEYCOMB ?
                android.R.layout.simple_list_item_activated_1 :
                android.R.layout.simple_list_item_1;

        setListAdapter(new ArrayAdapter<String>(getActivity(), layout,
            Strings.Items));
    }

    @Override
    public void onStart() {
        super.onStart();
        if (getFragmentManager().findFragmentById(R.id.item_fragment) != null) {
            getListView().setChoiceMode(ListView.CHOICE_MODE_SINGLE);
        }
    }

    @Override
    public void onAttach(Activity activity) {
        super.onAttach(activity);

        try {
            mCallback = (OnItemSelectedListener) activity;
        } catch (ClassCastException e) {
            throw new ClassCastException(activity.toString()
                + " must implement OnItemSelectedListener");
        }
    }

    @Override
    public void onListItemClick(ListView l, View v, int position, long id) {
        mCallback.onItemSelected(position);
        getListView().setItemChecked(position, true);
    }
}
```

代码清单 7-16 给出了 Fragment 元素的内容，所引用的文件为 **src/com/cookbook/fragments/TextFragment.java**。

**代码清单 7-16　src/com/cookbook/fragments/TextFragment.java**

```
package com.cookbook.fragments;

import com.cookbook.fragments.Strings;
import com.cookbook.fragments.R;

import android.os.Bundle;
import android.support.v4.app.Fragment;
import android.view.LayoutInflater;
import android.view.View;
import android.view.ViewGroup;
import android.widget.TextView;

public class TextFragment extends Fragment {
    final static String ARG_POSITION = "position";
    int mCurrentPosition = -1;

    @Override
    public View onCreateView(LayoutInflater inflater,
        ViewGroup container, Bundle savedInstanceState) {
        return inflater.inflate(R.layout.text_view, container, false);
    }
```

```
    @Override
    public void onStart() {
        super.onStart();

        Bundle args = getArguments();
        if (args != null) {
            updateTextView(args.getInt(ARG_POSITION));
        } else if (mCurrentPosition != -1) {
            updateTextView(mCurrentPosition);
        } else {
            TextView tv = (TextView) getActivity().findViewById(R.id.text);
            tv.setText("Select an item from the list.");
        }
    }

    public void updateTextView(int position){
        TextView tv = (TextView) getActivity().findViewById(R.id.text);
        tv.setText(Strings.Text[position]);
        mCurrentPosition = position;
    }

    @Override
    public void onSaveInstanceState(Bundle outState) {
        super.onSaveInstanceState(outState);
        outState.putInt(ARG_POSITION, mCurrentPosition);
    }
}
```

既然布局已经设定完成，Fragment 也被创建好了，就该轮到主 Activity。代码清单 7-17 给出了 **MainActivity.java**。Fragment 由 `FragmentManager` 来处理，后者又依靠 `FragmentTransaction`。`FragmentTransaction` 跟踪那些对视图可用的 Fragment。每当使用 `FragmentTransaction` 添加、删除或替换 Fragment 时，必然要调用 `commit()` 方法。

**代码清单 7-17　src/com/cookbook/fragments/MainActivity.java**

```
package com.cookbook.fragments;

import android.os.Bundle;
import android.support.v4.app.FragmentActivity;
import android.support.v4.app.FragmentTransaction;

// When using the support lib, use FragmentActivity
public class MainActivity extends FragmentActivity implements
    ItemFragment.OnItemSelectedListener {

    @Override
    protected void onCreate(Bundle savedInstanceState) {
        super.onCreate(savedInstanceState);
        setContentView(R.layout.activity_main);

        //If using large layout, use the Support Fragment Manager
        if (findViewById(R.id.fragment_container) != null) {
            if(savedInstanceState != null){
                return;
            }
            ItemFragment firstFragment = new ItemFragment();
            firstFragment.setArguments(getIntent().getExtras());

    getSupportFragmentManager().beginTransaction().add(R.id.fragment_container,
    firstFragment).commit();
        }
    }
```

```
    public void onItemSelected(int position) {
        TextFragment textFrag =
            (TextFragment) getSupportFragmentManager()
                .findFragmentById(R.id.text_fragment);
        if (textFrag != null) {
            textFrag.updateTextView(position);
        } else {
            TextFragment newFragment = new TextFragment();
            Bundle args = new Bundle();
            args.putInt(TextFragment.ARG_POSITION, position);
            newFragment.setArguments(args);

            FragmentTransaction transaction =
                getSupportFragmentManager().beginTransaction();
            transaction.replace(R.id.fragment_container, newFragment);
            transaction.addToBackStack(null);
            transaction.commit();
        }
    }
}
```

## 技巧 68：使用对话框 Fragment

除改变某个页面的布局之外，还可使用 DialogFragment 来显示一个包含 Fragment 的对话框窗体。顾名思义，DialogFragment 就是一个对话框包含在一个 Fragment 之中。推荐始终使用 DialogFragment 来创建对话框。借助支持库，可以将代码移植到早期的 Android 版本之上。代码清单 7-18 给出了如何使用 DialogFragment 来建立 Activity。注意因为要使用 Fragment，Activity 必须扩展 FragmentActivity。

**代码清单 7-18 src/com/cookbook/dialogfragment/MainActivity.java**

```
package com.cookbook.dialogfragment;

import android.os.Bundle;
import android.support.v4.app.FragmentActivity;
import android.view.View;
import android.widget.Button;
import android.widget.Toast;

public class MainActivity extends FragmentActivity {

    @Override
    public void onCreate(Bundle savedInstanceState) {
        super.onCreate(savedInstanceState);
        setContentView(R.layout.activity_main);

        Button buttonOpenDialog = (Button) findViewById(R.id.opendialog);
        buttonOpenDialog.setOnClickListener(new Button.OnClickListener() {

            @Override
            public void onClick(View arg0) {
                openDialog();
            }
        });

    }

    void openDialog() {
```

```
        MyDialogFragment myDialogFragment = MyDialogFragment.newInstance();
        myDialogFragment.show(getSupportFragmentManager(), "myDialogFragment");
    }

    public void protestClicked() {
        Toast.makeText(MainActivity.this, "Your protest has been recorded", Toast.
➥LENGTH_LONG).show();
    }

    public void forgetClicked() {
        Toast.makeText(MainActivity.this,
            "You have chosen to forget", Toast.LENGTH_LONG).show();
    }

}
```

代码清单 7-19 给出了 DialogFragment 的逻辑。在此，MyDialogFragment 扩展了 DialogFragment。onCreateDialog 方法被重写，而一个新 Dialog 和某些 onClick 逻辑一起在其中被建立。

**代码清单 7-19 src/com/cookbook/dialogfragment/MyDialogFragment.java**

```
package com.cookbook.dialogfragment;

import android.app.AlertDialog;
import android.app.Dialog;
import android.content.DialogInterface;
import android.os.Bundle;
import android.support.v4.app.DialogFragment;

public class MyDialogFragment extends DialogFragment {
  static MyDialogFragment newInstance() {
    MyDialogFragment mdf = new MyDialogFragment();
    Bundle args = new Bundle();
    args.putString("title", "Dialog Fragment");
    mdf.setArguments(args);
    return mdf;
  }

  @Override
  public Dialog onCreateDialog(Bundle savedInstanceState) {
    String title = getArguments().getString("title");
    Dialog myDialog = new AlertDialog.Builder(getActivity())
        .setIcon(R.drawable.ic_launcher)
        .setTitle(title)
        .setPositiveButton("Protest", new DialogInterface.OnClickListener() {

          @Override
          public void onClick(DialogInterface dialog, int which) {
            ((MainActivity) getActivity()).protestClicked();
          }
        })
        .setNegativeButton("Forget", new DialogInterface.OnClickListener() {

          @Override
          public void onClick(DialogInterface dialog, int which) {
            ((MainActivity) getActivity()).forgetClicked();
          }
        }).create();

    return myDialog;
  }
}
```

# 第 8 章

# 多媒体技术

Android 平台提供了全面的多媒体功能。本章会介绍操作图像、录制和播放声音、录制和播放视频等技术。大多数解码器都为 Android 所支持，用于读取多媒体，但只有一部分编码器可以用在 Android 上来创建多媒体。表 8-1 总结了 Android 4.1 所支持的基本媒体框架。供应商改版后的 Android 可能会支持比表中更多的格式，对于 Google TV 设备尤其如此。

表 8-1　Android 4.1 支持读写的媒体类型

| 格式/编码 | 编码器 | 解码器 | 详细说明 | 支持的文件类型/容器格式 |
|---|---|---|---|---|
| | | 图像 | | |
| JPEG | 支持 | 支持 | 标准型+渐进式 | JPEG（.jpg） |
| GIF | | 支持 | | GIF（.gif） |
| PNG | 支持 | 支持 | | PNG（.png） |
| BMP | | 支持 | | BMP（.bmp） |
| WEBP | 支持（Android 4.0 以上版本） | 支持（Android 4.0 以上版本） | | WebP（.webp） |
| | | 音频 | | |
| AAC LC | 支持 | 支持 | 支持标准采样率在 8～48 kHz 的单声道/立体声/5.0 声道/5.1 声道的内容 | ● 3GPP（.3gp）<br>● MPEG-4（.mp4、.m4a）<br>● ADTS 原始 ACC（.aac，在 Android 3.1 以上版本中可解码，Android 4.0 以上版本中可编码，不支持 ADIF）<br>● MPEG-TS（.ts，不可定位播放进度，需 Android 3.0 以上版本） |

续表

| 格式/编码 | 编码器 | 解码器 | 详细说明 | 支持的文件类型/容器格式 |
| --- | --- | --- | --- | --- |
| HE-AACv1（AAC+） | 支持（Android 4.1 以上版本） | 支持 | | |
| HE-AACv2（增强版 AAC+） | | 支持 | 支持标准采样率在 8～48 kHz 的立体声/5.0 声道/5.1 声道的内容 | |
| AAC ELD（增强版低延迟 AAC） | 支持（Android 4.1 以上版本） | 支持（Android 4.1 以上版本） | 支持标准采样率在 16～48 kHz 的单声道/立体声的内容 | |
| AMR-NB | 支持 | 支持 | 采样率 8 kHz，比特率 4.75～12 Kbps | 3GPP（.3gp） |
| AMR-WB | 支持 | 支持 | 采样率 16 kHz，比特率为 6.60～23.85 Kbps 的 9 种不同值 | 3GPP（.3gp） |
| FLAC | | 支持（Android 3.1 以上版本） | ● 单声道/立体声（无多声道）<br>● 采样率不大于 48 kHz（但建议在输出码率 44.1 kHz 的设备上不超过该值，因为 48～44.1 kHz 的降低采样频率采样器不含有低通滤波器）<br>● 建议采用 16 位<br>● 24 位下无高频振动 | 仅 FLAC（.flac） |
| MP3 | | 支持 | 单声道/立体声，8～320 Kbps 恒定码率（CBR）或可变码率（VBR） | MP3（.mp3） |
| MIDI | | 支持 | ● MIDI 类型 0 和 1<br>● DLS 版本 1 和版本 2<br>● XMF 及移动 XMF<br>● 支持铃声格式 RTTTL/RTX、OTA 和 iMelody | ● 类型 0 和 1（.mid、.xmf、.mxmf）<br>● RTTTL/RTX（.rtttl、.rtx）<br>● OTA（.ota）<br>● iMelody（.imy） |

续表

| 格式/编码 | 编码器 | 解码器 | 详细说明 | 支持的文件类型/容器格式 |
| --- | --- | --- | --- | --- |
| Vorbis | | 支持 | | • Ogg（.ogg）<br>• Matroska（.mkv，需 Android 4.0 以上版本） |
| PCM/WAVE | 支持（Android 4.1 以上版本） | 支持 | • 8 位及 16 位线性 PCM（比特率上限由硬件限制决定）<br>• 原始 PCM 录音的采样率可为 8 000 Hz、16 000 Hz 及 44 100 Hz | WAVE（.wav） |
| 视频 | | | | |
| H.263 | 支持 | 支持 | | • 3GPP（.3gp）<br>• MPEG-4（.mp4） |
| H.264 AVC | 支持（Android 3.0 以上版本） | 支持 | 基线协议（Baseline Profile，BP） | • 3GPP（.3gp）<br>• MPEG-4（.mp4）<br>• MPEG-TS（.ts，仅 AAC 音频，不可定位播放进度，需 Android 3.0 以上版本） |
| MPEG-4 SP | | 支持 | | 3GPP（.3gp） |
| VP8 | | 支持（Android 2.3.3 以上版本） | 仅在 Android 4.0 及以上版本中可流媒体化 | • WebM（.webm）<br>• Matroska（.mkv，需 Android 4.0 以上版本） |

录制任意类型媒体的应用程序需要在 **AndroidManifest.xml** 文件中设置适当的权限（即下面两行之一或二者兼用）：

```
<uses-permission android:name="android.permission.RECORD_AUDIO"/>
<uses-permission android:name="android.permission.RECORD_VIDEO"/>
```

## 8.1 图像

应用程序的本地图像一般放在 **res/drawable/** 目录中（这在第 5 章中谈到过），并与应用程序一起打包。可以通过适当的资源标识符来访问它们，如 `R.drawable.my_picture`。Android 设备文件系统上的图像可以用一般的 Java 类来访问，比如 `InputStream`。然而，在 Android 将图像读入内存以待操作的更好选择是使用内置类 `BitmapFactory`。

BitmapFactory 可以从文件、流或字节数组创建位图对象。可以像下面这样装载资源或者文件：

```
Bitmap myBitmap1 = BitmapFactory.decodeResource(getResources(),
                                              R.drawable.my_picture);
Bitmap myBitmap2 = BitmapFactory.decodeFile(filePath);
```

在图像被装入内存后，就可以用诸如 getPixel()和 setPixel()这样的位图方法对其进行操作。然而，大多数图像的尺寸对于嵌入式设备来说过大，不适宜作全尺寸的操作。反之，考虑对图像进行下采样：

```
Bitmap bm = Bitmap.createScaledBitmap(myBitmap2, 480, 320, false);
```

这有助于避免 OutOfMemory 运行时错误。接下来的技巧给出了装载大图像的一种优化方法。

## 技巧 69：装载和显示一幅可供操作的图像

本技巧给出了这样一个例子：将一幅图像切割成 4 块，并将其顺序打乱，然后显示在屏幕上。在技巧中还显示了如何创建可选择的图像列表。

设备拍下一张图片后，会将其放到 **DCIM/Camera/**目录中，本技巧将该目录作为示例图像目录。图像目录被传入名为 ListFiles 的 Activity，该 Activity 会列出目录中所有的文件，并返回用户选择的是哪一幅。ListFiles Activity 如代码清单 8-1 所示。

**代码清单 8-1　ListFiles.java**

```
public class ListFiles extends ListActivity {
    private List<String> directoryEntries = new ArrayList<String>();

    @Override
    public void onCreate(Bundle savedInstanceState) {
        super.onCreate(savedInstanceState);
        Intent i = getIntent();
        File directory = new File(i.getStringExtra("directory"));

        if (directory.isDirectory()){
            File[] files = directory.listFiles();

            //Sort in descending date order
            Arrays.sort(files, new Comparator<File>(){
                public int compare(File f1, File f2) {
                    returnLong.valueOf(
                        f1.lastModified()).compareTo(f2.lastModified()
                    );
                }
            });
```

```
                //Fill list with files
                this.directoryEntries.clear();
                for (File file : files){
                    this.directoryEntries.add(file.getPath());
                }

                ArrayAdapter<String> directoryList = new ArrayAdapter<String>(
                        this,
                        R.layout.file_row, this.directoryEntries);
                //Alphabetize entries
                //directoryList.sort(null);
                this.setListAdapter(directoryList);
            }
        }

        @Override
        protected void onListItemClick(ListView l, View v,
                                        int position, long id) {
            File clickedFile = new File(this.directoryEntries.get(position));
            Intent i = getIntent();
            i.putExtra("clickedFile", clickedFile.toString());
            setResult(RESULT_OK, i);
            finish();
        }
    }
```

基于传入 Activity 的目录字符串创建了一个 `File` 对象。如果字符串表示一个目录，则其中的文件会通过 `compare()` 方法，基于文件的 `lastModified()` 标志，按最近修改时间由晚到早进行排序。

如果想按字母顺序排序，则可以使用 `sort()` 方法（该方法也已在 `ListFile` Activity 中给出，只是被注释掉了）。随后，使用一个独立的布局文件 `R.layout.file_row` 来构建列表并将其显示到屏幕上，如代码清单 8-2 所示。

**代码清单 8-2　res/layout/file_row.xml**

```
<?xml version="1.0" encoding="utf-8"?>
<TextView
     xmlns:android="http://schemas.android.com/apk/res/android"
     android:layout_width="match_parent"
     android:layout_height="wrap_content"
     android:textSize="20sp"
     android:padding="3pt"
/>
```

`ListFiles` 会将被选择的文件路径返回给调用它的 Activity，后者可以通过其 `onActivityResult(..)` 方法中的 bundle 来读取该路径。

之后，将被选择的文件装入内存以待操作。如果文件过大，可在装载前对其进行下采样以节约内存；只需把代码清单 8-3 的 `onActivityResult` 中用粗体标记的一行语句用如下代码代替即可：

```
BitmapFactory.Options options = new BitmapFactory.Options();
options.inSampleSize = 4;
Bitmap imageToChange = BitmapFactory.decodeFile(tmp, options);
```

inSampleSize 值被设为 4，意味着将创建大小为原图 1/16 的图像（即在长宽两个像素维度上各缩为 1/4）。可以根据原图尺寸来设置合适的 inSampleSize 值。

另一个节约内存的方法是在进行操作之前重新调整内存中图像的大小，这可以通过 createScaledBitmap()方法来实现，本技巧也显示了这一点。代码清单 8-3 给出了主 Activity。

**代码清单 8-3 ImageManipulation.java**

```
package cc.dividebyzero.android.cookbook.chapter8.image;

import cc.dividebyzero.android.cookbook.chapter8.ListFiles;
import cc.dividebyzero.android.cookbook.chapter8.R;
import cc.dividebyzero.android.cookbook.chapter8.R.id;
import cc.dividebyzero.android.cookbook.chapter8.R.layout;
import android.app.Activity;
import android.content.Intent;
import android.graphics.Bitmap;
import android.graphics.BitmapFactory;
import android.os.Bundle;
import android.os.Environment;
import android.widget.ImageView;

public class ImageManipulation extends Activity {
     static final String CAMERA_PIC_DIR = "/DCIM/Camera/";
     ImageView iv;

     @Override
     public void onCreate(Bundle savedInstanceState) {
          super.onCreate(savedInstanceState);
          setContentView(R.layout.image_manipulation);
          iv = (ImageView) findViewById(R.id.my_image);

          String imageDir =
               Environment.getExternalStorageDirectory().getAbsolutePath()
               + CAMERA_PIC_DIR;

          Intent i = new Intent(this, ListFiles.class);
          i.putExtra("directory", imageDir);
          startActivityForResult(i,0);
     }

     @Override
     protected void onActivityResult(int requestCode,
               int resultCode, Intent data) {
          super.onActivityResult(requestCode, resultCode, data);
          if(requestCode == 0 && resultCode==RESULT_OK) {
               String tmp = data.getExtras().getString("clickedFile");
               Bitmap imageToChange= BitmapFactory.decodeFile(tmp);
               process_image(imageToChange);
          }
     }

     void process_image(Bitmap image) {
          Bitmap bm = Bitmap.createScaledBitmap(image, 480, 320, false);
          int width = bm.getWidth();
          int height = bm.getHeight();
          int x = width>>1;
          int y = height>>1;
          int[] pixels1 = new int[(width*height)];
```

```
        int[] pixels2 = new int[(width*height)];
        int[] pixels3 = new int[(width*height)];
        int[] pixels4 = new int[(width*height)];
        bm.getPixels(pixels1, 0, width, 0, 0, width>>1, height>>1);
        bm.getPixels(pixels2, 0, width, x, 0, width>>1, height>>1);
        bm.getPixels(pixels3, 0, width, 0, y, width>>1, height>>1);
        bm.getPixels(pixels4, 0, width, x, y, width>>1, height>>1);
        if(bm.isMutable()) {
            bm.setPixels(pixels2, 0, width, 0, 0, width>>1, height>>1);
            bm.setPixels(pixels4, 0, width, x, 0, width>>1, height>>1);
            bm.setPixels(pixels1, 0, width, 0, y, width>>1, height>>1);
            bm.setPixels(pixels3, 0, width, x, y, width>>1, height>>1);
        }
        iv.setImageBitmap(bm);
    }
}
```

与 Activity 关联的主布局如代码清单 8-4 所示。

**代码清单 8-4　image_manipulation.xml**

```
<?xml version="1.0" encoding="utf-8"?>
<LinearLayout
    xmlns:android="http://schemas.android.com/apk/res/android"
    android:orientation="vertical"
    android:layout_width="match_parent"
    android:layout_height="match_parent">
    <TextView android:layout_width="match_parent"
        android:layout_height="wrap_content"
        android:textSize="30sp"
        android:text="Scrambled Picture" />
    <ImageView android:id="@+id/my_image"
        android:layout_width="wrap_content"
        android:layout_height="wrap_content" />
</LinearLayout>
```

在 **AndroidManifest.xml** 文件中必须对两个 Activity 都进行声明，如代码清单 8-5 所示。图 8-1 则给出了输出结果的示例。

**代码清单 8-5　AndroidManifest.xml**

```
<manifest xmlns:android="http://schemas.android.com/apk/res/android"
    package="cc.dividebyzero.android.cookbook.chapter8"
    android:versionCode="1"
    android:versionName="1.0">
    <uses-sdk android:minSdkVersion="8" android:targetSdkVersion="15" />
    <application android:label="@string/app_name"
        android:icon="@drawable/ic_launcher"
        android:theme="@style/AppTheme">
        <activity android:name=".Chapter8">
            <intent-filter >
                <action android:name="android.intent.action.MAIN" />
                <category android:name="android.intent.category.LAUNCHER" />
            </intent-filter>
        </activity>
        <activity android:name=".ListFiles">
            <intent-filter >
              <action android:name="android.intent.action.PICK" />
              <category android:name="android.intent.category.DEFAULT" />
```

```
            </intent-filter>
        </activity>
        <activity android:name=".audio.AudioPlayback"/>
        <activity android:name=".audio.AudioRecording"/>
        <activity android:name=".audio.AudioSoundPool"/>

        <activity android:name=".video.VideoViewActivity"/>
        <activity android:name=".video.VideoPlayback"/>
        <activity android:name=".image.ImageManipulation"/>
    </application>
</manifest>
```

图 8-1　被切割打乱的图像

# 8.2　音频

录制和播放音频有两套不同的框架。选用哪一种取决于应用程序的需求。

MediaPlayer/MediaRecoder：这是操作音频的标准方法，但要求数据必须是基于文件或者基于流的。处理时会创建独立的线程。SoundPool 中使用这一框架。

AudioTrack/AudioRecorder：这个方法提供了对原始音频的直接访问，对于处理内存中的音频、在播放期间将数据写入缓冲区，或者其他不需文件或流的情形，该方法比较有用。处理时不会创建独立的线程。

上述方法在下面的技巧中均会用到。

## 技巧 70：选择和播放音频文件

MediaPlayer 和 MediaRecoder 类既可用来播放音频，也可用来播放视频。本技巧聚焦于前者，其使用过程是直截了当的。播放过程分为以下步骤。

（1）创建一个 MediaPlayer 实例：

```
MediaPlayer m_mediaPlayer = new MediaPlayer();
```

（2）指定源媒体。可以通过原始资源来创建源媒体：

```
m_mediaPlayer = MediaPlayer.create(this, R.raw.my_music);
```

另一种方式是从文件系统中选择文件（该方式还需要一条 prepare 语句）：

```
m_mediaPlayer.setDataSource(path);
m_mediaPlayer.prepare();
```

任何时候，这两条语句都需要被一个 try-catch 块包括起来，因为指定的资源不一定存在，可能产生异常。

（3）开始音频播放：

```
m_mediaPlayer.start();
```

（4）播放完成时，停止 MediaPlayer 并释放实例，以释放相关资源：

```
m_mediaPlayer.stop();
m_mediaPlayer.release();
```

技巧使用与代码清单 8-1 和代码清单 8-2 给出的同样的 ListFiles Activity 来创建一个待播放的可选音频文件列表。我们假设音频文件位于 Android 设备的**/sdcard/music/**目录中，当然这是可以调整的。

当 ListFiles Activity 返回一个文件后，该文件就被初始化为 MediaPlayer 的媒体源。随后调用 startMP()方法来启动 MediaPlayer，并将按钮文本设为“Pause”。类似地，pauseMP()方法会让 MediaPlayer 暂停，并将按钮文本变为“Play”。任意时刻，用户都可以点击按钮来暂停或继续音乐的播放。

通常，MediaPlayer 会创建它自己的后台线程，且并不随主 Activity 的中断而中断。对于音乐播放器这是个合理的行为，但通常开发者可能希望对其进行控制。因此，为方便说明，本技巧通过重写 onPause()和 onResume()方法，让音乐播放的中断和恢复与主 Activity 同步，如代码清单 8-6 所示。

**代码清单 8-6 AudioPlayback.java**

```
package cc.dividebyzero.android.cookbook.chapter8.audio;

import cc.dividebyzero.android.cookbook.chapter8.ListFiles;
import cc.dividebyzero.android.cookbook.chapter8.R;
import cc.dividebyzero.android.cookbook.chapter8.R.id;
import cc.dividebyzero.android.cookbook.chapter8.R.layout;
import android.app.Activity;
import android.content.Intent;
import android.media.MediaPlayer;
import android.os.Bundle;
import android.os.Environment;
import android.view.View;
import android.widget.Button;

public class AudioPlayback extends Activity {
    static final String MUSIC_DIR = "/music/";
    Button playPauseButton;

    private MediaPlayer m_mediaPlayer;

    @Override
    protected void onCreate(Bundle savedInstanceState) {
        super.onCreate(savedInstanceState);
```

```
        setContentView(R.layout.audio_playback);
        playPauseButton = (Button) findViewById(R.id.play_pause);

        m_mediaPlayer= new MediaPlayer();

        String musicDir = Environment.getExternalStorageDirectory()
        .getAbsolutePath() + MUSIC_DIR;

        //Show a list of music files to choose
        Intent i = new Intent(this, ListFiles.class);
        i.putExtra("directory", musicDir);
        startActivityForResult(i,0);

        playPauseButton.setOnClickListener(new View.OnClickListener() {
            public void onClick(View view) {
                if(m_mediaPlayer.isPlaying()) {
                    //Stop and give option to start again
                    pauseMP();
                } else {
                    startMP();
                }
            }
        });
    }

    @Override
    protected void onActivityResult(int requestCode, int resultCode,
                                    Intent data) {
        super.onActivityResult(requestCode, resultCode, data);
        if(requestCode == 0 && resultCode==RESULT_OK) {
            String tmp = data.getExtras().getString("clickedFile");
            try {
                m_mediaPlayer.setDataSource(tmp);
                m_mediaPlayer.prepare();
            } catch (Exception e) {
                e.printStackTrace();
            }
            startMP();
        }
    }

    void pauseMP() {
        playPauseButton.setText("Play");
        m_mediaPlayer.pause();
    }

    void startMP() {
        m_mediaPlayer.start();
        playPauseButton.setText("Pause");
    }

    boolean needToResume = false;
    @Override
    protected void onPause() {
        if(m_mediaPlayer != null && m_mediaPlayer.isPlaying()) {
            needToResume = true;
            pauseMP();
        }
        super.onPause();
    }
```

```
    @Override
    protected void onResume() {
        super.onResume();
        if(needToResume && m_mediaPlayer != null) {
            startMP();
        }
    }
}
```

与之相关的、带有 Play/Pause 按钮的主 XML 布局由代码清单 8-7 给出。

**代码清单 8-7 res/layout/audio_playback.xml**

```
<?xml version="1.0" encoding="utf-8"?>
<LinearLayout
    xmlns:android="http://schemas.android.com/apk/res/android"
    android:orientation="vertical"
    android:layout_width="match_parent"
    android:layout_height="match_parent">
    <Button android:id="@+id/play_pause"
        android:text="Play"
        android:textSize="20sp"
        android:layout_width="wrap_content"
        android:layout_height="wrap_content" />
</LinearLayout>
```

## 技巧 71：录制音频文件

使用 MediaRecorder 录制音频的过程与上一技巧中播放的过程近似，只是需要多指定一些东西（也可以使用 DEFAULT 默认设置，下面每个列表中的第一项为默认选项）。

- MediaRecorder.AudioSource
  - MIC：内置麦克风。
  - VOICE_UPLINK：语音通话时发送的音频。
  - VOICE_DOWNLINK：语音通话时接收的音频。
  - VOICE_CALL：同时包括语音通话时上行发送和下行接收的音频。
  - CAMCORDER：摄像状态下的麦克风。
  - VOICE_RECOGNITION：语音识别状态下的麦克风。
- MediaRecorder.OutputFormat
  - THREE_GPP：3GPP 媒体格式。
  - MPEG_4：MPEG4 媒体格式。
  - AMR_NB：自适应多速率窄频带文件格式。
- MediaRecorder.AudioEncoder
  - AMR_NB：自适应多速率窄频带声码器（vocoder）。

录制音频的具体步骤如下。

（1）创建一个 MediaRecorder 的实例：

```
MediaRecorder m_Recorder = new MediaRecorder();
```

（2）指定媒体源，比如麦克风：

```
m_Recorder.setAudioSource(MediaRecorder.AudioSource.MIC);
```

（3）设置输出文件格式及编码，比如：

```
m_Recorder.setOutputFormat(MediaRecorder.OutputFormat.THREE_GPP);
m_Recorder.setAudioEncoder(MediaRecorder.AudioEncoder.AMR_NB);
```

（4）设置保存文件的路径：

```
m_Recorder.setOutputFile(path);
```

（5）准备并开始录制：

```
m_Recorder.prepare();
m_Recorder.start();
```

以上录制音频的步骤的具体用法可以参照前一个技巧。

## 技巧 72：操作原始音频

MediaRecorder/MediaPlayer 框架对于大多数音频都是有用的，但要直接操作来自麦克风的原始音频、在不保存为文件的情况下进行处理，或播放原始音频，就要改用 AudioRecord/AudioTrack 来进行。首先要在 **AndroidManifest.xml** 文件中设置权限：

```
<uses-permission android:name="android.permission.RECORD_AUDIO" />
```

接下来，要采取如下步骤。

（1）创建一个 AudioRecord 实例，向构造函数中指定如下内容。

- 音频源：使用前面提到的 MediaRecorder.AudioSource 的可选项中的一个；例如 MediaRecorder.AudioSource.MIC。
- 采样频率，以 Hz（赫兹）为单位：对于 CD 音质的音频使用 44100 一值，也可以使用依次减半的 22050、11025 等值（对于语音而言这样的采样频率已经足够，而且系统也只确保支持这些频率）。
- 通道配置：使用 AudioFormat.CHANNEL_IN_STEREO 来录制立体声声音，或者用 CHANNEL_IN_MONO 来录制单声道声音。
- 音频编码：8 位量化编码时使用 AudioFormat.ENCODING_PCM_8BIT，16 位量化编码则使用 AudioFormat.ENCODING_PCM_16BIT。
- 缓冲区大小（以字节为单位）静态模式下为分配内存的总大小，流模式下则为所用的区块（chunk）的大小。至少要设为 getMinBufferSize()方法对应的值。

（2）用 AudioRecord 实例启动录制。

（3）使用下列方法之一，将音频数据读入内存中的 audioData[]数组：

```
read(short[] audioData, int offsetInShorts, int sizeInShorts)
read(byte[] audioData, int offsetInBytes, int sizeInBytes)
```

（4）停止录制。

例如，下面的代码可以从内置麦克风录制语音到内存缓冲区 RecordedAudio，后者可以被声明为 short[] [1]类型（例如，在每个样本为 16 位的时候）。使用 short[]型的好处是在将字节型的值转换为 short 型值时不需要担心字节顺序问题。注意，每秒 11 025 个样本以及 10 000 个样本的缓冲区大小意味着可录制的时长略小于 1 s。

```
short[] myRecordedAudio = new short[10000];
AudioRecord audioRecord = new AudioRecord(
          MediaRecorder.AudioSource.MIC, 11025,
          AudioFormat.CHANNEL_IN_MONO,
          AudioFormat.ENCODING_PCM_16BIT, 10000);
          audioRecord.startRecording();
          audioRecord.read(myRecordedAudio, 0, 10000);
          audioRecord.stop();
          audioRecord.release();
```

接下来是播放音频的步骤。

（1）创建一个 AudioTrack 实例，为其构造函数指定以下内容。

- 流类型：使用 AudioManager.STREAM_MUSIC 从麦克风捕获或从扬声器回放。其他选项还有 STREAM_VOICE_CALL、STREAM_SYSTEM、STREAM_RING 和 STREAM_ALARM。
- 采样频率（以赫兹为单位）：与录音时的含义相同。
- 通道配置：使用 AudioFormat.CHANNEL_OUT_STEREO 播放立体声声音。还有很多其他选项，例如 CHANNEL_OUT_MONO 和 CHANNEL_OUT_5POINT1（用于播放环绕声）。
- 音频编码：与录音时的含义相同。
- 缓冲区大小（以字节为单位）：一次播放的数据区块的大小。
- 缓冲模式：对于可以全部装入内存的、较短的声音，使用 AudioTrack.MODE_STATIC 避免传输开销。其他情况下，使用 AudioTrack.MODE_STREAM 将数据以区块形式写入硬件。

（2）通过 AudioTrack 实例开始播放。

（3）使用下列方法之一，将内存中的 audioData[]数组写入硬件：

```
write(short[] audioData, int offsetInShorts, int sizeInShorts)
write(byte[] audioData, int offsetInBytes, int sizeInBytes)
```

（4）停止播放（可选）。

例如，如下代码适用于播放前面的录音示例中生成的语音数据：

```
AudioTrack audioTrack = new AudioTrack(
           AudioManager.STREAM_MUSIC, 11025,
           AudioFormat.CHANNEL_OUT_MONO,
           AudioFormat.ENCODING_PCM_16BIT, 4096,
           AudioTrack.MODE_STREAM);
audioTrack.play();
audioTrack.write(myRecordedAudio, 0, 10000);
audioTrack.stop();
audioTrack.release();
```

---

① 即短整型（short）的数组。——译者注

本技巧使用了上面的两个选项将音频录制到内存后再将其播放出来。布局中为屏幕指定了两个按钮：一个用于录制音频，另一个用于播放录制的音频。其声明如代码清单 8-8 给出的主布局文件所示。

**代码清单 8-8 audio_recording.xml**

```
<?xml version="1.0" encoding="utf-8"?>
<LinearLayout xmlns:android="http://schemas.android.com/apk/res/android"
    android:orientation="vertical"
    android:layout_width="match_parent"
    android:layout_height="match_parent">
    <TextView android:id="@+id/status"
        android:text="Ready" android:textSize="20sp"
        android:layout_width="wrap_content"
        android:layout_height="wrap_content" />
    <Button android:id="@+id/record"
        android:text="Record for 5 seconds"
        android:textSize="20sp" android:layout_width="wrap_content"
        android:layout_height="wrap_content" />
    <Button android:id="@+id/play"
        android:text="Play" android:textSize="20sp"
        android:layout_width="wrap_content"
        android:layout_height="wrap_content" />
</LinearLayout>
```

由代码清单 8-9 给出的主 Activity 为上述两个按钮创建了 `OnClickListener`，对内存中音频缓冲区进行录制和播放操作。`onClick()` 回调方法创建了适当的背景线程，因为不论是 `AudioTrack` 还是 `AudioRecord` 都不应该在 UI 线程中运行。为方便说明，给出了两个不同的创建线程的方法：`record_thread()` 含有一个本地线程，其 UI 通过 `Handler` 来更新；而播放线程则使用了主 Activity 的 `run()` 方法。

缓冲区被保存在内存中。为便于说明，将录制时长设定为 5 秒。

**代码清单 8-9 AudioRecording.java**

```
package cc.dividebyzero.android.cookbook.chapter8.audio;

import cc.dividebyzero.android.cookbook.chapter8.R;
import cc.dividebyzero.android.cookbook.chapter8.R.id;
import cc.dividebyzero.android.cookbook.chapter8.R.layout;
import android.app.Activity;
import android.media.AudioFormat;
import android.media.AudioManager;
import android.media.AudioRecord;
import android.media.AudioTrack;
import android.media.MediaRecorder;
import android.os.Bundle;
import android.os.Handler;
import android.util.Log;
import android.view.View;
import android.widget.Button;
import android.widget.TextView;

public class AudioRecording extends Activity implements Runnable {
    private TextView statusText;
    public void onCreate(Bundle savedInstanceState) {
```

```
        super.onCreate(savedInstanceState);
        setContentView(R.layout.audio_recording);

        statusText = (TextView) findViewById(R.id.status);

        Button actionButton = (Button) findViewById(R.id.record);
        actionButton.setOnClickListener(new View.OnClickListener() {
            public void onClick(View view) {
                record_thread();
            }
        });

        Button replayButton = (Button) findViewById(R.id.play);
        replayButton.setOnClickListener(new View.OnClickListener() {
            public void onClick(View view) {
                Thread thread = new Thread(AudioRecording.this);
                thread.start();
            }
        });
    }

    String text_string;
    final Handler mHandler = new Handler();
    // Create runnable for posting
    final Runnable mUpdateResults = new Runnable() {
        public void run() {
            updateResultsInUi(text_string);
        }
    };

    private void updateResultsInUi(String update_txt) {
        statusText.setText(update_txt);
    }

    private void record_thread() {
        Thread thread = new Thread(new Runnable() {
            public void run() {
                text_string = "Starting";
                mHandler.post(mUpdateResults);

                record();

                text_string = "Done";
                mHandler.post(mUpdateResults);
            }
        });
        thread.start();
    }

    private int audioEncoding = AudioFormat.ENCODING_PCM_16BIT;
    int frequency = 11025; //hertz
    int bufferSize = 50*AudioTrack.getMinBufferSize(
            frequency,
            AudioFormat.CHANNEL_OUT_MONO,
            audioEncoding
            );
    // Create new AudioRecord object to record the audio
    public AudioRecord audioRecord = new AudioRecord(
            MediaRecorder.AudioSource.MIC,
            frequency,
            AudioFormat.CHANNEL_IN_MONO,
            audioEncoding,
```

```
            bufferSize
            );
    // Create new AudioTrack object w/same parameters as AudioRecord obj
    public AudioTrack audioTrack = new AudioTrack(
            AudioManager.STREAM_MUSIC,
            frequency,
            AudioFormat.CHANNEL_OUT_MONO,
            audioEncoding,
            4096,
            AudioTrack.MODE_STREAM
            );
    short[] buffer = new short[bufferSize];

    public void record() {
        try {
            audioRecord.startRecording();
            audioRecord.read(buffer, 0, bufferSize);
            audioRecord.stop();
            audioRecord.release();
        } catch (Throwable t) {
            Log.e("AudioExamplesRaw","Recording Failed");
        }
    }

    public void run() { //Play audio using runnable activity
        audioTrack.play();
        int i=0;
        while(i<bufferSize) {
            audioTrack.write(buffer, i++, 1);
        }
        return;
    }

    @Override
    protected void onPause() {
        if(audioTrack!=null) {
            if(audioTrack.getPlayState()==AudioTrack.PLAYSTATE_PLAYING) {
                audioTrack.pause();
            }
        }
        super.onPause();
    }
}
```

## 技巧 73：有效利用声音资源

若既要保持压缩音频文件的较小内存需求，又想获得原始音频低延迟播放的好处，可以使用 SoundPool 类。该类使用 MediaPlayer 服务来解码音频，同时提供重复播放声音缓存的方法，并且可以加速或减速播放。

该类的用法与之前的技巧里讲过的其他声音方法类似：先初始化，然后装载资源，播放，最后释放资源。然而，要注意，SoundPool 会启动一个后台线程，因此若在 load() 方法之后紧跟着使用 play() 方法，如果资源没有足够的时间被加载，就可能播放不出声音。类似地，紧跟在 play() 后调用 release() 方法，可能会在资源能够开始播放前就将其释放掉。因此，最好将 SoundPool 资源和 Activity 生命周期事件（如 onCreate 和 onPause）绑定，并将

SoundPool 资源的播放与一个用户生成的事件（如按下按钮或游戏过关）绑定。

本技巧的主 Activity 如代码清单 8-10 所示，它使用和代码清单 8-7 相同的布局。按钮按下会导致 SoundPool 将一种敲鼓的声音重复 8 次（初始的 1 次加上 7 次重复）。此外，多次按下按钮可以让播放速率在半速和两倍速之间切换。最多可以同时播放 10 个音频流，这意味着快速连续按下按钮 10 次可以让 10 个敲鼓声音流同时播放。

**代码清单 8-10 AudioSoundPool.java**

```
package cc.dividebyzero.android.cookbook.chapter8.audio;

import cc.dividebyzero.android.cookbook.chapter8.R;
import cc.dividebyzero.android.cookbook.chapter8.R.id;
import cc.dividebyzero.android.cookbook.chapter8.R.layout;
import cc.dividebyzero.android.cookbook.chapter8.R.raw;
import android.app.Activity;
import android.media.AudioManager;
import android.media.SoundPool;
import android.os.Bundle;
import android.view.View;
import android.widget.Button;

public class AudioSoundPool extends Activity {
    static float rate = 0.5f;
    @Override
    protected void onCreate(Bundle savedInstanceState) {
        super.onCreate(savedInstanceState);

        setContentView(R.layout.audio_soundpool);
        Button playDrumButton = (Button) findViewById(R.id.play_pause);

        final SoundPool mySP = new SoundPool(
                                    10,
                                    AudioManager.STREAM_MUSIC,
                                    0);
        final int soundId = mySP.load(this, R.raw.drum_beat, 1);

        playDrumButton.setOnClickListener(new View.OnClickListener() {
            public void onClick(View view) {
                rate = 1/rate;
                mySP.play(soundId, 1f, 1f, 1, 7, rate);
            }
        });
    }
}
```

## 技巧 74：添加媒体并更新路径

应用程序创建了一个新录音文件之后，就可以将其注册为系统可用的资源。这通过 MediaStore 类来实现。例如，代码清单 8-11 展示了如何将名为 myFile 的新保存的声音文件进行注册，使其可以用作铃声、通知提示音或闹钟铃声，但又让其对于 MP3 播放器不可见（因为 IS_MUSIC 标志被设为 false）。

**代码清单 8-11 将音频文件注册到系统中的示例**

```
//Reload MediaScanner to search for media and update paths
sendBroadcast(new Intent(Intent.ACTION_MEDIA_MOUNTED,
                Uri.parse("file://"
                            + Environment.getExternalStorageDirectory())));
ContentValues values = new ContentValues();
values.put(MediaStore.MediaColumns.DATA, myFile.getAbsolutePath());
values.put(MediaStore.MediaColumns.TITLE, myFile.getName());
values.put(MediaStore.MediaColumns.TIMESTAMP,
                                            System.currentTimeMillis());
values.put(MediaStore.MediaColumns.MIME_TYPE,
                                            recorder.getMimeContentType());
values.put(MediaStore.Audio.Media.ARTIST, SOME_ARTIST_HERE);
values.put(MediaStore.Audio.Media.IS_RINGTONE, true);
values.put(MediaStore.Audio.Media.IS_NOTIFICATION, true);
values.put(MediaStore.Audio.Media.IS_ALARM, true);
values.put(MediaStore.Audio.Media.IS_MUSIC, false);
ContentResolver contentResolver = new ContentResolver();
Uri base = MediaStore.Audio.INTERNAL_CONTENT_URI;
Uri newUri = contentResolver.insert(base, values);
String path = contentResolver.getDataFilePath(newUri);
```

在此，使用 ContentValues 为文件声明一些标准属性，比如 TITLE、TIMESTAMP 和 MIME_TYPE，而 ContentResolver 则用于在 MediaStore 内容数据库中创建一个条目，并自动添加文件路径。

# 8.3 视频

有两种不同的显示视频的方法。一种是使用 MediaPlayer 框架，与刚刚讨论过的用其播放音频的例子类似。另一种是使用 VideoView 类，该类会自动实现大部分工作，是较简单的应用情形下的推荐选择。

## 技巧 75：使用 VideoView

VideoView 的使用非常容易。在 XML 布局中将其声明，并加载该布局，剩下要做的就是为 VideoView 提供一个视频 URL，然后播放就会立即开始。如果视频格式不被框架所支持，或者发生了其他错误，VideoView 还会显示一个错误对话框。

为了更便于用户使用，还有一个名为 MediaController 的辅助类可以利用。该类会添加 Play/Pause（播放/暂停）、Forward（前进）、Rewind（倒带）按钮以及一个拖动条，我们只需将 MediaController 与 VideoView 相连接，使用前者的.setAnchorView 方法，使后者成为前者的锚（anchor）。这样，只需要几行代码就可以获得一个完整的视频播放器，如代码清单 8-12 所示。

**代码清单 8-12 VideoViewActivity.java**

```
public class VideoViewActivity extends Activity {

        private static final String VIDEO_DIR =
        File.separator+"DCIM"+File.separator+"Camera";
```

```
        private VideoView videoView;

        @Override
        public void onCreate(Bundle savedInstanceState){
            super.onCreate(savedInstanceState);

            setContentView(R.layout.video_view);
            videoView=(VideoView)findViewById(R.id.videoView1);
            MediaController controller=new MediaController(this);
            controller.setMediaPlayer(videoView);
            controller.setAnchorView(videoView);

            videoView.setMediaController(controller);

        String videoDir = Environment.getExternalStorageDirectory()
         .getAbsolutePath() + VIDEO_DIR;

         //Show a list of video files to choose
         Intent i = new Intent(this, ListFiles.class);
         i.putExtra("directory", videoDir);
         startActivityForResult(i,0);

         }

    @Override
    protected void onActivityResult(int requestCode,
                                    int resultCode, Intent data) {
        super.onActivityResult(requestCode, resultCode, data);
        if(requestCode == 0 && resultCode==RESULT_OK) {
            String path = data.getExtras().getString("clickedFile");

            videoView.setVideoPath(path);
            videoView.start();

        }
    }
}
```

与之配套的布局见代码清单 8-13。

**代码清单 8-13　video_view.xml**

```
<?xml version="1.0" encoding="utf-8"?>
<LinearLayout xmlns:android="http://schemas.android.com/apk/res/android"
    android:layout_width="match_parent"
    android:layout_height="match_parent"
    android:orientation="vertical" >

    <VideoView
        android:id="@+id/videoView1"
        android:layout_width="match_parent"
        android:layout_height="wrap_content"
    />

</LinearLayout>
```

## 技巧 76：使用 MediaPlayer 播放视频

MediaPlayer 框架也可以用来播放视频。与播放音频的主要区别在于必须提供一个用于渲染视频帧的表面（surface）。这可以使用 SurfaceView 类来实现，在代码清单 8-14 中，它被添加到紧邻 Play/Pause 按钮下方的位置。

**代码清单 8-14　video_playback.xml**

```
<?xml version="1.0" encoding="utf-8"?>
<LinearLayout
    xmlns:android="http://schemas.android.com/apk/res/android"
    android:orientation="vertical"
    android:layout_width="match_parent"
    android:layout_height="match_parent">
    <Button android:id="@+id/play_pause"
        android:text="Play"
        android:textSize="20sp"
        android:layout_width="wrap_content"
        android:layout_height="wrap_content" />
    <SurfaceView
                android:id="@+id/surface"
                android:layout_width="match_parent"
                android:layout_height="0dip"
                android:layout_weight="1"
                android:visibility="visible"
                />
</LinearLayout>
```

创建表面可能要花些时间。正因为如此，要使用 SurfaceHolder.Callback 方法将 MediaPlayer 的显示设置为在表面创建之后进行。完成这一步后，视频就可以播放了。如果开始播放前并未附加显示，或者用空参数调用 setDisplay 方法，就会只播放视频的音轨。

# 第 9 章

# 硬件接口

Android 设备有很多不同类型的内置硬件，且对开发者是可访问的。传感器（如摄像头、加速度计、磁力计、压力传感器、温度传感器和距离传感器等）在很多设备上都具有。电话、蓝牙、近场通信（NFC）以及其他类型的无线连接同样允许开发者通过某些形式来访问。本章介绍如何利用这些硬件 API 丰富应用程序的体验。注意，本章给出的例子最好在真实 Android 设备上运行，因为模拟器可能提供不了准确的或真实的硬件接口行为。

## 9.1 摄像头

摄像头是 Android 设备中最容易看到和最常用的传感器。对于大多数消费者来说，摄像头是卖点之一，其性能也逐代增强。图像处理应用程序通常在图像拍摄之后对其进行加工，但另一些应用程序，比如增强现实（augmented reality）类软件，会有覆盖（overlay）地实时使用摄像头。

从应用程序访问摄像头的方式有两种。一种是如第 2 章讲过的那样声明一个隐式 Intent，由隐式的 Intent 来启动默认的摄像接口：

```
Intent intent = new Intent("android.media.action.IMAGE_CAPTURE");
startActivity(intent);
```

第二种方式是利用 `Camera` 类，该类提供了更为灵活的设置。这种方式将创建一个自定义摄像接口，下面的技巧将着眼于这部分内容。对摄像头的硬件访问需要显式地在 **AndroidManifest.xml** 中添加权限：

```
<uses-permission android:name="android.permission.CAMERA" />
```

后面的技巧默认包含该权限设定。

### 技巧 77：自定义摄像头

在 Android 系统中，对摄像头的控制被抽象为多个组件。

- Camera 类：访问摄像头硬件。
- Camera.Parameters 类：指定摄像头参数，如图片大小、图片质量、闪光模式以及赋予 GPS 位置的方法。
- Camera Preview 方法：设置摄像头输出显示，开启和关闭流视频预览显示。
- SurfaceView 类：用于视图层级的最底层上的绘图表面，将其作为显示摄像头预览的预留位置。

在描述如何将上述组件整合到一起之前，要先介绍布局结构。主布局如代码清单 9-1 所示，它包含了一个 SurfaceView 类以保存摄像头输出。

**代码清单 9-1　res/layout/main.xml**

```
<LinearLayout
    xmlns:android="http://schemas.android.com/apk/res/android"
    android:layout_width="match_parent"
    android:layout_height="match_parent"
    android:orientation="vertical">

    <SurfaceView android:id="@+id/surface"
        android:layout_width="match_parent"
        android:layout_height="match_parent">
    </SurfaceView>
</LinearLayout>
```

控制接口可以用一个独立的布局添加到视图的上部，如代码清单 9-2 所示。该布局包含一个位于屏幕底部中央处的按钮，用于控制拍摄。

**代码清单 9-2　res/layout/cameraoverlay.xml**

```
<LinearLayout xmlns:android="http://schemas.android.com/apk/res/android"
    android:layout_width="match_parent"
    android:layout_height="match_parent"
    android:orientation="vertical"
    android:gravity="bottom"
    android:layout_gravity="bottom">

    <LinearLayout
    xmlns:android="http://schemas.android.com/apk/res/android"
        android:layout_width="match_parent"
        android:layout_height="wrap_content"
        android:orientation="horizontal"
        android:gravity="center_horizontal">

        <Button
        android:id="@+id/button"
        android:layout_width="wrap_content"
        android:layout_height="wrap_content"
        android:text="take picture"
        />
    </LinearLayout>
</LinearLayout>
```

主 Activity 包含多种功能。首先，布局的建立方式如下。

（1）将窗口设定为半透明并全屏显示。（在此情况下，标题和通知栏会被隐藏。）

（2）前面的布局中定义的 SurfaceView 类（R.id.surface）被摄像头预览填充。每个 SurfaceView 包含一个 SurfaceHolder 类，用于访问和控制表面。Activity 被添加为 SurfaceHolder 的回调，而 SurfaceHolder 的类型被设为 SURFACE_TYPE_PUSH_BUFFERS，这意味着创建了一个“push”型的表面。表面对象没有自己的缓冲区，这使得视频流更有效率。

（3）声明一个 LayoutInflater，在原有（**main.xml**）布局之上填充另一个布局（**cameraoverlay.xml**）。

接着，Activity 设置用于拍摄图片的一个触发器。

（1）将一个 OnClickListener 添加到 cameraOverlay 布局的按钮上，因此当按钮被按下后，就会拍摄一张图片（mCamera.takePicture()）。

（2）要使用 takePicture()方法，需要先实现下列接口。

- ShutterCallback()用于定义拍摄图片后需要的各种效果，比如播放拍照提示音。
- PictureCallback()用返回于原始图片数据，前提是硬件有足够的内存支持这一特性。（否则，返回数据可能为空。）
- 另一个 PictureCallback()方法用于压缩的图片数据。它将调用本地方法 done()来保存图片。

然后，Activity 会保存拍摄的图片。

（1）被压缩的图片字节数组被存储到名为 tempData 的本地变量，以备处理。BitmapFactory 用于将字节数组解码到一个 Bitmap 对象中。

（2）使用媒体内容提供器保存位图，并返回其 URL。如果主 Activity 是被其他 Activity 调用的，这个 URL 应当作为返回信息传给发起调用的 Activity，以保持对图像的跟踪。

（3）处理结束后，调用 finish()方法来杀死 Activity。

最后，Activity 创建响应（response），用于应对表面视图的变化。

（1）实现一个 SurfaceHolder.CallBack 接口。这需要重写以下三个方法。

- surfaceCreated()：当表面被首次创建时调用，方法中会初始化对象。
- surfaceChanged()：当表面被创建后或者表面发生变化时调用（例如，格式或大小发生变化）。
- surfaceDestroyed()：在把表面从用户视图中移除之后、销毁表面之前调用。用于清理内存。

（2）当表面发生改变时，摄像头的参数也相应改变（例如，PreviewSize 要根据表面的大小变化而变化）。

以上功能都包含在代码清单 9-3 给出的完整的 Activity 中。

**代码清单 9-3 src/com/cookbook/hardware/CameraApplication.java**

```
package com.cookbook.hardware;

import android.app.Activity;
import android.content.Intent;
import android.graphics.Bitmap;
```

```
import android.graphics.BitmapFactory;
import android.graphics.PixelFormat;
import android.hardware.Camera;
import android.hardware.Camera.PictureCallback;
import android.hardware.Camera.ShutterCallback;
import android.os.Bundle;
import android.provider.MediaStore.Images;
import android.util.Log;
import android.view.LayoutInflater;
import android.view.SurfaceHolder;
import android.view.SurfaceView;
import android.view.View;
import android.view.Window;
import android.view.WindowManager;
import android.view.View.OnClickListener;
import android.view.ViewGroup.LayoutParams;
import android.widget.Button;
import android.widget.Toast;

public class CameraApplication extends Activity
                              implements SurfaceHolder.Callback {
    private static final String TAG = "cookbook.hardware";
    private LayoutInflater mInflater = null;
    Camera mCamera;
    byte[] tempData;
    boolean mPreviewRunning = false;
    private SurfaceHolder mSurfaceHolder;
    private SurfaceView mSurfaceView;
    Button takepicture;
    @Override
    public void onCreate(Bundle savedInstanceState) {
        super.onCreate(savedInstanceState);

        getWindow().setFormat(PixelFormat.TRANSLUCENT);
        requestWindowFeature(Window.FEATURE_NO_TITLE);
        getWindow().setFlags(WindowManager.LayoutParams.FLAG_FULLSCREEN,
                WindowManager.LayoutParams.FLAG_FULLSCREEN);

        setContentView(R.layout.main);

        mSurfaceView = (SurfaceView)findViewById(R.id.surface);
        mSurfaceHolder = mSurfaceView.getHolder();
        mSurfaceHolder.addCallback(this);
        // Uncomment the following line if using less than Android 3.0 (API 11)
        // mSurfaceHolder.setType(SurfaceHolder.SURFACE_TYPE_PUSH_BUFFERS);

        mInflater = LayoutInflater.from(this);
        View overView = mInflater.inflate(R.layout.cameraoverlay, null);
        this.addContentView(overView,
                new LayoutParams(LayoutParams.MATCH_PARENT,
                        LayoutParams.MATCH_PARENT));
        takepicture = (Button) findViewById(R.id.button);
        takepicture.setOnClickListener(new OnClickListener(){
            public void onClick(View view){
                mCamera.takePicture(mShutterCallback,
                        mPictureCallback, mjpeg);
            }
        });
    }

    ShutterCallback mShutterCallback = new ShutterCallback(){
        @Override
        public void onShutter() {}
    };
```

```
PictureCallback mPictureCallback = new PictureCallback() {
    public void onPictureTaken(byte[] data, Camera c) {}
};
PictureCallback mjpeg = new PictureCallback() {
    public void onPictureTaken(byte[] data, Camera c) {
        if(data !=null) {
            tempdata=data;
            done();
        }
    }
};

void done() {
    Bitmap bm = BitmapFactory.decodeByteArray(tempdata,
                                            0, tempdata.length);
    String url = Images.Media.insertImage(getContentResolver(),
            bm, null, null);
    bm.recycle();
    Bundle bundle = new Bundle();
    if(url!=null) {
        bundle.putString("url", url);
        Intent mIntent = new Intent();
        mIntent.putExtras(bundle);
        setResult(RESULT_OK, mIntent);
    } else {
        Toast.makeText(this, "Picture cannot be saved",
                         Toast.LENGTH_SHORT).show();
    }
    finish();
}
@Override
public void surfaceChanged(SurfaceHolder holder, int format,
                           int w, int h) {
    Log.e(TAG, "surfaceChanged");
    try {
        if (mPreviewRunning) {
            mCamera.stopPreview();
            mPreviewRunning = false;
        }

        Camera.Parameters p = mCamera.getParameters();
        p.setPreviewSize(w, h);

        mCamera.setParameters(p);
        mCamera.setPreviewDisplay(holder);
        mCamera.startPreview();
        mPreviewRunning = true;
    } catch(Exception e) {
        Log.d("",e.toString());
    }
}

@Override
public void surfaceCreated(SurfaceHolder holder) {
    Log.e(TAG, "surfaceCreated");
    mCamera = Camera.open();
}

@Override
public void surfaceDestroyed(SurfaceHolder holder) {
    Log.e(TAG, "surfaceDestroyed");
    mCamera.stopPreview();
    mPreviewRunning = false;
    mCamera.release();
```

```
            mCamera=null;
        }
    }
```

注意，从摄像头硬件得到的摄像头的预览并不是标准化的，有的 Android 设备上显示的预览可能与屏幕的方向不一致。遇到这种情况，只要为 CameraPreview Activity 的 onCreate() 方法添加一行代码即可解决：

```
this.setRequestedOrientation(ActivityInfo.SCREEN_ORIENTATION_LANDSCAPE);
```

# 9.2　其他传感器

小型、低功耗的微机电系统（MEMS）的流行趋势越来越显著。智能手机成为了传感器的集合体，而智能手机制造商对传感器精度的提升也带动了对更高性能设备的需求。

我们在第 1 章讨论过，每款 Android 手机带有的传感器组合都是不同的。通常配备两个传感器：一个三轴加速度计，用于确定设备的倾斜度；一个三轴磁力计，用于确定罗盘方向。其他可能集成的传感器还有：温度传感器、距离传感器、光传感器以及陀螺仪。下面列出了当前 Android SDK 支持的传感器。

- TYPE_ACCELEROMETER：测量加速度，单位为 $m/s^2$。
- TYPE_AMBIENT_TEMPERATURE：测量温度，单位为摄氏度（℃）（在 API Level 14 以后替代了 TYPE_TEMPERATURE）。
- TYPE_GRAVITY：测量在三维坐标系中的运动，包括重力的大小。
- TYPE_GYROSCOPE：基于角动量测量方向。
- TYPE_LIGHT：测量环境光，以勒克斯（lux）为单位。
- TYPE_LINEAR_ACCELERATION：测量在三维坐标系中的运动，忽略重力的影响。
- TYPE_MAGNETIC_FIELD：测量磁场强度，以微特斯拉（microteslas）为单位。
- TYPE_PRESSURE：测量空气压力。
- TYPE_PROXIMITY：测量到某个目标对象的距离，以厘米为单位。
- TYPE_RELATIVE_HUMIDITY：测量湿度，结果为一个百分数[①]。
- TYPE_TEMPERATURE：测量温度，以摄氏度为单位。

getSensorList()方法能列出某个特定设备的所有可用传感器。SensorManager 负责管理所有的传感器，它通过 onSensorChanged()和 onAccuracyChanged()两个函数，提供了各种传感器的事件监听器，用于监听传感器值和精度的变化。

## 技巧 78：获取设备的旋转姿态

理想状况下，加速度计测量到的地球重力场的强度（即重力加速度）应为 $G=9.8\ m/s^2$，而磁力计测量到地球磁场的强度，取决于设备所处地理位置的不同，应为 $H=30\sim60\ \mu T$。有了这两

① 即相对湿度。——译者注

个向量，我们就可以对旋转姿态进行简单的、类似课本上那样的估计。我们在getRotationMatrix()方法中会用到这些。本技巧将展示如何实现。

设备的坐标系统（也称为“机身坐标系”）框架像下面这样定义。

- $x$ 轴定义为屏幕较短边的方向（即与菜单键的排列方向一致）。
- $y$ 轴定义为屏幕较长边的方向。
- $z$ 轴定义为贯穿屏幕前后的方向。

真实世界的坐标系统（也被称为“惯性坐标系”）框架则像这样定义：

- $x$ 轴方向与 $z$ 轴方向（向量）的叉乘[1]方向
- $y$ 轴是与底面相切并指向北极的方向。
- $z$ 轴方向是垂直于底面并指向天空的方向。

当设备平放在一个水平桌面，屏幕朝上并指向北方时，上述两个坐标系是一致的。此时，加速度计在 $x$、$y$、$z$ 方向的测量值为(0, 0, G)。在大多数地点，即便设备指向北方，地球的磁场也会以一个小角度 $\theta$ 微微指向地面，其值为$(0, H\cos(\theta), -H\sin(\theta))$。

一旦设备发生倾斜和旋转，SensorManager.getRotationMatrix()提供了3×3的旋转矩阵R[]，其值赋为从设备坐标系统到世界坐标系统的差距；还提供了3×3的倾斜矩阵I[]（绕 $x$ 轴的旋转量），其值赋为理想情况下的磁场值(0, H, 0)与真实磁场值的差距。

注意，如果设备正在加速，或者处于某个强磁场附近，我们得到的值未必能反映真实的地球参考系框架。

另一种表示旋转量的方法是利用SensorManager.getOrientation()，该方法会给出旋转矩阵R[]及姿态向量attitude[]。

- attitude[0]：方位角（azimuth，以弧度为单位）是围绕世界坐标系 $z$ 轴的旋转角，要求设备必须朝北。其值域为$-\pi$～$\pi$（−PI～PI），其中0表示朝北，$\pi/2$ 表示朝东。
- attitude[1]：倾斜度（pitch，以弧度为单位）是围绕世界坐标系 $x$ 轴的旋转角，要求设备沿着设备屏幕较长边方向竖直向上。其值域亦为$-\pi$～$\pi$（−PI～PI），其中0表示朝上，而 $\pi/2$ 意味着设备朝向地面。
- attitude[2]：旁向倾角（roll，以弧度为单位）是围绕世界坐标系 $y$ 轴的旋转角，需要设备沿其较短边方向竖直向上。其值域也是$-\pi$～$\pi$（−PI～PI），其中0表示朝上，而 $\pi/2$ 意味着指向右方。

本技巧将姿态向量信息显示到屏幕上。布局带有一个ID为attitude的文本框，如代码清单9-4所示。

**代码清单 9-4 res/layout/main.xml**

```
<?xml version="1.0" encoding="utf-8"?>
<LinearLayout xmlns:android="http://schemas.android.com/apk/res/android"
```

① 该运算又称向量的外积或向量积。结果方向可用右手法则判断，对于a、b两个向量，伸出右手，先让手掌张开，大拇指与四指方向垂直；然后保持大拇指不动，弯曲四指，让四指的方向与从a旋转到b的方向一致，此时大拇指所指方向就是a、b向量叉乘的方向（该方向与a、b的方向一定都是垂直的）。——译者注

```
    android:orientation="vertical"
    android:layout_width="match_parent"
    android:layout_height="match_parent"
    >
<TextView android:id="@+id/attitude"
    android:layout_width="match_parent"
    android:layout_height="wrap_content"
    android:text="Azimuth, Pitch, Roll"
    />
</LinearLayout>
```

主 Activity 在代码清单 9-5 中给出，其中对加速度计和磁力计进行了注册，让其向传感器监听器返回数据。SensorEventListener 会确定是哪个传感器触发了回调，并进行赋值。姿态信息基于旋转矩阵来决定，其单位会从弧度转换为角度，并显示到屏幕上。注意，传感器的刷新率可采用下列值之一。

- SENSOR_DELAY_FASTEST：能取到的最快刷新率。
- SENSOR_DELAY_GAME：适合游戏的刷新率。
- SENSOR_DELAY_NORMAL：适合应对屏幕方向改变的默认刷新率。
- SENSOR_DELAY_UI：适合用户界面的刷新率。

**代码清单 9-5 src/com/cookbook/orientation/OrientationMeasurements.java**

```
package com.cookbook.orientation;

import android.app.Activity;
import android.hardware.Sensor;
import android.hardware.SensorEvent;
import android.hardware.SensorEventListener;
import android.hardware.SensorManager;
import android.os.Bundle;
import android.widget.TextView;

public class OrientationMeasurements extends Activity {
    private SensorManager myManager = null;
    TextView tv;

    @Override
    public void onCreate(Bundle savedInstanceState) {
        super.onCreate(savedInstanceState);
        setContentView(R.layout.main);
        tv = (TextView) findViewById(R.id.attitude);
        // Set Sensor Manager
        myManager = (SensorManager)getSystemService(SENSOR_SERVICE);
        myManager.registerListener(mySensorListener,
                myManager.getDefaultSensor(Sensor.TYPE_ACCELEROMETER),
                SensorManager.SENSOR_DELAY_GAME);
        myManager.registerListener(mySensorListener,
                myManager.getDefaultSensor(Sensor.TYPE_MAGNETIC_FIELD),
                SensorManager.SENSOR_DELAY_GAME);
    }

    float[] mags = new float[3];
    float[] accels = new float[3];
    float[] rotationMat = new float[9];
    float[] inclinationMat = new float[9];
    float[] attitude = new float[3];
```

```
    final static double RAD2DEG = 180/Math.PI;
    private final SensorEventListener mySensorListener
                                    = new SensorEventListener() {
        @Override
        public void onSensorChanged(SensorEvent event)
        {
            int type = event.sensor.getType();

            if(type == Sensor.TYPE_MAGNETIC_FIELD) {
                mags = event.values;
            }
            if(type == Sensor.TYPE_ACCELEROMETER) {
                accels = event.values;
            }

            SensorManager.getRotationMatrix(rotationMat,
                    inclinationMat, accels, mags);
            SensorManager.getOrientation(rotationMat, attitude);
            tv.setText("Azimuth, Pitch, Roll:\n"
                    + attitude[0]*RAD2DEG + "\n"
                    + attitude[1]*RAD2DEG + "\n"
                    + attitude[2]*RAD2DEG);
        }

        public void onAccuracyChanged(Sensor sensor, int accuracy) {}
    };
}
```

要获得一致的数据，最好避免在 onSensorChanged()方法中放入高计算密度的代码。还要注意，SensorEvent 要为后续的传感器数据所重用。因此，作为高精度数据，最好使用 clone()方法来处理事件值，例如：

```
accels = event.values.clone();
```

这样确保了如果在类中的其他地方重用了 accels，它不会随着传感器持续采样而改变。

## 技巧 79：使用温度传感器和光传感器

温度传感器探测手机温度，用于内部硬件的校准。光传感器测量环境光，用于自动调整屏幕亮度。

这些传感器并不是在所有手机上都可用，但如果手机带有它们，开发者就可以用它们做一些其他的事。从传感器读取值的代码如代码清单 9-6 所示。可以将这段代码添加到上一个技巧的 Activity 中，并查看效果。

**代码清单 9-6 访问温度传感器和光传感器**

```
private final SensorEventListener mTListener
                                = new SensorEventListener(){
    @Override
    public void onAccuracyChanged(Sensor sensor, int accuracy) {}

    @Override
    public void onSensorChanged(SensorEvent event) {
        Log.v("test Temperature",
```

```
                "onSensorChanged:"+event.sensor.getName());
        if(event.sensor.getType()==Sensor.TYPE_AMBIENT_TEMPERATURE){
            tv2.setText("Temperature:"+event.values[0]);
        }
    }
};

private final SensorEventListener mLListener
                                    = new SensorEventListener(){
    @Override
    public void onAccuracyChanged(Sensor sensor, int accuracy) {}

    @Override
    public void onSensorChanged(SensorEvent event) {
        Log.v("test Light",
              "onSensorChanged:"+event.sensor.getName());
        if(event.sensor.getType()==Sensor.TYPE_LIGHT){
            tv3.setText("Light:"+event.values[0]);
        }
    }
};

myManager.registerListener(mTListener, sensorManager
                    .getDefaultSensor(Sensor.TYPE_TEMPERATURE),
                    SensorManager.SENSOR_DELAY_FASTEST);
myManager.registerListener(mLListener, sensorManager
                    .getDefaultSensor(Sensor.TYPE_LIGHT),
                    SensorManager.SENSOR_DELAY_FASTEST);
```

# 9.3 电话

Android API 提供了一种检测手机基本信息（比如网络类型、连接状态以及所提供的操作电话号码字符串）的工具。

## 技巧 80：使用电话管理器

电话 API 中有一个名为 `TelephonyManager` 的类，它是 Android 系统服务，负责访问设备上电话服务的信息。某些电话信息是有权限保护的，所以必须在 **AndroidManifest.xml** 文件中对访问权限加以声明：

```
<uses-permission android:name="android.permission.READ_PHONE_STATE" />
```

主 Activity 如代码清单 9-7 所示。

**代码清单 9-7 src/com/cookbook/hardware.telephony/TelephonyApp.java**

```
package com.cookbook.hardware.telephony;

import android.app.Activity;
import android.os.Bundle;
import android.telephony.TelephonyManager;
import android.widget.TextView;

public class TelephonyApp extends Activity {
```

```
    TextView tv1;
    TelephonyManager telManager;
    @Override
    public void onCreate(Bundle savedInstanceState) {
        super.onCreate(savedInstanceState);
        setContentView(R.layout.main);
        tv1 =(TextView) findViewById(R.id.tv1);
        telManager = (TelephonyManager)
                    getSystemService(TELEPHONY_SERVICE);

        StringBuilder sb = new StringBuilder();
        sb.append("deviceid:")
          .append(telManager.getDeviceId()).append("\n");
        sb.append("device Software Ver:")
          .append(telManager.getDeviceSoftwareVersion()).append("\n");
        sb.append("Line number:")
          .append(telManager.getLine1Number()).append("\n");
        sb.append("Network Country ISO:")
          .append(telManager.getNetworkCountryIso()).append("\n");
        sb.append("Network Operator:")
          .append(telManager.getNetworkOperator()).append("\n");
        sb.append("Network Operator Name:")
          .append(telManager.getNetworkOperatorName()).append("\n");
        sb.append("Sim Country ISO:")
          .append(telManager.getSimCountryIso()).append("\n");
        sb.append("Sim Operator:")
          .append(telManager.getSimOperator()).append("\n");
        sb.append("Sim Operator Name:")
          .append(telManager.getSimOperatorName()).append("\n");
        sb.append("Sim Serial Number:")
          .append(telManager.getSimSerialNumber()).append("\n");
        sb.append("Subscriber Id:")
          .append(telManager.getSubscriberId()).append("\n");
        sb.append("Voice Mail Alpha Tag:")
          .append(telManager.getVoiceMailAlphaTag()).append("\n");
        sb.append("Voice Mail Number:")
          .append(telManager.getVoiceMailNumber()).append("\n");
        tv1.setText(sb.toString());
    }
}
```

主布局 XML 文件在代码清单 9-8 中给出，屏幕输出结果见图 9-1。

### 代码清单 9-8 res/layout/main.xml

```
<?xml version="1.0" encoding="utf-8"?>
<LinearLayout xmlns:android="http://schemas.android.com/apk/res/android"
    android:orientation="vertical"
    android:layout_width="match_parent"
    android:layout_height="match_parent"
    >
<TextView
    android:id="@+id/tv1"
    android:layout_width="match_parent"
    android:layout_height="wrap_content"
    android:text="@string/hello"
    />
</LinearLayout>
```

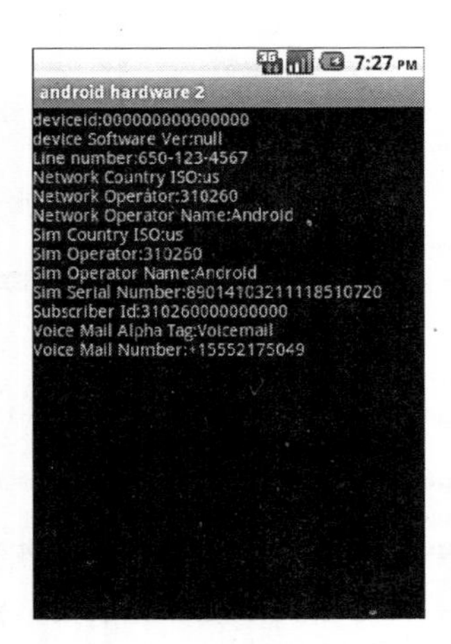


图 9-1　使用 TelephonyManager 类的输出结果

# 技巧 81：监听电话状态

PhoneStateListener 提供了设备上不同电话状态的信息，包括网络服务状态、信号强度、信息等待指示器（语音信息）等。某些信息需要显式地声明权限，如表 9-1 所示。

表 9-1　可能的电话状态监听器事件及其所需权限

| 电话状态监听器 | 描　述 | 权　限 |
|---|---|---|
| LISTEN_CALL_FORWARDING_INDICATOR | 监听呼叫转移指示器的变化 | READ_PHONE_STATE |
| LISTEN_CALL_STATE | 监听通话状态变化 | READ_PHONE_STATE |
| LISTEN_CELL_INFO | 监听观察到的手机信息变化 | 无 |
| LISTEN_CELL_LOCATION | 监听手机位置变化 | ACCESS_COARSE_LOCATION |
| LISTEN_DATA_ACTIVITY | 监听手机数据传输上下行方向变化 | READ_PHONE_STATE |
| LISTEN_DATA_CONNECTION_STATE | 监听数据连接状态变化 | 无 |
| LISTEN_MESSAGE_WAITING_INDICATOR | 监听信息等待指示器变化 | READ_PHONE_STATE |
| LISTEN_NONE | 移除监听器 | 无 |
| LISTEN_SERVICE_STATE | 监听网络服务状态变化 | 无 |
| LISTEN_SIGNAL_STRENGTHS | 监听网络信号强度变化 | 无 |

例如，要接听来电，TelephonyManager 需要针对 PhoneStateListener.LISTEN_CALL_STATE 事件注册一个监听器。可能的呼叫状态有下列三种。

- CALL_STATE_IDLE：设备未处于通话状态。
- CALL_STATE_RINGING：设备收到呼叫请求。
- CALL_STATE_OFFHOOK：通话中。

本技巧会在手机通话状态发生改变时将状态列举出来。借助于 LogCat 工具（将在第 16 章中探讨），可以在有呼入或呼出电话时看到相应的不同状态。

主 Activity 如代码清单 9-9 所示，它创建了一个扩展了 PhoneStateListener 的内部类，该类重写了 onCallStateChanged 方法，用于捕捉手机通话状态的变化。其他可以重写的方法还有 onCallForwardingIndicator()、onCellLocationChanged()和 onDataActivity()等。

**代码清单 9-9　src/com/cookbook/hardware.telephony/HardwareTelephony.java**

```
package com.cookbook.hardware.telephony;

import android.app.Activity;
import android.os.Bundle;
import android.telephony.PhoneStateListener;
import android.telephony.TelephonyManager;
import android.util.Log;
import android.widget.TextView;

public class HardwareTelephony extends Activity {
    TextView tv1;
    TelephonyManager telManager;
    @Override
    public void onCreate(Bundle savedInstanceState) {
        super.onCreate(savedInstanceState);
        setContentView(R.layout.main);
        tv1 =(TextView) findViewById(R.id.tv1);
        telManager = (TelephonyManager)
                    getSystemService(TELEPHONY_SERVICE);

        telManager.listen(new TelListener(),
                    PhoneStateListener.LISTEN_CALL_STATE);
    }
{
    private class TelListener extends PhoneStateListener {
        public void onCallStateChanged(int state, String incomingNumber)
            super.onCallStateChanged(state, incomingNumber);

            Log.v("Phone State", "state:"+state);
            switch (state) {
                case TelephonyManager.CALL_STATE_IDLE:
                    Log.v("Phone State",
                          "incomingNumber:"+incomingNumber+" ended");
                    break;
                case TelephonyManager.CALL_STATE_OFFHOOK:
                    Log.v("Phone State",
                          "incomingNumber:"+incomingNumber+" picked up");
                    break;
                case TelephonyManager.CALL_STATE_RINGING:
                    Log.v("Phone State",
                          "incomingNumber:"+incomingNumber+" received");
                    break;
                default:
                    break;
            }
```

```
        }
    }
}
```

## 技巧82：拨叫一个号码

要在应用程序中拨打电话，需要将下面的权限添加到 **AndroidManifest.xml** 文件中：

```
<uses-permission android:name="android.permission.CALL_PHONE" />
```

拨打电话的动作可以使用 ACTION_CALL 或者 ACTION_DIALER 这两个隐式 Intent 之一来实现。使用 ACTION_DIALER Intent 时，电话拨号盘用户界面会与指定的待拨电话号码一起显示给用户。这通过下列代码实现：

```
startActivity(new Intent(Intent.ACTION_CALL,
  Uri.parse("tel:15102345678")));
```

使用 ACTION_CALL Intent 时，不会显示电话拨号盘，而是直接拨打指定的号码。这通过下列代码实现：

```
startActivity(new Intent(Intent.ACTION_DIAL,
  Uri.parse("tel:15102345678")));
```

# 9.4　蓝牙

蓝牙来自于 IEEE 802.15.1 标准，是一种在设备间近距离交换数据的开放的无线协议。常见的例子如手机和耳机之间的通信，而另一些应用程序可用蓝牙进行近距离追踪。要使用蓝牙完成设备间通信，需要完成以下4个步骤。

（1）为设备打开蓝牙。

（2）在有效范围内找到已配对的或可用的设备。

（3）连接到设备。

（4）在设备间传输数据。

要使用蓝牙服务，应用程序必须拥有 BLUETOOTH 权限才能实现接收和发送；拥有 BLUETOOTH_ADMIN 权限才能控制蓝牙设置或初始化设备发现功能。以上这些可以通过在 **AndroidManifest.xml** 中加入下面的几行代码来实现：

```
<uses-permission android:name="android.permission.BLUETOOTH" />
<uses-permission android:name="android.permission.BLUETOOTH_ADMIN" />
```

所有的蓝牙 API 功能都包含在 android.bluetooth 包中。一共有以下5个主要的提供蓝牙特性的类。

- BluetoothAdapter：代表蓝牙无线接口，该接口用于发现设备并实例化蓝牙连接。
- BluetoothClass：用于描述蓝牙设备的一般特征。
- BluetoothDevice：代表某个远程蓝牙设备。

- BluetoothSocket：代表用于与另一个蓝牙设备进行数据交换的套接字或连接点。
- BluetoothServerSocket：代表一个监听到来的请求的开放套接字。

我们会在后面几个技巧中详细讨论上述类。

## 技巧 83：开启蓝牙

蓝牙通过 BluetoothAdapter 类来初始化。getDefaultAdapter()方法检索蓝牙无线接口的相关信息。如果该方法返回 null，说明设备并不支持蓝牙：

```
BluetoothAdapter myBluetooth = BluetoothAdapter.getDefaultAdapter();
```

激活蓝牙时，要使用这个 BluetoothAdapter 实例来查询状态。如果蓝牙未开启，就可以用 Android 内建的 ACTION_REQUEST_ENABLE 类来请求用户开启蓝牙：

```
if(!myBluetooth.isEnabled()) {
     Intent enableIntent = new Intent(BluetoothAdapter
                                    .ACTION_REQUEST_ENABLE);
     startActivity(enableIntent);
}
```

## 技巧 84：发现蓝牙设备

在激活蓝牙后，要发现已配对或可用的蓝牙设备，需要异步调用 BluetoothAdapter 实例的 startDiscovery()方法。这需要注册一个 BroadcastReceiver 来监听 ACTION_FOUND 事件，这类事件会在某个远程蓝牙设备被发现时告知应用程序。代码清单 9-10 中的示例代码给出了具体实现。

**代码清单 9-10 发现蓝牙设备**

```
private final BroadcastReceiver mReceiver = new BroadcastReceiver() {
     public void onReceive(Context context, Intent intent) {
          String action = intent.getAction();
          // When discovery finds a device
          if (BluetoothDevice.ACTION_FOUND.equals(action)) {
               // Get the BluetoothDevice object from the intent
               BluetoothDevice device = intent.getParcelableExtra(
                                   BluetoothDevice.EXTRA_DEVICE);
               Log.v("BlueTooth Testing",device.getName() + "\n"
                    + device.getAddress());
          }
     }
};

IntentFilter filter = new IntentFilter(BluetoothDevice.ACTION_FOUND);
registerReceiver(mReceiver, filter);
myBluetooth.startDiscovery();
```

BroadcastReceiver 还可以监听 ACTION_DISCOVERY_STARTED 和 ACTION_DISCOVERY_FINISHED 事件，这两类事件会在发现过程开始和结束时通知应用程序。

要让其他蓝牙设备能发现当前设备，应用程序要使用 ACTION_REQUEST_DISCOVERABLE Intent 来开启设备的可见性。该 Activity 会在应用程序顶部显示另一个对话框，来询问用户是否要让当前设备对其他设备可见。

```
Intent discoverableIntent
      = new Intent(BluetoothAdapter.ACTION_REQUEST_DISCOVERABLE);
startActivity(discoverableIntent);
```

## 技巧 85：与已绑定的蓝牙设备配对

已绑定的蓝牙设备是指那些在过去已经与当前设备配过对的设备。在为两个设备配对时，要使用 BluetoothSocket 和 BluetoothServerSocket 类，其中一个设备被当作服务器，另一个则作为客户端。要获得已绑定的设备，可以使用 BluetoothAdapter 实例的 getBondedDevices() 方法：

```
Set<BluetoothDevice> pairedDevices = mBluetoothAdapter.getBondedDevices();
```

## 技巧 86：打开蓝牙套接字

要与另一个设备建立蓝牙连接，应用程序应要么实现客户端套接字，要么实现服务器端套接字。在服务器与客户端绑定之后，每个设备都会有一个基于 RFCOMM（蓝牙传输协议）的已连接蓝牙套接字。然而，客户端设备和服务器设备获取蓝牙套接字的方式不同。服务器在接受一个到来的请求时得到蓝牙套接字；而客户端则在打开一个连接服务器的 RFCOMM 通道时得到套接字实例。

服务器端的初始化使用通用的客户端-服务器编程模式，应用程序需要一个开放的套接字来接受到来的请求（这与 TCP 类似）。应当使用 BluetoothServerSocket 接口来创建一个服务器监听端口。在连接被接受之后，会返回一个 BluetoothSocket 类，利用它可以管理连接。

BluetoothServerSocket 可通过 BluetoothAdapter 实例的 listenUsingRFcommWithServiceRecord() 方法来获取。获取了套接字之后，accept() 方法开始监听请求，并仅当一个连接被接受或者发生了异常时返回。接着，当 accept() 返回一个有效值时，BluetoothSocket 类随之返回。最后，应当调用 close() 方法来释放服务器套接字及其资源，因为 RFCOMM 只允许一个通道一次连接一个客户端。这样做并不会关闭已连接的 BluetoothSocket。下面的代码片段显示了如何完成上述步骤：

```
BluetoothServerSocket myServerSocket
    = myBluetoothAdapter.listenUsingRfcommWithServiceRecord(name, uuid);
myServerSocket.accept();
myServerSocket.close();
```

注意，accept() 方法是一个会引起阻塞的调用，不应在主线程内实现，更好地办法是将其实现到一个工作线程中，如代码清单 9-11 所示。

**代码清单 9-11　建立蓝牙套接字**

```
private class AcceptThread extends Thread {
    private final BluetoothServerSocket mmServerSocket;

    public AcceptThread() {
        // Use a temporary object that is later assigned
```

```
        // to mmServerSocket, because mmServerSocket is final
        BluetoothServerSocket tmp = null;
        try {
            // MY_UUID is the app's UUID string, also used by the client
            tmp = mAdapter.listenUsingRfcommWithServiceRecord(NAME,MY_UUID);
            } catch (IOException e) { }
        mmServerSocket = tmp;
    }

    public void run() {
        BluetoothSocket socket = null;
        // Keep listening until an exception occurs or a socket is returned
        while (true) {
            try {
                socket = mmServerSocket.accept();
            } catch (IOException e) {
                break;
            }
            // If a connection was accepted
            if (socket != null) {
                // Do work to manage the connection (in a separate thread)
                manageConnectedSocket(socket);
                mmServerSocket.close();
                break;
            }
        }
    }

    /** will cancel the listening socket and cause thread to finish */
    public void cancel() {
        try {
            mmServerSocket.close();
        } catch (IOException e) { }
    }
}
```

要实现客户端设备的相关机制，需要从远程设备获取 BluetoothDevice，然后要检索套接字以建立连接。要检索 BluetoothSocket 类，可使用 BluetoothDevice 的 createRfcommSocketToServiceRecord(UUID) 方法，其中的 UUID 来自 listenUsingRfcommWithServiceRecord 中。检索到套接字之后，可使用 connect() 方法来初始化连接。这个方法同样会引起阻塞，也需要在单独的线程中实现，如代码清单 9-12 所示。其中 UUID 存放在 BluetoothDevice 对象中，该对象在设备发现过程中被获取。

**代码清单 9-12　连接到蓝牙套接字**

```
private class ConnectThread extends Thread {
    private final BluetoothSocket mmSocket;
    private final BluetoothDevice mmDevice;

    public ConnectThread(BluetoothDevice device) {
        // Use a temporary object that is later assigned to mmSocket,
        // because mmSocket is final
        BluetoothSocket tmp = null;
        mmDevice = device;

        // Get a BluetoothSocket to connect with the given BluetoothDevice
        try {
        // MY_UUID is the app's UUID string, also used by the server code
        tmp = device.createRfcommSocketToServiceRecord(MY_UUID);
        } catch (IOException e) { }
```

```
            mmSocket = tmp;
        }

        public void run() {
            // Cancel discovery because it will slow down the connection
            mAdapter.cancelDiscovery();

            try {
                // Connect the device through the socket. This will block
                // until it succeeds or throws an exception.
                mmSocket.connect();
            } catch (IOException connectException) {
                // Unable to connect; close the socket and get out
                try {
                    mmSocket.close();
                } catch (IOException closeException) { }
                return;
            }

            // Do work to manage the connection (in a separate thread)
            manageConnectedSocket(mmSocket);
        }

        /** will cancel an in-progress connection and close the socket */
        public void cancel() {
            try {
                mmSocket.close();
            } catch (IOException e) { }
        }
    }
```

建立连接后，就可以使用常见的 InputStream 和 OutputStream 在蓝牙设备间读取和发送数据了。

## 技巧 87：使用设备振动功能

设备振动是所有手机共有的一个特性。要控制 Android 设备上的振动功能，须在 **AndroidManifest.xml** 文件中定义下面的权限：

```
<uses-permission android:name="android.permission.VIBRATE" />
```

接着，对设备振动器的使用不过是由 Android 框架提供的又一项系统服务而已。可以使用 Vibrator 类来访问该服务：

```
Vibrator myVib = (Vibrator) getSystemService(Context.VIBRATOR_SERVICE);
```

有了 Vibrator 的实例之后，只需调用 vibrate()方法即可启动设备振动功能：

```
myVib.vibrate(3000); //Vibrate for 3 seconds
```

如果需要，可以用 cancel()方法在振动自行停止之前将其立即停止：

```
myVib.cancel(); //Cancel the vibration
```

还可以实现某种节奏模式下的振动，可以通过指定一个振动-暂停序列来实现。例如：

```
long[] pattern = {2000,1000,5000};
 myVib.vibrate(pattern,1);
```

这两行代码会让设备先等待 2s，之后开启这样的模式：振动 1 s，暂停 5 s，如此往复。vibrate()方法的第二个参数指定 pattern 数组中表示重复起始点的元素下标。如将其设为 -1，表示不对模式做任何重复。

## 技巧 88：访问无线网络

许多应用程序都会用到 Android 设备的无线网络连接。为更好地理解如何依照网络变化来处理应用程序的行为，Android 设备对底层网络状态的访问途径。其实现方式是，通过广播 Intent 来将网络连接的变化通知给应用程序组件，并提供对网络设定和连接的控制。

Android 通过 ConnectivityManager 类提供了一个系统服务，让开发者能够监视连接状态、设定首选的网络连接，以及管理连接失效转移（failover）等。该服务的初始化如下：

```
ConnectivityManager myNetworkManager
 = (ConnectivityManager) getSystemService(Context.CONNECTIVITY_SERVICE);
```

要使用连接管理器（connective manager），需要在 **AndroidManifest.xml** 中为应用程序设置如下权限：

```
<uses-permission android:name="android.permission.ACCESS_NETWORK_STATE" />
```

连接管理器提供了 getNetworkInfo()和 getActiveNetworkInfo()两个方法，用以在一个 NetworkInfo 类中获取当前网络状况的细节。然而，更好的监视网络变化的方法是创建一个广播接收器，如下面的例子所示：

```
private BroadcastReceiver mNetworkReceiver = new BroadcastReceiver(){
    public void onReceive(Context c, Intent i){
        Bundle b = i.getExtras();
        NetworkInfo ni = (NetworkInfo)
                        b.get(ConnectivityManager.EXTRA_NETWORK_INFO);
        if(ni.isConnected()){
            //Do the operation
        }else{
            //Announce to the user the network problem
        }
    }
};
```

定义了广播接收器之后，可以将其注册，用于监听 ConnectivityManager.CONNECTIVITY_ACTION 类型的 Intent。

```
this.registerReceiver(mNetworkReceiver,
           new IntentFilter(ConnectivityManager.CONNECTIVITY_ACTION));
```

前面定义的 mNetworkReceiver 类只会从 ConnectivityManager.EXTRA_NETWORK_INFO 中提取 NetworkInfo。然而，连接管理器还拥有其他可显示的信息。下面给出不同类型的可用信息。

- EXTRA_EXTRA_INFO：包含关于网络状态的附加信息。
- EXTRA_IS_FAILOVER：返回一个布尔值，表示当前连接是否是失效转移网络的结果。

- EXTRA_NETWORK_INFO：返回一个 NetworkInfo 对象。
- EXTRA_NETWORK_TYPE：触发一个 CONNECTIVITY_ACTION 广播。
- EXTRA_NO_CONNECTIVITY：返回一个布尔值，表示是否没有网络连接。
- EXTRA_OTHER_NETWORK_INFO：返回一个指示网络断开时，可用于失效转移的网络的 NetworkInfo 对象。
- EXTRA_REASON：返回一个描述连接失败原因的字符串。

ConnectivityManager 还提供对网络硬件和失效转移的偏好设置。可以用 setNetworkPreference()方法来选择网络类型。要变更网络，需要在 **AndroidManifest.xml** 文件中为应用程序设定另一个权限：

```
<uses-permission android:name="android.permission.CHANGE_NETWORK_STATE" />
```

## 9.5 近场通信（NFC）

NFC 是已被植入很多 Android 设备的一种无线技术。当有两台支持 NFC 的设备，并有少量的数据，比如播放列表、网络地址、通讯录信息等需要被转移的时候，NFC 将是一个绝妙的通信中介。它之所以是理想选择，是因为它不需要复杂的密码，也不需要发现机制或设备配对机制。如果使用了 Android Beam，只需在一个设备上点击一下，并在另一个设备的接受对话框上再点击一下，就可以利用 NFC 在两个设备间传输数据。

NFC 通过经编码的 NFC 数据交换格式（NDEF）信息，在设备间读取和传送少量数据。每条 NDEF 包含至少一条 NDEF 记录，该记录包含如下的字段：

- 3 位类型名称格式（TNF）;
- 可变长度类型；
- 可变长度 ID（可选）；
- 可变长度有效本体（payload）。

3 位 TNF 字段可能包含许多不同的值，这些值被 Android 用作标签分配系统的一部分，来决定如何将一个 MIME 类型或 URI 映射到正被读取的 NDEF 信息上。如果标签被识别，就要用到 ACTION_NDEF_DISCOVERED Intent，会启动一个 Activity 来处理这个 Intent。如果当标签被扫描时没有已注册的 Activity，或数据未被识别，就启动 ACTION_TECH_DISCOVERED Intent，并提示用户选择打开一个用于处理的程序。开发应用程序时，开发者可能会利用前台分配系统，让 Android 系统不要退出当前应用，同时打开另一个程序来处理 NFC 数据。

在应用程序中使用 NFC 需要访问 NFC 硬件的权限。可以将下面的代码添加到 **AndroidManifest.xml** 文件中：

```
<uses-permission android:name="android.permission.NFC" />
```

在 SDK 级别 10 及以上的版本中，对 NFC 读写的支持以及增强的 NDEF 选项对于应用程序都是可用的。要想使用 Android Beam，需要的最低版本为级别 14。

## 技巧 89：读取 NFC 标签

使用NFC通常涉及的就是读和写。代码清单9-13中的代码可以用来建立一个能够返回NFC标签中存储的某些数据的读取器。这些数据随后会被显示在一个 `TextView` 中。

**代码清单 9-13　src/com/cookbook/nfcreader/MainActivity.java**

```
package com.cookbook.nfcreader;

import com.cookbook.nfcreader.R;
import android.app.Activity;
import android.app.PendingIntent;
import android.content.Intent;
import android.content.IntentFilter;
import android.nfc.NdefMessage;
import android.nfc.NdefRecord;
import android.nfc.NfcAdapter;
import android.os.Bundle;
import android.os.Parcelable;
import android.util.Log;
import android.widget.TextView;

public class MainActivity extends Activity {

  protected NfcAdapter nfcAdapter;
  protected PendingIntent nfcPendingIntent;

  private static final String TAG = MainActivity.class.getSimpleName();

    @Override
    protected void onCreate(Bundle savedInstanceState) {
        super.onCreate(savedInstanceState);
        setContentView(R.layout.activity_main);

        nfcAdapter = NfcAdapter.getDefaultAdapter(this);
    nfcPendingIntent = PendingIntent.getActivity(this, 0,
        new Intent(this, this.getClass()).addFlags(Intent.FLAG_ACTIVITY_SINGLE_TOP),
        0);
    }

    public void enableForegroundMode() {
    Log.d(TAG, "enableForegroundMode");

    IntentFilter tagDetected = new IntentFilter(NfcAdapter.ACTION_TAG_DISCOVERED);
    IntentFilter[] writeTagFilters = new IntentFilter[] {tagDetected};
    nfcAdapter.enableForegroundDispatch(this, nfcPendingIntent,
        writeTagFilters, null);
  }

    public void disableForegroundMode() {
    Log.d(TAG, "disableForegroundMode");

    nfcAdapter.disableForegroundDispatch(this);
  }

    @Override
  public void onNewIntent(Intent intent) {
    Log.d(TAG, "onNewIntent");
    String stringOut = "";

    if (NfcAdapter.ACTION_TAG_DISCOVERED.equals(intent.getAction())) {
      TextView textView = (TextView) findViewById(R.id.main_tv);
```

```
        Parcelable[] messages =
intent.getParcelableArrayExtra(NfcAdapter.EXTRA_NDEF_MESSAGES);
        if (messages != null) {
          for (int i = 0; i < messages.length; i++) {
            NdefMessage message = (NdefMessage)messages[i];
            NdefRecord[] records = message.getRecords();

            for (int j = 0; j < records.length; j++) {
              NdefRecord record = records[j];
              stringOut += "TNF: " + record.getTnf() + "\n";
              stringOut += "MIME Type: " + new String(record.getType()) + "\n";
              stringOut += "Payload: " + new String(record.getPayload()) + "\n\n";
              textView.setText(stringOut);
            }
            }
          }
      }
    }

    @Override
    protected void onResume() {
      Log.d(TAG, "onResume");

      super.onResume();

      enableForegroundMode();
    }

    @Override
    protected void onPause() {
      Log.d(TAG, "onPause");

      super.onPause();

      disableForegroundMode();
    }
  }
```

代码清单 9-13 建立的应用程序使用了前台分配系统来确保应用程序自己被用来处理任意扫描到的 NFC 标签。注意，在 onPause()方法中调用了 disableForegroundMode()方法使应用程序不再继续作为 NFC 标签的默认处理者。onResume()方法则能恢复这一能力。

## 技巧 90：写入 NFC 标签

支持 NFC 的 Android 设备还能够对未受保护的 NFC 标签进行写入。代码清单 9-14 给出了一个能向 NFC 标签写入信息的示例应用程序。注意，在 NFC 标签写入期间，存储卡上的信息会被擦除，由写入的信息取而代之。

**代码清单 9-14　src/com/cookbook/nfcwriter/MainActivity.java**

```
package com.cookbook.nfcwriter;

import java.io.IOException;
import java.nio.charset.Charset;

import android.app.Activity;
import android.app.PendingIntent;
import android.content.Intent;
import android.content.IntentFilter;
```

```
import android.nfc.NdefMessage;
import android.nfc.NdefRecord;
import android.nfc.NfcAdapter;
import android.nfc.Tag;
import android.nfc.tech.Ndef;
import android.nfc.tech.NdefFormatable;
import android.os.Bundle;
import android.view.View;
import android.view.View.OnClickListener;
import android.widget.Button;
import android.widget.Toast;

public class MainActivity extends Activity implements OnClickListener {

  protected NfcAdapter nfcAdapter;
  private Button mainButton;
  private boolean mInWriteMode;

  @Override
  protected void onCreate(Bundle savedInstanceState) {
    super.onCreate(savedInstanceState);
    setContentView(R.layout.activity_main);

    nfcAdapter = NfcAdapter.getDefaultAdapter(this);

    mainButton = (Button)findViewById(R.id.main_button);
        mainButton.setOnClickListener(this);
  }

  public void onClick(View v) {
    displayMessage("Touch and hold tag against phone to write.");
    beginWrite();
  }

  @Override
  protected void onPause() {
    super.onPause();
    stopWrite();
  }

  @Override
    public void onNewIntent(Intent intent) {
    if(mInWriteMode) {
      mInWriteMode = false;
      Tag tag = intent.getParcelableExtra(NfcAdapter.EXTRA_TAG);
      writeTag(tag);
      }
      }

    private void beginWrite() {
      mInWriteMode = true;

          PendingIntent pendingIntent = PendingIntent.getActivity(this, 0,
new Intent(this, getClass()).addFlags(Intent.FLAG_ACTIVITY_SINGLE_TOP), 0);
        IntentFilter tagDetected =
new IntentFilter(NfcAdapter.ACTION_TAG_DISCOVERED);
        IntentFilter[] filters = new IntentFilter[] { tagDetected };

    nfcAdapter.enableForegroundDispatch(this, pendingIntent, filters, null);
  }

  private void stopWrite() {
    nfcAdapter.disableForegroundDispatch(this);
  }
```

```
    private boolean writeTag(Tag tag) {
      byte[] payload = "Text stored in an NFC tag".getBytes();
      byte[] mimeBytes = "text/plain".getBytes(Charset.forName("US-ASCII"));
           NdefRecord cardRecord = new NdefRecord(NdefRecord.TNF_MIME_MEDIA,
           mimeBytes, new byte[0], payload);
           NdefMessage message = new NdefMessage(new NdefRecord[] { cardRecord });

      try {
        Ndef ndef = Ndef.get(tag);
        if (ndef != null) {
           ndef.connect();

           if (!ndef.isWritable()) {
             displayMessage("This is a read-only tag.");
             return false;
           }

           int size = message.toByteArray().length;
           if (ndef.getMaxSize() < size) {
              displayMessage("There is not enough space to write.");
              return false;
           }

           ndef.writeNdefMessage(message);
           displayMessage("Write successful.");
           return true;
         } else {
           NdefFormatable format = NdefFormatable.get(tag);
           if (format != null) {
              try {
                format.connect();
                format.format(message);
                displayMessage("Write successful\nLaunch a scanning app or scan and
➥choose to read.");
                return true;
              } catch (IOException e) {
                displayMessage("Unable to format tag to NDEF.");
                return false;
              }
           } else {
             displayMessage("Tag doesn't appear to support NDEF format.");
             return false;
           }
         }
      } catch (Exception e) {
        displayMessage("Write failed.");
      }
        return false;
      }

    private void displayMessage(String message) {
      Toast.makeText(MainActivity.this, message, Toast.LENGTH_LONG).show();
    }
  }
```

代码清单9-14建立的应用程序会将“Text stored in an NFC tag”这串文本作为信息的有效净荷进行写入，而信息的MIME类型则为“text/plain”。可以看到这些值在 `writeTag()` 方法中被置入。通过改变这些值，可以改变Android系统处理NFC标签的方式。

## 9.6 通用串行总线（USB）

从Android 3.1开始，可以以主机模式或附件模式使用USB设备。运行了Gingerbread 2.3.4d

的较早的 Android 设备可以在引用了 Google 支持 API 的情况下使用附件模式。进行附件模式开发时会用到下面两个类。

- `UsbManager`：允许与已连接的 USB 附件通信。
- `UsbAccessory`：代表一个 USB 设备，带有可以检索其信息的方法。

主机模式允许 Android 设备控制连接到其上的其他设备。这还意味着 Android 设备需要为其他 USB 设备供电。主机模式下可能用到的设备的例子有键盘、鼠标和其他输入设备。Android 3.1 及以上版本可以使用主机模式。下面是进行主机模式开发时会用到的类。

- `UsbManager`：访问 USB 连接的状态，并与 USB 设备通信。
- `UsbDevice`：代表一个插入的 USB 设备。
- `UsbInterface`：`UsbDevice` 上的一个接口。
- `UsbEndpoint`：`UsbInterface` 上的一个端点。
- `UsbDeviceConnection`：对 USB 设备进行信息的发送和接收。
- `UsbRequest`：代表一个 USB 请求数据包。
- `UsbConstants`：USB 协议要用到的常量。

附件模式使得其他 USB 设备可以为 Android 设备供电。使用附件模式的 USB 设备必须遵循 Android 附件开发工具包（ADK）中的规则。能在附件模式下使用的 USB 设备有外部识别设备、音乐控制器、插接站（docking station）和其他类似的设备。

对一个插入的 USB 设备进行开发时，开发者可能无法使用 USB 调试功能。在此情况下，LogCat 仍然可用，但要经由无线连接，这通过在使用 TCP 上的 ADB（Android 调试桥）来完成。要实现这一点，需在调试模式下启动 Android 设备，并将其连接到计算机上。打开设备的无线连接并查找当前 IP 地址。打开（计算机的）终端或命令行，进到 **SDK 安装**目录中的 **platform-tools** 子目录下。然后在命令行或终端下输入如下命令：

**adb tcpip 5555**

这样就可将连接模式从 USB 改为 TCP/IP。如果不进行这一改变，经由网络连接设备的尝试将会失败。要连接到设备，键入以下命令，其中 **DEVICEIPADDRESS** 是要连接的设备的 IP 地址：

**adb connect DEVICEIPADDRESS:5555**

注意对于某些操作系统，并不需要在命令的最后指定端口号（**:5555**）。端口号并不限于 5555，任意开放的端口都可以使用。如果连接命令使用 5555 端口时失败，尝试再次执行不指定任何端口的连接命令。

连接之后，通常在 ADB 下可以运行的命令就都能执行。当然，`adb logcat` 可能是最重要的一个。

使用网络对应用程序进行调试之后，如果要回到 USB 调试模式，可在控制台下键入 **adb usb** 命令，ADB 就会重启，并寻找 USB 连接。

要获得 ADB 及其用法的更多信息，可访问 http://developer.android.com/tools/help/adb.html。

# 第 10 章

# 网络

基于网络的应用程序为用户提供了更多的价值，因为其中的内容可以是动态和交互的。网络使得从社交网络到云计算等很多东西变为可能。

本章着眼于网络状态、短信息服务（SMS）、基于因特网资源的应用和社交网络应用程序。对于通过网络连接来获取或更新可用信息的应用程序而言，连接状态是很重要的。SMS 是允许手机设备之间进行短文本信息交换的一种通信服务。基于因特网资源的应用程序依赖于诸如 HTML（超文本标记语言）、XML（可扩展标记语言）和 JSON（JavaScript 对象标记）等网络内容。社交网络应用程序（比如 Twitter）则是人们联系彼此的一种重要方式。

## 10.1 响应网络状态

了解设备如何连接和是否连接到某个网络对于 Android 开发而言是十分重要的。从网络服务器传输数据流的应用程序需要针对可能会造成金钱开销的大量数据，对用户提出警告。应用程序延迟问题可能也要纳入考虑。做几个简单的查询就可以使用户确认他们当前是否是通过网络设备进行连接，以及当连接状态改变时如何反应。

### 技巧 91：检查网络连接

`ConnectivityManager` 用于判定设备的连接状况。本技巧可用来判定有哪些网络接口被连接到了网络上。代码清单 10-1 使用 `ConnectivityManager` 显示设备是否通过 Wi-Fi 或蓝牙进行了连接。

**代码清单 10-1　src/com/cookbook/connectivitycheck/MainActivity.java**

```
package com.cookbook.connectivitycheck;

import android.app.Activity;
```

```
import android.content.Context;
import android.net.ConnectivityManager;
import android.net.NetworkInfo;
import android.os.Bundle;
import android.widget.TextView;

public class MainActivity extends Activity {
    TextView tv;

    @Override
    protected void onCreate(Bundle savedInstanceState) {
        super.onCreate(savedInstanceState);
        setContentView(R.layout.activity_main);
        tv = (TextView) findViewById(R.id.tv_main);
        try {
            String service = Context.CONNECTIVITY_SERVICE;
            ConnectivityManager cm = (ConnectivityManager)getSystemService(service);
            NetworkInfo activeNetwork = cm.getActiveNetworkInfo();

            boolean isWiFi = activeNetwork.getType() == ConnectivityManager.TYPE_WIFI;
            boolean isBT = activeNetwork.getType() == ConnectivityManager.TYPE_BLUETOOTH;

            tv.setText("WiFi connected: "+isWiFi+"\nBluetooth connected: "+isBT);
        } catch(Exception nullPointerException) {
            tv.setText("No connected networks found");
        }
    }
}
```

代码清单 10-1 使用了 TYPE_WIFI 和 TYPE_BLUETOOTH 常量来检查相应网络的连接情况。除 TYPE_WIFI 和 TYPE_BLUETOOTH 之外，判定连接情况的常量还有下面这些。

- TYPE_DUMMY：用于虚拟数据连接。
- TYPE_ETHERNET：用于默认的以太网（Ethernet）连接。
- TYPE_MOBILE：用于默认的移动数据连接。
- TYPE_MOBILE_DUN：用于特定于 DUN 的移动数据连接。
- TYPE_MOBILE_HIPRI：用于高优先级的移动数据连接。
- TYPE_MOBILE_MMS：用于特定于 MMS 的移动数据连接。
- TYPE_MOBILE_SUPL：用于特定于 SUPL 的移动数据连接。
- TYPE_WIMAX：用于默认的 WiMAX 数据连接。

图 10-1 显示了一个运行了代码清单 10-1 中代码的显示效果。虽然已经启用了蓝牙，由于当前没有活动的连接，所以报告的连接状态依然是 false。

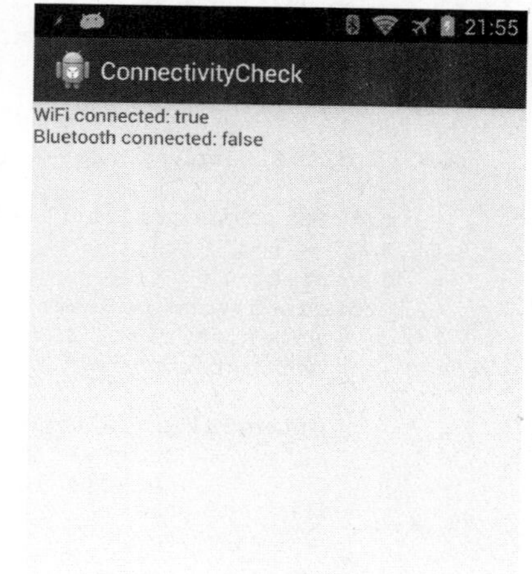

图 10-1 检查设备连接状况

# 技巧 92：接收连接变化信息

在需要对连接状态变化做出反应时，可以用广播接收器来检查网络连接的状态。

广播接收器可以在应用程序的 manifest 中声明，也可以作为主 Activity 内的一个子类。二者都是可访问的，这个技巧使用子类的方式，并结合 onCreate()和 onDestory()方法对接收器注册和取消注册。

因为本技巧要检查连接状况，需要在应用程序的 manifest 中添加下列权限：

```
<uses-permission android:name="android.permission.INTERNET" />
<uses-permission android:name="android.permission.ACCESS_NETWORK_STATE" />
```

代码清单 10-2 显示了检查连接状态变化所需的代码。当检查到变化时，应用程序会显示一条 Toast 信息，把变化告知给用户。

**代码清单 10-2　src/com/cookbook/connectivitychange/MainActivity.java**

```
package com.cookbook.connectivitychange;

import android.app.Activity;
import android.content.BroadcastReceiver;
import android.content.Context;
import android.content.Intent;
import android.content.IntentFilter;
import android.net.ConnectivityManager;
import android.net.NetworkInfo;
import android.os.Bundle;
import android.widget.Toast;

public class MainActivity extends Activity {

     private ConnectivityReceiver receiver = new ConnectivityReceiver();

     @Override
     protected void onCreate(Bundle savedInstanceState) {
          super.onCreate(savedInstanceState);
          setContentView(R.layout.activity_main);

          IntentFilter filter = new IntentFilter(ConnectivityManager.CONNECTIVITY_ACTION);
                    receiver = new ConnectivityReceiver();
                    this.registerReceiver(receiver, filter);
     }

     @Override
          public void onDestroy() {
                    super.onDestroy();
                    if (receiver != null) {
                              this.unregisterReceiver(receiver);
                    }
          }

     public class ConnectivityReceiver extends BroadcastReceiver {

          @Override
          public void onReceive(Context context, Intent intent) {
```

```
                ConnectivityManager conn =
(ConnectivityManager)context.getSystemService(Context.CONNECTIVITY_SERVICE);
                NetworkInfo networkInfo = conn.getActiveNetworkInfo();

                if (networkInfo != null && networkInfo.getType() == ConnectivityManager.
➥TYPE_WIFI) {
                  Toast.makeText(context, "WiFi is connected", Toast.LENGTH_SHORT).show();
                } else if (networkInfo != null) {
                  Toast.makeText(context, "WiFi is disconnected", Toast.LENGTH_SHORT).show();
                } else {
                  Toast.makeText(context, "No active connection", Toast.LENGTH_SHORT).show();
                }
                }
        }
    }
```

图 10-2 显示了当连接了 Wi-Fi 时出现的信息。图 10-3 则显示了当 Wi-Fi 和移动数据都被断开时出现的信息。

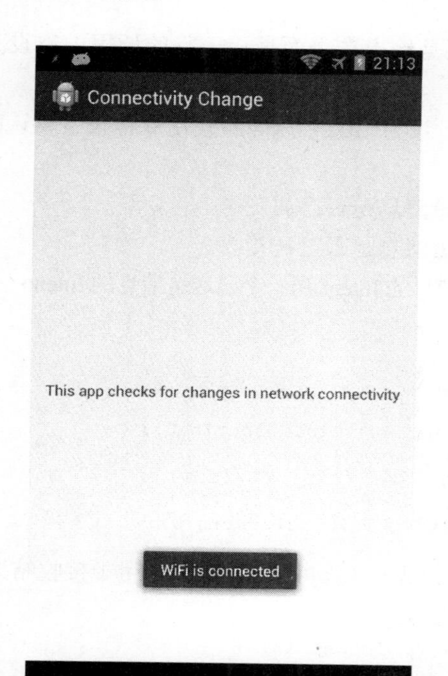


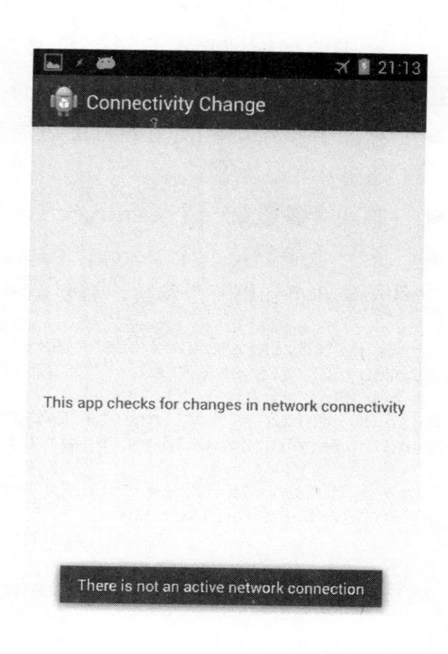

图 10-2　当 Wi-Fi 启用，会出现一条 Toast 信息，将连接告知用户

图 10-3　Wi-Fi 和移动数据都禁用时，会显示一条 Toast 信息，通知用户缺少网络连接

## 10.2　使用短消息

Android 框架通过 SmsManager 类提供了对短消息功能的完全访问。Android 的早期版本将 SmsManager 放在 android.telephony.gsm 包中。从 Android 1.5 起，SmsManager 同

时支持GSM和CDMA移动电话标准，现在SmsManager被移入android.telephony包中。

通过SmsManager类发送短消息非常简便。

（1）在**AndroidManifest.xml**中设置发送短消息的权限：

```
<uses-permission android:name="android.permission.SEND_SMS" />
```

（2）使用SmsManager.getDefault()静态方法获取短消息管理器（SMS manager）实例：

```
SmsManager mySMS = SmsManager.getDefault();
```

（3）定义目标手机号码以及待发送消息的内容。使用sendTextMessage()方法将短消息发送到另一台设备上。

```
String destination = "16501234567";
String msg = "Sending my first message";
mySMS.sendTextMessage(destination, null, msg, null, null);
```

以上对于发送短消息而言已经足够。不过我们还是给出在上面最后一行的调用中被设为null的三个参数的用法。

- 第二个参数是要使用的特定短消息服务中心。设为null意味着使用载体设备默认的服务中心。
- 第四个参数是一个PendingIntent，跟踪短消息是否已被发送。
- 第五个参数是一个PendingIntent，跟踪短消息是否已被接收。

要使用第四个和第五个参数，需要声明一个已发送消息的Intent和一个已送达消息的Intent：

```
String SENT_SMS_FLAG = "SENT_SMS";
String DELIVER_SMS_FLAG = "DELIVER_SMS";

Intent sentIn = new Intent(SENT_SMS_FLAG);
PendingIntent sentPIn = PendingIntent.getBroadcast(this,0,sentIn,0);

Intent deliverIn = new Intent(SENT_SMS_FLAG);
PendingIntent deliverPIn
                = PendingIntent.getBroadcast(this,0,deliverIn,0);
```

然后需要为每一个PendingIntent注册一个BroadcastReceiver类，用于接收结果：

```
BroadcastReceiver sentReceiver = new BroadcastReceiver(){
      @Override public void onReceive(Context c, Intent in) {
           switch(getResultCode()){
                case Activity.RESULT_OK:
                     //sent SMS message successfully;
                     break;
                default:
                     //sent SMS message failed
                     break;
           }
      }
};
BroadcastReceiver deliverReceiver = new BroadcastReceiver(){
      @Override public void onReceive(Context c, Intent in) {
           //SMS delivered actions
      }
```

```
};

registerReceiver(sentReceiver, new IntentFilter(SENT_SMS_FLAG));
registerReceiver(deliverReceiver, new IntentFilter(DELIVER_SMS_FLAG));
```

多数短消息的长度上限为 140 个字符。为确保短消息长度在限制范围内，可使用 divideMessage()方法将文本拆分成长度不超过短消息字数上限的片段，然后使用 sendMultipartTextMessage()方法代替 sendTextMessage()方法。两者的区别在于 sendMultipartTextMessage()的信息使用 ArrayList 存放，而且两者用到的 pending Intent 也不同：

```
ArrayList<String> multiSMS = mySMS.divideMessage(msg);
ArrayList<PendingIntent> sentIns = new ArrayList<PendingIntent>();
ArrayList<PendingIntent> deliverIns = new ArrayList<PendingIntent>();

for(int i=0; i< multiSMS.size(); i++){
    sentIns.add(sentIn);
    deliverIns.add(deliverIn);
}

mySMS.sendMultipartTextMessage(destination, null,
                               multiSMS, sentIns, deliverIns);
```

## 技巧 93：收到短消息后自动回复

大多数短消息都不会被接收者立即查阅，本技巧将介绍如何在收到短消息时发送自动回复信息。实现方式是在后台创建一个能够接收到来信息的 Android 服务。还有一种方法是在 **AndroidManifest.xml** 文件中注册一个广播接收器。

应用程序必须在 **AndroidManifest.xml** 文件中声明发送和接收短消息的权限，如代码清单 10-3 所示。清单中还声明了一个名为 SMSResponder 的主 Activity，用于创建自动回复；以及一个名为 ResponderService 的服务，用于在收到 SMS 时发送回复。

**代码清单 10-3 AndroidManifest.xml**

```
<?xml version="1.0" encoding="utf-8"?>
<manifest xmlns:android="http://schemas.android.com/apk/res/android"
      package="com.cookbook.SMSResponder"
      android:versionCode="1"
      android:versionName="1.0">
    <application android:icon="@drawable/icon"
                 android:label="@string/app_name">
        <activity android:name=".SMSResponder"
                  android:label="@string/app_name">
            <intent-filter>
              <action android:name="android.intent.action.MAIN" />
              <category android:name="android.intent.category.LAUNCHER" />
            </intent-filter>
        </activity>
        <service android:enabled="true" android:name=".ResponderService">
        </service>
    </application>

    <uses-permission android:name="android.permission.RECEIVE_SMS"/>
```

```
    <uses-permission android:name="android.permission.SEND_SMS"/>
</manifest>
```

代码清单 10-4 所示的主布局文件中包含一个 LinearLayout，它带有三个视图：用于显示自动回复信息的 TextView、用于提交在应用程序内对回复信息的修改的按钮，以及供用户键入回复信息的 EditText。

### 代码清单 10-4 res/layout/main.xml

```
<?xml version="1.0" encoding="utf-8"?>
<LinearLayout xmlns:android="http://schemas.android.com/apk/res/android"
    android:orientation="vertical"
    android:layout_width="match_parent"
    android:layout_height="match_parent">
        <TextView android:id="@+id/display"
        android:layout_width="match_parent"
        android:layout_height="wrap_content"
        android:text="@string/hello"
        android:textSize="18dp"
        />
        <Button android:id="@+id/submit"
        android:layout_width="wrap_content"
        android:layout_height="wrap_content"
        android:text="Change my response"
        />
        <EditText android:id="@+id/editText"
        android:layout_width="match_parent"
        android:layout_height="match_parent"
      />
</LinearLayout>
```

主 Activity 如代码清单 10-5 所示。它启动了用于监听并自动回复短消息的服务，同时还允许用户修改回复信息并将其保存到 SharedPreferences 中供今后使用。

### 代码清单 10-5 src/com/cookbook/SMSresponder/SMSResponder.java

```
package com.cookbook.SMSresponder;

import android.app.Activity;
import android.content.Intent;
import android.content.SharedPreferences;
import android.content.SharedPreferences.Editor;
import android.os.Bundle;
import android.preference.PreferenceManager;
import android.util.Log;
import android.view.View;
import android.view.View.OnClickListener;
import android.widget.Button;
import android.widget.EditText;
import android.widget.TextView;

public class SMSResponder extends Activity {
    TextView tv1;
    EditText ed1;
    Button bt1;
    SharedPreferences myprefs;
    Editor updater;
    String reply=null;
```

```
    @Override
    public void onCreate(Bundle savedInstanceState) {
        super.onCreate(savedInstanceState);
        setContentView(R.layout.main);

        myprefs = PreferenceManager.getDefaultSharedPreferences(this);
        tv1 = (TextView) this.findViewById(R.id.display);
        ed1 = (EditText) this.findViewById(R.id.editText);
        bt1 = (Button) this.findViewById(R.id.submit);

        reply = myprefs.getString("reply",
                "Thank you for your message. I am busy now."
                + "I will call you later");
        tv1.setText(reply);

        updater = myprefs.edit();
        ed1.setHint(reply);
        bt1.setOnClickListener(new OnClickListener() {
            public void onClick(View view) {
                updater.putString("reply", ed1.getText().toString());
                updater.commit();
                SMSResponder.this.finish();
            }
        });

        try {
            // Start service
            Intent svc = new Intent(this, ResponderService.class);
            startService(svc);
        }
        catch (Exception e) {
            Log.e("onCreate", "service creation problem", e);
        }
    }
}
```

本技巧代码的大部分位于代码清单 10-6 所示的服务中。该服务首先为应用程序检索 SharedPreferences。然后，注册两个分别用于监听到来短信息和发出短消息的广播接收器。用于发出短消息的广播接收器在本例中实际并未用到，只是出于完整性考虑才给出的。

到来短消息的广播接收器使用一个 bundle 来检索协议描述单元（protocol desciption unit，PDU），后者包含短消息文本以及其他短消息的元数据。广播接收器将 PNU 解析为一个 Object 数组。createFromPdu()方法将 Object 数组转换为一个 SmsMessage。之后，可使用 getOriginatingAddress()方法获取发送者的电话号码，还可使用 getMessageBody()方法获取文本信息。

本技巧中，在检索到发送者地址之后，会调用 respond()方法。这一方法尝试获取存储在 SharedPreferences 中的有关自动回复信息的数据。如果 SharedPreferences 中并未保存相关数据，则会使用一个默认值。随后，respond()方法分别为发送状态和接收状态各创建一个 PendingIntent。divideMessage()方法用于确保信息大小没有超过限制。所有数据在处理完成之后，会通过 sendMultiTextMessage()方法发送。

**代码清单 10-6 src/com/cookbook/SMSresponder/ResponderService.java**

```
package com.cookbook.SMSresponder;

import java.util.ArrayList;

import android.app.Activity;
import android.app.PendingIntent;
import android.app.Service;
import android.content.BroadcastReceiver;
import android.content.Context;
import android.content.Intent;
import android.content.IntentFilter;
import android.content.SharedPreferences;
import android.os.Bundle;
import android.os.IBinder;
import android.preference.PreferenceManager;
import android.telephony.SmsManager;
import android.telephony.SmsMessage;
import android.util.Log;
import android.widget.Toast;

public class ResponderService extends Service {
    //the action fired by the Android system when an SMS was received
    private static final String RECEIVED_ACTION =
                                "android.provider.Telephony.SMS_RECEIVED";
    private static final String SENT_ACTION="SENT_SMS";
    private static final String DELIVERED_ACTION="DELIVERED_SMS";

    String requester;
    String reply="";
    SharedPreferences myprefs;

    @Override
    public void onCreate() {
        super.onCreate();
        myprefs = PreferenceManager.getDefaultSharedPreferences(this);

        registerReceiver(sentReceiver, new IntentFilter(SENT_ACTION));
        registerReceiver(deliverReceiver,
                             new IntentFilter(DELIVERED_ACTION));

        IntentFilter filter = new IntentFilter(RECEIVED_ACTION);
        registerReceiver(receiver, filter);

        IntentFilter attemptedfilter = new IntentFilter(SENT_ACTION);
        registerReceiver(sender,attemptedfilter);
    }

    private BroadcastReceiver sender = new BroadcastReceiver(){
        @Override
        public void onReceive(Context c, Intent i) {
            if(i.getAction().equals(SENT_ACTION)) {
                if(getResultCode() != Activity.RESULT_OK) {
                    String recipient = i.getStringExtra("recipient");
                    requestReceived(recipient);
                }
            }
        }
    };
```

```
BroadcastReceiver sentReceiver = new BroadcastReceiver() {
    @Override public void onReceive(Context c, Intent in) {
        switch(getResultCode()) {
            case Activity.RESULT_OK:
                //sent SMS message successfully;
                smsSent();
                break;
            default:
                //sent SMS message failed
                smsFailed();
                break;
        }
    }
};

public void smsSent() {
    Toast.makeText(this, "SMS sent", Toast.LENGTH_SHORT);
}
public void smsFailed() {
    Toast.makeText(this, "SMS sent failed", Toast.LENGTH_SHORT);
}
public void smsDelivered() {
    Toast.makeText(this, "SMS delivered", Toast.LENGTH_SHORT);
}

BroadcastReceiver deliverReceiver = new BroadcastReceiver() {
    @Override public void onReceive(Context c, Intent in) {
        //SMS delivered actions
        smsDelivered();
    }
};

public void requestReceived(String f) {
    Log.v("ResponderService","In requestReceived");
    requester=f;
}

BroadcastReceiver receiver = new BroadcastReceiver() {
    @Override
    public void onReceive(Context c, Intent in) {
        Log.v("ResponderService","On Receive");
        reply="";
        if(in.getAction().equals(RECEIVED_ACTION)) {
            Log.v("ResponderService","On SMS RECEIVE");

            Bundle bundle = in.getExtras();
            if(bundle!=null) {
                Object[] pdus = (Object[])bundle.get("pdus");
                SmsMessage[] messages = new SmsMessage[pdus.length];
                for(int i = 0; i<pdus.length; i++) {
                    Log.v("ResponderService","FOUND MESSAGE");
                    messages[i] =
                            SmsMessage.createFromPdu((byte[])pdus[i]);
                }
                for(SmsMessage message: messages) {
                    requestReceived(message.getOriginatingAddress());
                }
                respond();
            }
        }
    }
};
```

```
    @Override
    public void onStart(Intent intent, int startId) {
        super.onStart(intent, startId);
    }

    public void respond() {
        Log.v("ResponderService","Responding to " + requester);
        reply = myprefs.getString("reply",
                                  "Thank you for your message. I am busy now."
                                  + "I will call you later.");
        SmsManager sms = SmsManager.getDefault();
        Intent sentIn = new Intent(SENT_ACTION);
        PendingIntent sentPIn = PendingIntent.getBroadcast(this,
                                                             0,sentIn,0);
        Intent deliverIn = new Intent(DELIVERED_ACTION);
        PendingIntent deliverPIn = PendingIntent.getBroadcast(this,
                                                             0,deliverIn,0);
        ArrayList<String> Msgs = sms.divideMessage(reply);
        ArrayList<PendingIntent> sentIns = new ArrayList<PendingIntent>();
        ArrayList<PendingIntent> deliverIns =
                                   new ArrayList<PendingIntent>();
        for(int i=0; i< Msgs.size(); i++) {
            sentIns.add(sentPIn);
            deliverIns.add(deliverPIn);
        }

        sms.sendMultipartTextMessage(requester, null,
                                   Msgs, sentIns, deliverIns);
    }

    @Override
    public void onDestroy() {
        super.onDestroy();
        unregisterReceiver(receiver);
        unregisterReceiver(sender);
    }

    @Override
    public IBinder onBind(Intent arg0) {
        return null;
    }
}
```

## 10.3 使用 Web 内容

要启动因特网浏览器来显示 Web 内容，可使用第 2 章提到过的隐式的 Intent ACTION_VIEW，例如：

```
Intent i = new Intent(Intent.ACTION_VIEW);
i.setData(Uri.parse("http://www.google.com"));
startActivity(i);
```

开发者还可以通过 WebView 来创建自己的 Web 浏览器。WebView 是一个用来显示 Web 内容的 View。与其他视图一样，在 Activity 中，它既可以占据整个屏幕，也可以仅仅占据布局的一部分。WebView 使用 WebKit 来渲染网页，WebKit 者是一个开源的浏览器引擎，被苹果公司的 Safari 用于渲染网页。

## 技巧 94：自定义 Web 浏览器

有两种获取 WebView 对象的方法。一种是通过构造函数对其进行实例化：

```
WebView webview = new WebView(this);
```

另一种是在布局中使用并在 Activity 中进行声明：

```
WebView webView = (WebView) findViewById(R.id.webview);
```

在获取了 WebView 对象之后，可以使用 loadURL() 方法显示网页：

```
webview.loadUrl("http://www.google.com/");
```

可以使用 WebSettings 类来定义浏览器的特性。例如，可以使用 setBlockNetworkImage() 方法屏蔽网络图片，以减少加载的数据量。显示网页内容的字体大小可以用 setDefaultFontSize() 来设置。其他一些常用设置如下面的例子所示：

```
WebSettings webSettings = webView.getSettings();
webSettings.setSaveFormData(false);
webSettings.setJavaScriptEnabled(true);
webSettings.setSavePassword(false);
webSettings.setSupportZoom(true);
```

## 技巧 95：使用 HTTP GET 请求

除了启动浏览器或在 Activity 中使用 Webview 微件来包含基于 WebKit 的浏览器控件外，开发者也许还想创建基于因特网的原生应用程序。这意味着应用程序只依赖来自因特网的原始数据（例如图像、媒体文件和 XML 数据），且只载入与程序功能有关的数据。这对于创建社交网络应用程序非常重要。在 Android 中有两个包可用于处理网络通信：java.net 和 android.net。

本技巧中，使用 HTTP GET 来检索 XML 或 JSON 数据（关于 JSON，在 www.json.org/上可找到其概览）。特别地，本技巧演示了 Google 搜索的表述性状态转移（REST）API，并使用了如下查询：

```
http://ajax.googleapis.com/ajax/services/search/web?v=1.0&q=
```

要搜索任何主题，只需将其添加到查询语句末尾。例如，要搜索有关美国职业篮球联赛（NBA）的信息，可使用下面的查询，之后会得到 JSON 数据：

```
http://ajax.googleapis.com/ajax/services/search/web?v=1.0&q=NBA
```

Activity 需要具有访问因特网的权限才能运行，因此要将下面的代码添加到 **AndroidManifest.xml** 文件中：

```
<uses-permission android:name="android.permission.INTERNET"/>
```

主布局如代码清单 10-7 所示。它拥有三个视图：EditText 供用户输入搜索主题、Button

用于触发检索、TextView 用来显示搜索结果。

**代码清单 10-7　res/layout/main.xml**

```
<?xml version="1.0" encoding="utf-8"?>
<LinearLayout xmlns:android="http://schemas.android.com/apk/res/android"
    android:orientation="vertical"
    android:layout_width="match_parent"
    android:layout_height="match_parent"
    >
            <EditText
            android:id="@+id/editText"
            android:layout_width="match_parent"
            android:layout_height="wrap_content"
            android:singleLine="true"
             />
             <Button
               android:id="@+id/submit"
             android:layout_width="wrap_content"
             android:layout_height="wrap_content"
             android:text="Search"
             />
             <TextView
             android:id="@+id/display"
             android:layout_width="match_parent"
             android:layout_height="match_parent"
             android:text="@string/hello"
             android:textSize="18dp"
             />
</LinearLayout>
```

主 Activity 如代码清单 10-8 所示。它在 onCreate() 方法中对三个布局元素进行了初始化。在按钮的 OnClickListener 类中调用了 searchRequest()，该方法使用 Google REST API URL 生成搜索项目，然后初始化一个 URL 类的实例。之后，使用 URL 类实例获得一个 HttpURLConnection 实例。

HttpURLConnection 实例能够检索连接状态。当 HttpURLConnection 返回的结果代码为 HTTP_OK 时，意味着整个 HTTP 事务的完成。接着，从 HTTP 事务返回的 JSON 数据可以被转存到一个字符串中。这一步的实现要利用一个传递给 BufferReader 的 InputStream-Reader，后者会读取数据并创建一个 String 实例。从 HTTP 获取了结果之后，Activity 会使用 processResponse() 方法对 JSON 数据进行解析。

**代码清单 10-8　src/com/cookbook/internet/search/GoogleSearch.java**

```
package com.cookbook.internet.search;

import java.io.BufferedReader;
import java.io.IOException;
import java.io.InputStreamReader;
import java.net.HttpURLConnection;
import java.net.MalformedURLException;
import java.net.URL;
import java.security.NoSuchAlgorithmException;

import org.json.JSONArray;
```

```
import org.json.JSONException;
import org.json.JSONObject;

import android.app.Activity;
import android.os.Bundle;
import android.util.Log;
import android.view.View;
import android.view.View.OnClickListener;
import android.widget.Button;
import android.widget.EditText;
import android.widget.TextView;

public class GoogleSearch extends Activity {
    /** called when the activity is first created */
    TextView tv1;
    EditText ed1;
    Button bt1;
    static String url =
"http://ajax.googleapis.com/ajax/services/search/web?v=1.0&q=";

    @Override
    public void onCreate(Bundle savedInstanceState) {
        super.onCreate(savedInstanceState);
        setContentView(R.layout.main);
        tv1 = (TextView) this.findViewById(R.id.display);
        ed1 = (EditText) this.findViewById(R.id.editText);
        bt1 = (Button) this.findViewById(R.id.submit);

        bt1.setOnClickListener(new OnClickListener() {
            public void onClick(View view) {
                if(ed1.getText().toString()!=null) {
                    try{
                        processResponse(
                          searchRequest(ed1.getText().toString()));
                    } catch(Exception e) {
                        Log.v("Exception Google search",
                              "Exception:"+e.getMessage());
                    }
                }
                ed1.setText("");
            }
        });
    }

    public String searchRequest(String searchString)
                            throws MalformedURLException, IOException {
        String newFeed=url+searchString;
        StringBuilder response = new StringBuilder();
        Log.v("gsearch","gsearch url:"+newFeed);
        URL url = new URL(newFeed);

        HttpURLConnection httpconn
                          = (HttpURLConnection) url.openConnection();
        if(httpconn.getResponseCode()==HttpURLConnection.HTTP_OK) {
            BufferedReader input = new BufferedReader(
                    new InputStreamReader(httpconn.getInputStream()),
                    8192);
            String strLine = null;
            while ((strLine = input.readLine()) != null) {
                response.append(strLine);
            }
            input.close();
```

```
        }
        return response.toString();
    }

    public void processResponse(String resp) throws IllegalStateException,
                    IOException, JSONException, NoSuchAlgorithmException {
        StringBuilder sb = new StringBuilder();
        Log.v("gsearch","gsearch result:"+resp);
        JSONObject mResponseObject = new JSONObject(resp);
        JSONObject responObject
                        = mResponseObject.getJSONObject("responseData");
        JSONArray array = responObject.getJSONArray("results");
        Log.v("gsearch","number of results:"+array.length());
        for(int i = 0; i<array.length(); i++) {
            Log.v("result",i+"] "+array.get(i).toString());
            String title = array.getJSONObject(i).getString("title");
            String urllink = array.getJSONObject(i)
                                .getString("visibleUrl");
            sb.append(title);
            sb.append("\n");
            sb.append(urllink);
            sb.append("\n");
        }
        tv1.setText(sb.toString());
    }
}
```

要理解解析工作的具体机制，需要了解 JSON 数据结构。在本例中，Google REST API 通过 `JSONArray` 提供了所有的搜索结果。图 10-4 显示了对 NBA 的搜索结果。

图 10-4　通过 Google REST API 查询得到的搜索结果

注意，本技巧只能在早于 API Level 11 的 Android 项目上运行，这要归因于在主线程上运行的网络请求。在下个技巧中，会使用 AsyncTask 来修正抛出的 NetworkOnMainThreadException。

## 技巧 96：使用 HTTP POST 请求

有时需要从因特网检索原始二进制数据，比如图像、视频和音频文件。这可以使用 setRequestMethod() 方法，通过 HTTP POST 协议来实现，例如：

```
httpconn.setRequestMethod(POST);
```

通过因特网访问数据可能很费时，且有不可预测性。因此，只要想获取网络数据，就应当生成一个独立的线程。

除第 3 章中给出的方法外，还有一个名为 AsyncTask 的 Android 内置类可以执行后台操作并将结果发布到 UI 线程上，而不需要直接操作线程或 Handler。基于 AsyncTask，可以异步地实现 POST 方法，所需代码如下：

```
private class MyGoogleSearch extends AsyncTask<String, Integer, String> {
    protected String doInBackground(String... searchKey) {

        String key = searchKey[0];

        try {
            return searchRequest(key);
        } catch(Exception e) {
            Log.v("Exception Google search",
                  "Exception:"+e.getMessage());
            return "";
        }
    }

    protected void onPostExecute(String result) {
        try {
            processResponse(result);
        } catch(Exception e) {
            Log.v("Exception Google search",
                  "Exception:"+e.getMessage());
        }
    }
}
```

可以把这段代码添加到代码清单 10-8 给出的 **GoogleSearch.java** 的末尾。只要对按钮的 OnClickListener 中的代码进行如下改动，即可以让修改后的程序产生与原先相同的结果：

```
new MyGoogleSearch().execute(ed1.getText().toString());
```

## 技巧 97：使用 WebView

WebView 对于显示半定期性变化的内容，或者在修改时不需强制应用程序进行更新的数据很有用处。WebView 还为 Web 应用程序访问某些客户端一侧 Android 系统特性提供了可能，比如使用 Toast 消息系统。

要为应用程序添加 WebView，需要把下面的代码添加进布局 XML：

```
<WebView xmlns:android="http://schemas.android.com/apk/res/android"
  android:id="@+id/webview"
  android:layout_width="match_parent"
  android:layout_height="match_parent" />
```

还必须向应用程序的 manifest 添加如下权限：

```
<uses-permission android:name="android.permission.INTERNET" />
```

要创建一个不带任何用户交互的简单页面，可向主 Activity 的 `onCreate()` 方法中添加如下代码：

```
WebView myWebView = (WebView) findViewById(R.id.webview);
myWebView.loadUrl("http://www.example.com/");
```

若要对 WebView 内的页面启用 JavaScript，需要修改 WebSettings。可使用下面的代码来实现：

```
WebSettings webSettings = myWebView.getSettings();
webSettings.setJavaScriptEnabled(true);
```

要在 JavaScript 中触发本地方法，需要创建一个可以作为接口的类。代码清单 10-9 给出了整合了前面各部分功能的完整的 Activity。

**代码清单 10-9 src/com/cookbook/viewtoaweb/MainActivity.java**

```
package com.cookbook.viewtoaweb;

import android.app.Activity;
import android.content.Context;
import android.os.Bundle;
import android.webkit.JavascriptInterface;
import android.webkit.WebSettings;
import android.webkit.WebView;
import android.widget.Toast;

public class MainActivity extends Activity {

    @Override
    protected void onCreate(Bundle savedInstanceState) {
      super.onCreate(savedInstanceState);
      setContentView(R.layout.activity_main);

      WebView myWebView = (WebView) findViewById(R.id.webview);
      WebSettings webSettings = myWebView.getSettings();
      webSettings.setJavaScriptEnabled(true);
      myWebView.addJavascriptInterface(new WebAppInterface(this), "Android");
      myWebView.loadUrl("http://www.devcannon.com/androidcookbook/chapter10/webview/");
    }

    public class WebAppInterface {
        Context context;

        WebAppInterface(Context c) {
            context = c;
        }

        @JavascriptInterface
        public void triggerToast(String toast) {
```

```
            Toast.makeText(context, toast, Toast.LENGTH_SHORT).show();
        }
    }
}
```

下面的 HTML 用来触发代码清单 10-9 中的相关代码：

```
<input type="text" name="toastText" id="toastText" />
<button id="btn" onClick="androidToast()">Toast it</button>
```

而下面的 JavaScript 也是用来触发代码的：

```
function androidToast() {
  var input = document.getElementById('toastText');
  Android.triggerToast(input.value);
}
```

图 10-5 显示了 WebView 的运行效果，其中包含一个被浏览的页面开启的 Toast。

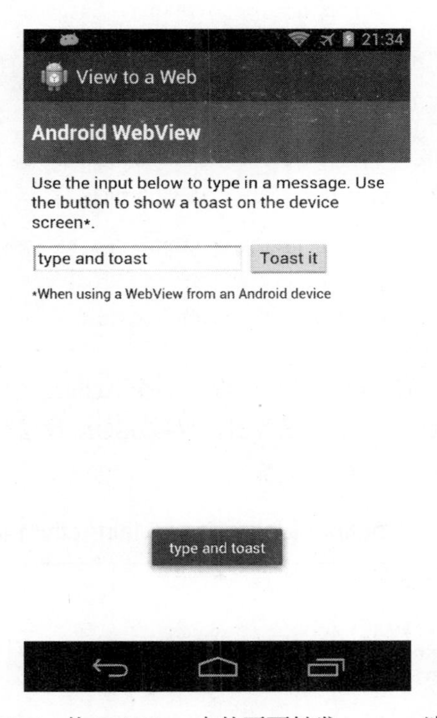


图 10-5　从 WebView 中的页面触发 Toast 消息

# 技巧 98：解析 JSON

JSON 是非常流行的数据传输格式，特别是对 Web 服务而言。Android 在 org.json 包中包含有一系列的类，把它们导入到程序中就可以操作 JSON 数据。

要实现解析，首先要创建一个 JSON 对象，可以像这样来实现：

```
private JSONObject jsonObject;
```

当然还需要 JSON 格式的数据。下面的代码创建了一个包含 JSON 数据的字符串：

```
private String jsonString =
"{\"item\":{\"name\":\"myName\",\"numbers\":[{\"id\":\"1\"},{\"id\":\"2\"}]}}";
```

由于字符串并不是一个 JSON 对象，因此还要创建一个 JSON 对象来转存字符串中的值。实现如下：

```
jsonObject = new JSONObject(jsonString);
```

这样就有一个可以操作的对象，可以从中取得数据。如果试图使用 `getString()` 方法从 `jsonObject` 中的“对象”里抽取数据，则会抛出 `JSONException` 异常。这是因为 `jsonObject` 的内容并非字符串形式。要抽取特定的值，需要建立另外一个对象来包含所需的值，就像这样：

```
JSONObject itemObject = jsonObject.getJSONObject("item");
```

之后就可以获取`"name"`的值了，语句如下：

```
String jsonName = itemObject.getString("name");
```

可以使用一个循环来获取存储在 `jsonObject` 的`"numbers"`字段中的信息。具体的实现是创建一个 `JSONArray` 对象，对其进行循环，就像下面这样：

```
JSONArray numbersArray = itemObject.getJSONArray("numbers");

for(int i = 0;i < numbersArray.length();i++){
  numbersArray.getJSONObject(i).getString("id");
}
```

代码清单 10-10 显示了如何把解析过程整合进一个 Activity 中，并将结果输出到 `TextView` 中。注意，从远程位置，比如一个 Web 服务器，抽取 JSON 数据时必须使用一个独立的类或者一个 `AsyncTask`，以避免主 UI 线程被阻塞。

**代码清单 10-10　src/com/cookbook/parsejson/MainActivity.java**

```
package com.cookbook.parsejson;

import org.json.JSONArray;
import org.json.JSONException;
import org.json.JSONObject;

import android.app.Activity;
import android.os.Bundle;
import android.widget.TextView;

public class MainActivity extends Activity {

  TextView tv;
  private JSONObject jsonObject;
  private String jsonString =
  "{\"item\":{\"name\":\"myName\",\"numbers\":[{\"id\":\"1\"},{\"id\":\"2\"}]}}";

  @Override
```

```
    protected void onCreate(Bundle savedInstanceState) {
      super.onCreate(savedInstanceState);
      setContentView(R.layout.activity_main);
      tv = (TextView) findViewById(R.id.tv_main);

      try {
        jsonObject = new JSONObject(jsonString);
        JSONObject itemObject = jsonObject.getJSONObject("item");
        String jsonName = "name: " +itemObject.getString("name");
        JSONArray numbersArray = itemObject.getJSONArray("numbers");
        String jsonIds = "";

        for(int i = 0;i < numbersArray.length();i++){
          jsonIds += "id: " +
              numbersArray.getJSONObject(i).getString("id").toString() + "\n";
        }

          tv.setText(jsonName+"\n"+jsonIds);

        } catch (JSONException e) {
          e.printStackTrace();
        }
      }
  }
```

## 技巧 99：解析 XML

Android 的官方文档建议使用 XmlPullParser 来解析 XML 数据。你可以使用任何喜欢的方法来获取 XML 数据，但在本技巧中只使用一个简单的单节点 XML 字符串。代码清单 10-11 给出了一个 Activity，能够把 XML 文档的读取过程（包括节点和文本值）显示到一个 TextView 上。

XML 数据要被逐行处理，通过 next() 方法移动到下一行。要解析 XML 数据中特定的节点，需要在 while 循环中为其添加专门的 if else 语句。

**代码清单 10-11　src/com/cookbook/parsexml/MainActivity.java**

```
package com.cookbook.parsexml;

import java.io.IOException;
import java.io.StringReader;

import org.xmlpull.v1.XmlPullParser;
import org.xmlpull.v1.XmlPullParserException;
import org.xmlpull.v1.XmlPullParserFactory;

import android.app.Activity;
import android.os.Bundle;
import android.widget.TextView;

public class MainActivity extends Activity {

  TextView tv;

  @Override
  protected void onCreate(Bundle savedInstanceState) {
    super.onCreate(savedInstanceState);
    setContentView(R.layout.activity_main);
```

```
        tv = (TextView) findViewById(R.id.tv_main);
        String xmlOut = "";
        XmlPullParserFactory factory = null;
        try {
          factory = XmlPullParserFactory.newInstance();
        } catch (XmlPullParserException e) {
          e.printStackTrace();
        }
        factory.setNamespaceAware(true);
        XmlPullParser xpp = null;
        try {
          xpp = factory.newPullParser();
        } catch (XmlPullParserException e) {
          e.printStackTrace();
        }

        try {
          xpp.setInput(new StringReader("<node>This is some text</node>"));
        } catch (XmlPullParserException e) {
          e.printStackTrace();
        }

        int eventType = 0;
        try {
          eventType = xpp.getEventType();
        } catch (XmlPullParserException e) {
          e.printStackTrace();
        }

        while (eventType != XmlPullParser.END_DOCUMENT) {
          if(eventType == XmlPullParser.START_DOCUMENT) {
            xmlOut += "Start of XML Document";
          } else if (eventType == XmlPullParser.START_TAG) {
            xmlOut += "\nStart of tag: "+xpp.getName();
          } else if (eventType == XmlPullParser.END_TAG) {
            xmlOut += "\nEnd of tag: "+xpp.getName();
          } else if (eventType == XmlPullParser.TEXT) {
            xmlOut += "\nText: "+xpp.getText();
          }
          try {
            eventType = xpp.next();
          } catch (XmlPullParserException e) {
            e.printStackTrace();
          } catch (IOException e) {
            e.printStackTrace();
          }
        }
        xmlOut += "\nEnd of XML Document";

        tv.setText(xmlOut);
      }
    }
```

## 10.4 社交网络

Twitter 是一个社交网络和微博服务，用户用它可以发送和阅读被称为 tweet 的消息。Twitter 被描述为“因特网上的短消息”，的确，每条 tweet 不能超过 140 个字符（不过其中的链接会被转换为更短的形式，且并不计算在 140 个字符的限制内）。Twitter 用户可以关注（follow）其他人的 tweet，也可以被其他人所关注。

## 技巧 100：读取所有者设定档

从 API Level 14（Ice Cream Sandwich）起，开发者能够访问所有者设定档。这是一种特殊的通讯录，其中存有 RawContent 数据。要读取某个设备的所有者设定档，必须为 **AndroidManifest.xml** 添加以下权限：

```
<uses-permission android:name="android.permission.READ_PROFILE" />
```

下面的代码能够访问设定档数据：

```
// sets the columns to retrieve for the owner profile - RawContact data
String[] mProjection = new String[]
    {
        Profile._ID,
        Profile.DISPLAY_NAME_PRIMARY,
        Profile.LOOKUP_KEY,
        Profile.PHOTO_THUMBNAIL_URI
    };

// retrieves the profile from the Contacts Provider
Cursor mProfileCursor =
    getContentResolver().query(Profile.CONTENT_URI,mProjection,null,null,null);
// Set the cursor to the first entry (instead of -1)
boolean b = mProfileCursor.moveToFirst();
for(int i = 0, length = mProjection.length;i < length;i++) {
    System.out.println("*** " +
        mProfileCursor.getString(mProfileCursor.getColumnIndex(mProjection[i])));
}
```

注意，代码中用到 System.out.println()的地方就是可以插入处理设定档信息逻辑的地方。另一点值得注意的是，虽然代码中并没有来自 Log.*的方法，仍然可以用 LogCat 来显示输出。

## 技巧 101：与 Twitter 集成

有一些来自第三方的库，能够帮助我们将 Twitter 集成到 Android 应用程序中（在 http://dev.twitter.com/pages/libraries#java 上可以找到）。

- Yusuke Yamamoto 开发的 Twitter4J：这是个针对 Twitter API 的、开源的、Maven[①]化的、并可安全使用 Google App 引擎的 Java 库，在 BSD 许可证下发布。
- Pablo Fernandez 开发的 Scribe：面向 Java 的 OAuth[②]模块，已 Maven 化，可用在 Facebook、LinkedIn、Twitter、Evernote 和 Vimeo 等多个平台上。

本技巧使用的是 Yusuke Yamamoto 开发的 Twitter4J 库，该库的文档可以在 http://twitter4j.org/en/javadoc/overview-summary.html 上看到。本技巧允许用户借助 OAuth 登录到 Twitter，并生成一条 tweet。

① Maven 是一种采用 Java 编写的开源项目管理工具，第 6 章曾经提到过。——译者注

② OAuth 是一种开放式协议，使用户可以通过一种简单和标准化的方法，从 Web、移动或桌面应用程序那里进行安全认证（译自 OAuth 官网）。——译者注

Twitter 对其验证系统进行了修改，现在需要应用程序进行注册，以访问公共提要（public feed）。首先，应用程序需要在 https://dev.twitter.com/apps/new 进行注册。注册过程中会生成 OAuth 公钥和私钥，本技巧要用到它们，所以请把它们记下来。

因为本应用程序要访问因特网，所以需要 INTERNET 权限。另外还要检查确保设备连接到了网络，所以 ACCESS_NETWORK_STATE 权限也是需要的。编辑 **AndroidManifest.xml** 添加这两个权限，如代码清单 10-12 所示。

**代码清单 10-12　AndroidManifest.xml**

```
<?xml version="1.0" encoding="utf-8"?>
<manifest xmlns:android="http://schemas.android.com/apk/res/android"
    package="com.cookbook.tcookbook"
    android:versionCode="1"
    android:versionName="1.0" >

    <uses-sdk
        android:minSdkVersion="9"
        android:targetSdkVersion="17" />

    <uses-permission android:name="android.permission.INTERNET" />
    <uses-permission android:name="android.permission.ACCESS_NETWORK_STATE" />

    <application
        android:allowBackup="true"
        android:icon="@drawable/ic_launcher"
        android:label="@string/app_name"
        android:theme="@style/AppTheme" >
        <activity
            android:name="com.cookbook.tcookbook.MainActivity"
            android:label="@string/app_name" >
            <intent-filter>
                <action android:name="android.intent.action.MAIN" />

                <category android:name="android.intent.category.LAUNCHER" />
            </intent-filter>
            <intent-filter>
                <action android:name="android.intent.action.VIEW" />
                <category android:name="android.intent.category.DEFAULT" />
                <category android:name="android.intent.category.BROWSABLE" />
                <data android:scheme="oauth" android:host="tcookbook"/>
            </intent-filter>
        </activity>
    </application>
</manifest>
```

至于应用程序的布局，有关的一切都被放进了 **activity_main.xml** 文件中。该文件包含一个页面装载时可见的按钮以及若干其他按钮、TextView 及 EditText 微件。注意，其中的某些要通过 android:visibility="gone"属性进行隐藏。代码清单 10-13 给出了 **activity_main.xml** 文件的内容。

**代码清单 10-13　res/layout/activity_main.xml**

```
<LinearLayout xmlns:android="http://schemas.android.com/apk/res/android"
    xmlns:tools="http://schemas.android.com/tools"
```

```
        android:layout_width="match_parent"
        android:layout_height="match_parent"
        android:orientation="vertical" >
        tools:context=".MainActivity" >

        <Button android:id="@+id/btnLoginTwitter"
            android:layout_width="match_parent"
            android:layout_height="wrap_content"
            android:text="Login with OAuth"
            android:layout_marginLeft="10dip"
            android:layout_marginRight="10dip"
            android:layout_marginTop="30dip"/>

        <TextView android:id="@+id/lblUserName"
            android:layout_width="match_parent"
            android:layout_height="wrap_content"
            android:padding="10dip"
            android:layout_marginTop="30dip"/>

        <TextView android:id="@+id/lblUpdate"
            android:text="Enter Your Tweet:"
            android:layout_width="match_parent"
            android:layout_height="wrap_content"
            android:layout_marginLeft="10dip"
            android:layout_marginRight="10dip"
            android:visibility="gone"/>

        <EditText android:id="@+id/txtUpdateStatus"
            android:layout_width="match_parent"
            android:layout_height="wrap_content"
            android:layout_margin="10dip"
            android:visibility="gone"/>

        <Button android:id="@+id/btnUpdateStatus"
            android:layout_width="match_parent"
            android:layout_height="wrap_content"
            android:text="Tweet it!"
            android:layout_marginLeft="10dip"
            android:layout_marginRight="10dip"
            android:visibility="gone"/>

        <Button android:id="@+id/btnLogoutTwitter"
            android:layout_width="match_parent"
            android:layout_height="wrap_content"
            android:text="Logout/invalidate OAuth"
            android:layout_marginLeft="10dip"
            android:layout_marginRight="10dip"
            android:layout_marginTop="50dip"
            android:visibility="gone"/>
</LinearLayout>
```

应用程序使用了一个 Activity，其中用到了两个类：一个用于连接检测，另一个会在使用错误的应用程序 OAuth 密钥时显示警告信息。

在主 Activity 中建立了若干常量。其中包含 OAuth 使用者关键字（consumer key）和使用者密码（consumer secret）。运行一个连接检查以确定用户是否能连上 Twitter。还注册了若干 `OnClickListener` 类，用于在按钮被点击时触发诸如登入、登出、更新一类的逻辑。

由于 Twitter 会为用户处理验证，传回的信息会被存储在应用程序偏好中，并在用户试图登录到应用程序时被再次检查。还要使用 AsyncTask 把所有创建的 tweet 移到一个后台线程中。

代码清单 10-14 给出了 Activity 的完整内容。

**代码清单 10-14　src/com/cookbook/tcookbook/MainActivity.java**

```java
package com.cookbook.tcookbook;

import twitter4j.Twitter;
import twitter4j.TwitterException;
import twitter4j.TwitterFactory;
import twitter4j.User;
import twitter4j.auth.AccessToken;
import twitter4j.auth.RequestToken;
import twitter4j.conf.Configuration;
import twitter4j.conf.ConfigurationBuilder;
import android.app.Activity;
import android.app.ProgressDialog;
import android.content.Intent;
import android.content.SharedPreferences;
import android.content.SharedPreferences.Editor;
import android.content.pm.ActivityInfo;
import android.net.Uri;
import android.os.AsyncTask;
import android.os.Build;
import android.os.Bundle;
import android.os.StrictMode;
import android.text.Html;
import android.util.Log;
import android.view.View;
import android.widget.Button;
import android.widget.EditText;
import android.widget.TextView;
import android.widget.Toast;

public class MainActivity extends Activity {

    // Replace the following value with the Consumer key
    static String TWITTER_CONSUMER_KEY = "01189998819991197253";
    // Replace the following value with the Consumer secret
    static String TWITTER_CONSUMER_SECRET =
        "616C6C20796F757220626173652061726520626656C6F6E6720746F207573";

    static String PREFERENCE_NAME = "twitter _ oauth";
    static final String PREF_KEY_OAUTH_TOKEN = "oauth_token";
    static final String PREF_KEY_OAUTH_SECRET = "oauth_token_secret";
    static final String PREF_KEY_TWITTER_LOGIN = "isTwitterLoggedIn";

    static final String TWITTER_CALLBACK_URL = "oauth://tcookbook";

    static final String URL_TWITTER_AUTH = "auth_url";
    static final String URL_TWITTER_OAUTH_VERIFIER = "oauth_verifier";
    static final String URL_TWITTER_OAUTH_TOKEN = "oauth_token";

    Button btnLoginTwitter;
    Button btnUpdateStatus;
    Button btnLogoutTwitter;
    EditText txtUpdate;
    TextView lblUpdate;
    TextView lblUserName;
```

```
ProgressDialog pDialog;

private static Twitter twitter;
private static RequestToken requestToken;

private static SharedPreferences mSharedPreferences;

private ConnectionDetector cd;

AlertDialogManager adm = new AlertDialogManager();

@Override
public void onCreate(Bundle savedInstanceState) {
  super.onCreate(savedInstanceState);
  setContentView(R.layout.activity_main);
  // used for Android 2.3+
  if (Build.VERSION.SDK_INT > Build.VERSION_CODES_GINGERBREAD) {
    StrictMode.ThreadPolicy policy =
        new StrictMode.ThreadPolicy.Builder().permitAll().build();
    StrictMode.setThreadPolicy(policy);
  }

  setRequestedOrientation(ActivityInfo.SCREEN_ORIENTATION_PORTRAIT);

  cd = new ConnectionDetector(getApplicationContext());

  if (!cd.isConnectingToInternet()) {
    adm.showAlertDialog(MainActivity.this, "Internet Connection Error",
        "Please connect to working Internet connection", false);
    return;
  }

  if(TWITTER_CONSUMER_KEY.trim().length() == 0 ||
      TWITTER_CONSUMER_SECRET.trim().length() == 0){
    adm.showAlertDialog(MainActivity.this,
        "Twitter OAuth tokens",
        "Please set your Twitter OAuth tokens first!", false);
    return;
  }

  btnLoginTwitter = (Button) findViewById(R.id.btnLoginTwitter);
  btnUpdateStatus = (Button) findViewById(R.id.btnUpdateStatus);
  btnLogoutTwitter = (Button) findViewById(R.id.btnLogoutTwitter);
  txtUpdate = (EditText) findViewById(R.id.txtUpdateStatus);
  lblUpdate = (TextView) findViewById(R.id.lblUpdate);
  lblUserName = (TextView) findViewById(R.id.lblUserName);

  mSharedPreferences = getApplicationContext().getSharedPreferences("MyPref", 0);

  btnLoginTwitter.setOnClickListener(new View.OnClickListener() {
    @Override
    public void onClick(View arg0) {
      // Call login Twitter function
      loginToTwitter();
    }
  });

  btnUpdateStatus.setOnClickListener(new View.OnClickListener() {
    @Override
    public void onClick(View v) {
      String status = txtUpdate.getText().toString();
```

```
            if (status.trim().length() > 0) {
                new updateTwitterStatus().execute(status);
            } else {
                Toast.makeText(getApplicationContext(),
                    "Please enter status message", Toast.LENGTH_SHORT).show();
            }
        }
    });

    btnLogoutTwitter.setOnClickListener(new View.OnClickListener() {
        @Override
        public void onClick(View arg0) {
            // Call logout Twitter function
            logoutFromTwitter();
        }
    });

    if (!isTwitterLoggedInAlready()) {
        Uri uri = getIntent().getData();
        if (uri != null && uri.toString().startsWith(TWITTER_CALLBACK_URL)) {
            String verifier = uri.getQueryParameter(URL_TWITTER_OAUTH_VERIFIER);

            try {
                AccessToken accessToken = twitter.getOAuthAccessToken(requestToken,
➥verifier);
                Editor e = mSharedPreferences.edit();

                e.putString(PREF_KEY_OAUTH_TOKEN, accessToken.getToken());
                e.putString(PREF_KEY_OAUTH_SECRET,accessToken.getTokenSecret());
                e.putBoolean(PREF_KEY_TWITTER_LOGIN, true);
                e.commit();

//              Log.e("Twitter OAuth Token", "> " + accessToken.getToken());

                btnLoginTwitter.setVisibility(View.GONE);

                lblUpdate.setVisibility(View.VISIBLE);
                txtUpdate.setVisibility(View.VISIBLE);
                btnUpdateStatus.setVisibility(View.VISIBLE);
                btnLogoutTwitter.setVisibility(View.VISIBLE);

                long userID = accessToken.getUserId();
                User user = twitter.showUser(userID);
                String username = user.getName();

                lblUserName.setText(Html.fromHtml("<b>Welcome " + username + "</b>"));
            } catch (Exception e) {
                Log.e("***Twitter Login Error: ",e.getMessage());
            }
        }
    }
}

private void loginToTwitter() {
    if (!isTwitterLoggedInAlready()) {
        ConfigurationBuilder builder = new ConfigurationBuilder();
        builder.setOAuthConsumerKey(TWITTER_CONSUMER_KEY);
        builder.setOAuthConsumerSecret(TWITTER_CONSUMER_SECRET);
        Configuration configuration = builder.build();

        TwitterFactory factory = new TwitterFactory(configuration);
        twitter = factory.getInstance();
```

```
        if(!(Build.VERSION.SDK_INT >= Build.VERSION_CODES.HONEYCOMB)) {
          try {
             requestToken = twitter.getOAuthRequestToken(TWITTER_CALLBACK_URL);
             this.startActivity(new Intent(Intent.ACTION_VIEW,
                  Uri.parse(requestToken.getAuthenticationURL())));
          } catch (TwitterException e) {
               e.printStackTrace();
             }
             } else {
             new Thread(new Runnable() {
             public void run() {
               try {
                  requestToken = twitter.getOAuthRequestToken(TWITTER_CALLBACK_URL);
                  MainActivity.this.startActivity(new Intent(Intent.ACTION_VIEW,
                      Uri.parse(requestToken.getAuthenticationURL())));
               } catch (TwitterException e) {
                  e.printStackTrace();
               }
             }
          }).start();
        }
      } else {
        Toast.makeText(getApplicationContext(),"Already logged into Twitter",
            Toast.LENGTH_LONG).show();
      }
    }

    class updateTwitterStatus extends AsyncTask<String, String, String> {
      @Override
      protected void onPreExecute() {
        super.onPreExecute();
        pDialog = new ProgressDialog(MainActivity.this);
        pDialog.setMessage("Updating to Twitter...");
        pDialog.setIndeterminate(false);
        pDialog.setCancelable(false);
        pDialog.show();
      }
      protected String doInBackground(String... args) {
//      Log.d("*** Text Value of Tweet: ",args[0]);
        String status = args[0];
        try {
        ConfigurationBuilder builder = new ConfigurationBuilder();
        builder.setOAuthConsumerKey(TWITTER_CONSUMER_KEY);
        builder.setOAuthConsumerSecret(TWITTER_CONSUMER_SECRET);

        String access_token =
           mSharedPreferences.getString(PREF_KEY_OAUTH_TOKEN, "");
        String access_token_secret =
          mSharedPreferences.getString(PREF_KEY_OAUTH_SECRET, "");

        AccessToken accessToken =
          new AccessToken(access_token, access_token_secret);
        Twitter twitter =
          new TwitterFactory(builder.build()).getInstance(accessToken);

        twitter4j.Status response = twitter.updateStatus(status);

//      Log.d("*** Update Status: ",response.getText());
        } catch (TwitterException e) {
        Log.d("*** Twitter Update Error: ", e.getMessage());
      }
```

```
            return null;
        }

        protected void onPostExecute(String file_url) {
            pDialog.dismiss();
            runOnUiThread(new Runnable() {
                @Override
                public void run() {
                    Toast.makeText(getApplicationContext(),
                            "Status tweeted successfully", Toast.LENGTH_SHORT).show();
                    txtUpdate.setText("");
                }
            });
        }
    }

    private void logoutFromTwitter() {
        Editor e = mSharedPreferences.edit();
        e.remove(PREF_KEY_OAUTH_TOKEN);
        e.remove(PREF_KEY_OAUTH_SECRET);
        e.remove(PREF_KEY_TWITTER_LOGIN);
        e.commit();

        btnLogoutTwitter.setVisibility(View.GONE);
        btnUpdateStatus.setVisibility(View.GONE);
        txtUpdate.setVisibility(View.GONE);
        lblUpdate.setVisibility(View.GONE);
        lblUserName.setText("");
        lblUserName.setVisibility(View.GONE);

        btnLoginTwitter.setVisibility(View.VISIBLE);
    }

    private boolean isTwitterLoggedInAlready() {
        return mSharedPreferences.getBoolean(PREF_KEY_TWITTER_LOGIN, false);
    }

    protected void onResume() {
        super.onResume();
    }
}
```

关于 Twitter4j 使用方法的更多信息，可以在下面的资源中找到。

- www.androidhive.info/2012/09/android-twitter-oauth-connect-tutorial/，创建者 Ravi Tamada。
- http://blog.doityourselfandroid.com/2011/08/08/improved-twitter-oauth-android/，创建者 Do-it-yourself Android。
- http://davidcrowley.me/?p=410，创建者 David Crowley。
- https://tutsplus.com/tutorials/?q=true&filter_topic=90，创建者 Sue Smith。
- http://blog.blundell-apps.com/sending-a-tweet/，创建者 Blundell。

## 技巧 102：与 Facebook 集成

Facebook 近几年日新月异，始终是社交网站领域的领跑者之一。Facebook 团队最近做了一件好事，修整了他们的文档以便开发者使用。官方文档可以在 https://developers.facebook.com/docs/getting-started/facebook-sdk-for-android/3.0/上找到。

要进行 Facebook 开发，首先要下载 Facebook SDK 和 Facebook Android 包（APK），地址为 https://developers.facebook.com/resources/facebook-android-sdk-3.0.zip。APK 提供了一种不使用 WebView 就能验证的手段。如果 Facebook 应用程序已经被安装到手机中，就不需要再安装 APK 文件了。

下一步，把 Facebook SDK 作为一个库项目添加到 Eclipse 安装中去。做法是先选择 **File→Import**，再选择 **General→Existing Projects into Workspace**。注意，Facebook 警告用户不要使用“Copy projects into work space”选项，因为这可能会建立不正确的文件系统，并导致 SDK 功能异常。

导入了 Facebook SDK 之后，就可以试验其中的范例项目。注意，大部分项目需要生成一个用于为应用程序签名的 key hash（哈希值），开发者也可以将其添加到自己的 Facebook 开发者设定档中，从而能快速访问 SDK 项目。

密钥通过 Java 所带的 `keytool` 工具来生成。打开终端或命令行，键入如下命令来生成密钥。

对于 OS X：

```
keytool -exportcert -alias androiddebugkey -keystore ~/.android/debug.keystore |
[ccc]openssl sha1 -binary | openssl base64
```

对于 Windows：

```
keytool -exportcert -alias androiddebugkey -keystore %HOMEPATH%\.android\debug.
keystore [ccc]| openssl sha1 -binary | openssl base64
```

上面的命令应当作为完整的一行键入（尽管终端或命令行显示时可能会分成多行）。命令执行后，会显示输入密码的提示。键入的密码是 **android**。成功生成的密钥会显示在屏幕上。注意如果出现“`'keytool' is not recognized as an internal or external command...`”错误，请将当前目录跳转到 JRE 安装目录下的 **bin** 子目录中，并再次尝试。还有一种错误信息与上述类似，只是关键词变为“openssl”，如果出现该错误，请到 http://code.google.com/p/openssl-for-windows/下载 OpenSSL。如果仍有其他错误，请确保 **bin** 目录已被添加到系统路径之中，或者已使用正确路径来代替 **%HOMEPATH%**。

如果用于开发的电脑不止一台，需要分别为每台电脑生成一个哈希值，并添加到 https://developers.facebook.com/上的开发者设定档中。

完成上述过程之后，请研究范例应用程序，并用它们登录到 Facebook。其中名为 **HelloFacebookSample** 的示例项目演示了如何访问设定档、更新状态，以及上传照片。

创建于 Facebook 集成的应用程序的最后一步是创建一个 Facebook 应用，并通过已生成的哈希值将其与 Android 应用程序捆绑起来。这样就能实现集成，并使得用户在使用应用程序时能够完成身份验证。

开发者网站上给出了所有所需事项极好的分解说明。请一定要阅读官方 Scrumptious 教程，你可以在 http://developers.facebook.com/docs/tutorials/androidsdk/3.0/scrumptious/上找到它。

# 第 11 章

# 数据存储方法

复杂而健壮的 Android 应用程序通常需要某种类型的数据存储。开发者可以根据实际情形选择不同的数据存储方法。

- 对于轻量级应用场合（比如保存应用程序设置和 UI 状态）可选择 SharedPreferences。
- 对于更复杂的应用场合，比如保存应用程序记录，可使用内建 SQLite 数据库。
- 还可使用标准的 Java 平面（flat）文件[①]存储方法：InputFileStream 和 OutputFileStream。

以上这些内容本章都将涉及。本章还会讨论 ContentProvider，该 Android 组件用于在应用程序间分享数据。需要注意的是，我们在第 2 章中还讨论过另外一种基本的由 Android 系统管理的数据存储方法，即 onSaveInstanceState()和 onRestoreInstanceState()方法对。所有这些方法哪个最佳取决于实际情况，在本章的每个技巧中我们都将讨论此问题。

## 11.1 shared preference

SharedPreferences 是一个接口，应用程序可以使用它来快速高效地将数据存储为名字-值对（name-value pair）的形式，该形式与 bundle 类似。信息被存储在 Android 设备的一个 XML 文件中。例如，如果应用程序 com.cookbook.datastorage 创建了一个 shared preference，Android 系统会在**/data/data/com.cookbook.datastorage/shared_prefs** 目录下创建一个新的 XML 文件。

shared preference 通常用于保存应用程序设置，比如用户设置、主题及其他通用应用程序属性等。它也可以保存登录信息，比如用户名、密码、自动登录标志、记住用户标志等。shared preference 数据可以被创建它的应用程序的所有组件访问。

① 平面文件是去除了所有特定应用程序格式的电子记录，从而使数据元素可以迁移到其他的应用上进行处理。这种去除电子数据格式的模式可以避免因为硬件和专有软件的过时而导致数据丢失。——译者注

## 技巧 103：创建和检索 shared preference

Activity 中可以用 getPreferences()方法访问 shared preference，该方法指定了对默认 preference 文件的操作模式。如果需要多个 preference 文件，则可以用 getSharedPreferences()方法分别指定每个文件的操作模式。如果在数据目录下存在 shared preference XML 文件，则会将其打开；否则就会创建一个新文件。操作模式提供了对 preference 的不同类型的访问权限的控制。

- MODE_PRIVATE：只有发起调用的应用程序能够访问 XML 文件。
- MODE_WORLD_READABLE：所有应用程序都能读取 XML 文件。该设定在 API Level 17 后就不提倡使用了；可以使用 ContentProvider、BroadcastReceiver 或一个服务来代替。
- MODE_WORLD_WRITEABLE：所有应用程序都可以写入 XML 文件；可以使用 ContentProvider、BroadcastReceiver 或一个服务来代替。

检索到 SharedPreferences 对象之后，需要一个 Editor 对象，该对象使用 put()方法向 XML 文件中写入名字-值对。目前支持的数据类型有 5 种：int、long、float、String 和 boolean。以下代码显示了如何创建和存储 shared preference 数据：

```
SharedPreferences prefs = getSharedPreferences("myDataStorage",
                                               MODE_PRIVATE);
Editor mEditor = prefs.edit();
mEditor.putString("username","datastorageuser1");
mEditor.putString("password","password1234");
mEditor.apply();
```

注意，在使用 Android 2.3（Level 9）或以上版本上开发时，需要使用 apply()方法提交变更，这将触发一个 async 请求更新文件。Android 的早期版本则要使用 commit()方法。

下面的代码显示了如何检索 shared preference 数据：

```
SharedPreferences prefs = getSharedPreferences("myDataStorage",
                                               MODE_PRIVATE);
String username = prefs.getString("username", "");
String password = prefs.getString("password", "");
```

## 技巧 104：使用 preference 框架

Android 提供了一个标准化的框架，用于设置跨应用程序的 preference。该框架使用分类 preference 和不同屏幕将相关的设定分组。PreferenceCategory 用于将一组 preference 声明到特定分类下。PreferenceScreen 在一个新的屏幕上展现一组 preference。

本技巧使用了代码清单 11-1 中 XML 文件定义的 preference。PreferenceScreen 是根元素，包含两个 EditTextPreference 元素，分别用于存储用户名（username）和密码（password）。其他可用的元素还有 CheckBoxPreference、RingtonePreference 和 DialogPreference。接下来，Android 系统生成用于一个操作 preferenc 的 UI，如图 11-1 所示。这些 preference 被存储在 shared preference 中，这意味着可以通过调用 getPreferences()检索到它们。

**代码清单 11-1　res/xml/preferences.xml**

```
<?xml version="1.0" encoding="utf-8"?>
<PreferenceScreen xmlns:android="http://schemas.android.com/apk/res/android">
  <EditTextPreference
    android:title="User Name"
    android:key="username"
     android:summary="Please provide user name">
  </EditTextPreference>
  <EditTextPreference
    android:title="Password"
    android:password="true"
    android:key="password"
    android:summary="Please enter your password">
  </EditTextPreference>
</PreferenceScreen>
```

图 11-1　Android 系统通过 preference XML 文件生成的 preference UI

之后，程序中扩展了 PreferenceActivity 的 Activity 会调用 addPreferencesFromResource()方法，将上述 preference 包含进 Activity 中，如代码清单 11-2 所示。注意，如果在 API 级别 11 或更高版本下开发，必须使用 PreferenceFragment 来调用 addPreferencesFromResource()方法。

**代码清单 11-2　src/com/cookbook/datastorage/MyPreferences.java**

```
package com.cookbook.datastorage;

import android.os.Bundle;
import android.preference.PreferenceActivity;

public class MyPreferences extends PreferenceActivity {

  @Override
  public void onCreate(Bundle savedInstanceState) {
    super.onCreate(savedInstanceState);

    addPreferencesFromResource(R.xml.preferences);
  }
}
```

主 Activity 仅仅在需要时调用 PreferenceActivity（例如，当按下菜单键的时候）。代码清单 11-3 给出了一个简单的例子，在 Activity 启动时显示 preference。

**代码清单 11-3 src/com/cookbook/datastorage/DataStorage.java**

```
package com.cookbook.datastorage;

import android.app.Activity;
import android.content.Intent;
import android.os.Bundle;

public class DataStorage extends Activity {
    /** called when the activity is first created */
    @Override
    public void onCreate(Bundle savedInstanceState) {
        super.onCreate(savedInstanceState);
        setContentView(R.layout.main);
        Intent i = new Intent(this, MyPreferences.class);
        startActivity(i);
    }
}
```

**AndroidManifest.xml** 文件需要包含所有的 Activity，包括新建立的 PreferenceActivity，如代码清单 11-4 所示。这样就能够创建图 11-1 所示的 preference 界面了。

**代码清单 11-4 AndroidManifest.xml**

```
<?xml version="1.0" encoding="utf-8"?>
<manifest xmlns:android="http://schemas.android.com/apk/res/android"
      package="com.cookbook.datastorage"
      android:versionCode="1"
      android:versionName="1.0">
    <application android:icon="@drawable/icon" android:label="@string/app_name">
        <activity android:name=".DataStorage"
                  android:label="@string/app_name">
            <intent-filter>
                <action android:name="android.intent.action.MAIN" />
                <category android:name="android.intent.category.LAUNCHER"
            />
            </intent-filter>
        </activity>
        <activity android:name=".MyPreferences" />
    </application>
    <uses-sdk android:minSdkVersion="7" />
</manifest>
```

## 技巧 105：基于存储的数据改变用户界面

我们可以扩展上一技巧中的 DataStorage Activity，使其能在装载时检查 shared preference，并对程序行为进行相应的调整。在本技巧中，如果用户名和密码已经被保存在 SharedPreferences 文件中，则会显示一个登录界面。成功登录后，Activity 也可以成功地继续。如果文件中没有登录信息，则 Activity 不执行前述步骤而是直接继续。

可以将 **main.xml** 布局文件修改为如代码清单 11-5 所示的登录页面。该页面使用两个 EditText 对象，用于输入用户名和密码，这在第 5 章中曾涉及过。

**代码清单 11-5 res/layout/main.xml**

```
<?xml version="1.0" encoding="utf-8"?>
<LinearLayout xmlns:android="http://schemas.android.com/apk/res/android"
    android:orientation="vertical"
    android:layout_width="match_parent"
    android:layout_height="match_parent">
    <TextView
        android:layout_width="match_parent"
        android:layout_height="wrap_content"
        android:text="username"
    />
    <EditText
        android:id="@+id/usertext"
        android:layout_width="match_parent"
        android:layout_height="wrap_content"
    />
    <TextView
        android:layout_width="match_parent"
        android:layout_height="wrap_content"
        android:text="password"
    />
    <EditText
        android:id="@+id/passwordtext"
        android:layout_width="match_parent"
        android:layout_height="wrap_content"
        android:password="true"
    />
    <Button
        android:id="@+id/loginbutton"
        android:layout_width="wrap_content"
        android:layout_height="wrap_content"
        android:text="login"
        android:textSize="20dp"
    />
</LinearLayout>
```

作为主 Activity 的 DataStorage（如代码清单 11-6 所示）已被修改成首先从 SharedPreferences 实例中读取 username 和 password 数据。如果数据未设置，应用程序会直接启动 MyPreferences Activity（代码清单 11-2）来设置 preference。如果数据已设置，应用程序则会显示由 **main.xml** 定义的登录布局，效果见图 11-2。

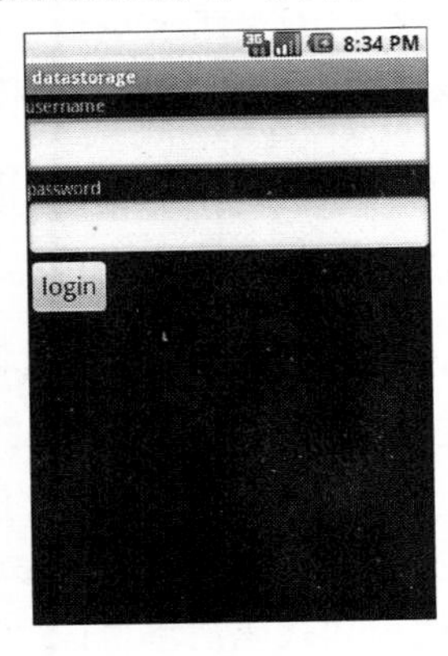


图 11-2 代码清单 11-5 描述的登录屏幕

图中的按钮拥有一个 onClickListener，用于检查登录信息是否与 SharedPreferences 中的用户名和密码相匹配。如果登录成功，应用程序就能够继续下一步，启动 MyPreference Activity。每次登录尝试后都会显示一条 Toast 信息，向用户表明登录是成功还是失败。

**代码清单 11-6 src/com/cookbook/datastorage/DataStorage.java**

```
package com.cookbook.datastorage;

import android.app.Activity;
import android.content.Intent;
import android.content.SharedPreferences;
import android.os.Bundle;
import android.preference.PreferenceManager;
import android.view.View;
import android.view.View.OnClickListener;
import android.widget.Button;
import android.widget.EditText;
import android.widget.Toast;

public class DataStorage extends Activity {
    SharedPreferences myprefs;
    EditText userET, passwordET;
    Button loginBT;
    @Override
    public void onCreate(Bundle savedInstanceState) {
        super.onCreate(savedInstanceState);
        myprefs = PreferenceManager.getDefaultSharedPreferences(this);
        final String username = myprefs.getString("username", null);
        final String password = myprefs.getString("password", null);
        if (username != null && password != null){
            setContentView(R.layout.main);
            userET = (EditText)findViewById(R.id.usertext);
            passwordET = (EditText)findViewById(R.id.passwordtext);
            loginBT = (Button)findViewById(R.id.loginbutton);
            loginBT.setOnClickListener(new OnClickListener() {
                public void onClick(View v) {
                    try {
                        if(username.equals(userET.getText().toString())
                            && password.equals(
                                    passwordET.getText().toString())) {
                            Toast.makeText(DataStorage.this,
                                        "login passed!!",
                                        Toast.LENGTH_SHORT).show();
                            Intent i = new Intent(DataStorage.this,
                                            MyPreferences.class);
                            startActivity(i);
                        } else {
                            Toast.makeText(DataStorage.this,
                                        "login failed!!",
                                        Toast.LENGTH_SHORT).show();
                        }
                    } catch (Exception e) {
                        e.printStackTrace();
                    }
                }
            });
        } else {
            Intent i = new Intent(this, MyPreferences.class);
            startActivity(i);
```

```
        }
    }
}
```

## 技巧 106：添加最终用户许可协议

我们在第 1 章曾提到过，在用户首次安装和运行应用时，往往有必要显示一个最终用户许可协议（EULA）。如果用户不接受该协议，下载的应用程序就不会运行。如果用户接受了协议，该协议就不会再次显示。

EULA 功能已经在代码清单 11-7 中实现为 `Eula` 类，并可在 Apache 许可证下公开使用。该类使用 `SharedPreferences`，通过 `PREFERENCE_EULA_ACCEPTED` 布尔量来判定之前是否接受或拒绝过 EULA。

**代码清单 11-7 src/com/cookbook/eula_example/Eula.java**

```
/*
 * Copyright (C) 2008 The Android Open Source Project
 *
 * Licensed under the Apache License, Version 2.0 (the "License");
 * you may not use this file except in compliance with the License.
 * You may obtain a copy of the License at
 *
 *      http://www.apache.org/licenses/LICENSE-2.0
 *
 * Unless required by applicable law or agreed to in writing, software
 * distributed under the License is distributed on an "AS IS" BASIS,
 * WITHOUT WARRANTIES OR CONDITIONS OF ANY KIND, either expressed or implied.
 * See the License for the specific language governing permissions and
 * limitations under the License.
 */

package com.cookbook.eula_example;

import android.app.Activity;
import android.app.AlertDialog;
import android.content.DialogInterface;
import android.content.SharedPreferences;

import java.io.IOException;
import java.io.BufferedReader;
import java.io.InputStreamReader;
import java.io.Closeable;

/**
 * displays a EULA ("End User License Agreement") that the user has to accept before
 * using the application
 */
class Eula {
    private static final String ASSET_EULA = "EULA";
    private static final String PREFERENCE_EULA_ACCEPTED = "eula.accepted";
    private static final String PREFERENCES_EULA = "eula";

    /**
     * callback to let the activity know when the user accepts the EULA
     */
```

```
static interface OnEulaAgreedTo {
    void onEulaAgreedTo();
}

/**
 * displays the EULA if necessary
 */
static boolean show(final Activity activity) {

    final SharedPreferences preferences =
                        activity.getSharedPreferences(
                                PREFERENCES_EULA, Activity.MODE_PRIVATE);
    //to test:
    //  preferences.edit()
    //      .putBoolean(PREFERENCE_EULA_ACCEPTED, false).commit();

    if (!preferences.getBoolean(PREFERENCE_EULA_ACCEPTED, false)) {
        final AlertDialog.Builder builder =
                        new AlertDialog.Builder(activity);
        builder.setTitle(R.string.eula_title);
        builder.setCancelable(true);
        builder.setPositiveButton(R.string.eula_accept,
                                new DialogInterface.OnClickListener() {
            public void onClick(DialogInterface dialog, int which) {
                accept(preferences);
                if (activity instanceof OnEulaAgreedTo) {
                    ((OnEulaAgreedTo) activity).onEulaAgreedTo();
                }
            }
        });
        builder.setNegativeButton(R.string.eula_refuse,
                                new DialogInterface.OnClickListener() {
            public void onClick(DialogInterface dialog, int which) {
                refuse(activity);
            }
        });
        builder.setOnCancelListener(
                                new DialogInterface.OnCancelListener() {
            public void onCancel(DialogInterface dialog) {
                refuse(activity);
            }
        });
        builder.setMessage(readEula(activity));
        builder.create().show();
        return false;
    }
    return true;
}

private static void accept(SharedPreferences preferences) {
    preferences.edit().putBoolean(PREFERENCE_EULA_ACCEPTED,
                                true).commit();
}

private static void refuse(Activity activity) {
    activity.finish();
}

private static CharSequence readEula(Activity activity) {
    BufferedReader in = null;
    try {
        in = new BufferedReader(new InputStreamReader(activity.getAssets().
```

```
➥open(ASSET_EULA)));
            String line;
            StringBuilder buffer = new StringBuilder();
            while ((line = in.readLine()) != null)
                buffer.append(line).append('\n');
            return buffer;
        } catch (IOException e) {
            return "";
        } finally {
            closeStream(in);
        }
    }

    /**
     * closes the specified stream
     */
    private static void closeStream(Closeable stream) {
        if (stream != null) {
            try {
                stream.close();
            } catch (IOException e) {
                // Ignore
            }
        }
    }
}
```

Eula 类需要按照下面的要求来自定义。

（1）EULA 包含的实际文本需要放入一个名为 **EULA** 的文本文件（在代码清单 11-7 中通过 ASSET_EULA 变量指定了它）中，并放到 Android 项目的 **assets/**目录下。该文件会被 Eula 类的 readEula()方法载入。

（2）需要为“接受协议”对话框指定几个字符串。这些字符串可以被收集到字符串资源文件中。代码清单 11-8 给出了字符串措辞的范例。

**代码清单 11-8　res/values/strings.xml**

```
<?xml version="1.0" encoding="utf-8"?>
<resources>
    <string name="hello">Welcome to MyApp</string>
    <string name="app_name">MyApp</string>
    <string name="eula_title">License Agreement</string>
    <string name="eula_accept">Accept</string>
    <string name="eula_refuse">Don\'t Accept</string>
</resources>
```

这样一来，只要在主 Activity 的 onCreate()方法中加入下面一行代码，任何应用程序都能够自动拥有 EULA 功能：

```
Eula.show(this);
```

## 11.2　SQLite 数据库

对于更复杂的数据结构，数据库提供了比平面文件更快捷和灵活的访问方法。Android 提供了名为 SQLite 的内置数据库，它提供了完整的关系型数据库的功能，可使用 SQL 命令。每个使用 SQLite 的应用程序都可拥有自己的数据库实例，该实例默认情况下只能被应用程序自己访问。数据库被储存在 Android 设备的**/data/data/<package_name>/databases** 文件夹下。可使用内容提供器在应用程序间分享数据库信息。使用 SQLite 的步骤如下。

（1）创建数据库。

（2）打开数据库。

（3）创建数据库表。

（4）创建数据集的插入接口。

（5）创建数据集的查询接口。

（6）关闭数据库。

下面的技巧给出了实现上述步骤的一般方法。

## 技巧 107：创建一个独立的数据库包

好的类模块结构对于较复杂的 Android 项目来说是必要的。在此，我们让数据库类自成一个包，名为 `com.cookbook.data`，这样重用起来就比较容易。包中包含三个类：`MyDB`、`MyDBhelper` 和 `Constants`。

`MyDB` 类如代码清单 11-9 所示。它包含一个 `SQLiteDatabase` 实例和一个 `MyDBhelper` 类（将在下文说明），并带有下列方法。

- `MyDB()`：初始化 `MyDBhelper` 实例（构造函数）。
- `open()`：使用 `MyDBhelper` 初始化 `SQLiteDatabase` 实例。这将会打开一个可写的数据连接。如果 SQLite 抛出了异常，将尝试获取一个可读的数据库来替代。
- `close()`：关闭数据库连接。
- `insertDiary()`：将日记（diary）条目保存到数据库中。首先将其保存为一个 `ContentValues` 实例的名字-值对，再将数据传递给 `SQLiteDatabase` 实例进行插入。
- `getDiaries()`：从数据库读取日记条目，把它们保存到 `Cursor` 类中，然后将它们返回。

**代码清单 11-9 src/com/cookbook/data/MyDB.java**

```
package com.cookbook.data;

import android.content.ContentValues;
import android.content.Context;
import android.database.Cursor;
import android.database.sqlite.SQLiteDatabase;
import android.database.sqlite.SQLiteException;
import android.util.Log;

public class MyDB {
    private SQLiteDatabase db;
    private final Context context;
    private final MyDBhelper dbhelper;
    public MyDB(Context c){
        context = c;
        dbhelper = new MyDBhelper(context, Constants.DATABASE_NAME, null,
```

```
                                   Constants.DATABASE_VERSION);
    }
    public void close()
    {
        db.close();
    }
    public void open() throws SQLiteException
    {
        try {
            db = dbhelper.getWriteableDatabase();
        } catch(SQLiteException ex) {
            Log.v("Open database exception caught", ex.getMessage());
            db = dbhelper.getReadableDatabase();
        }
    }
    public long insertDiary(String title, String content)
    {
        try{
            ContentValues newTaskValue = new ContentValues();
            newTaskValue.put(Constants.TITLE_NAME, title);
            newTaskValue.put(Constants.CONTENT_NAME, content);
            newTaskValue.put(Constants.DATE_NAME,
                            java.lang.System.currentTimeMillis());
            return db.insert(Constants.TABLE_NAME, null, newTaskValue);
        } catch(SQLiteException ex) {
            Log.v("Insert into database exception caught",
                ex.getMessage());
            return -1;
        }
    }
    public Cursor getDiaries()
    {
        Cursor c = db.query(Constants.TABLE_NAME, null, null, null, null, null, null);
        return c;
    }
}
```

MyDBhelper 类，如代码清单 11-10 所示，扩展了 SQLiteOpenHelper。SQLiteOpenHelper 框架提供了管理数据库的创建和升级的方法。数据库在构造方法 MyDBhelper() 中被初始化。初始化过程需要指定上下文和数据库名，用于在 **/data/data/com.cookbook.datastorage/databases** 目录下创建数据库文件；还需要指定数据库模式的版本，用于判定被调用的是 onCreate() 还是 onUpgrade() 方法。

数据库表可以在 onCreate() 方法中使用自定义的 SQL 命令添加，比如：

```
create table MyTable (key_id integer primary key autoincrement,
                     title text not null, content text not null,
                     recordDate long);
```

一旦数据库需要升级（比如用户下载了应用程序的一个新版本），数据库版本号的变化会导致 onUpgrade() 方法被调用。该方法可用于根据需要改变或删除表格，将数据库表格更新到新的模式。

**代码清单 11-10　src/com/cookbook/data/MyDBhelper.java**

```
package com.cookbook.data;

import android.content.Context;
import android.database.sqlite.SQLiteDatabase;
import android.database.sqlite.SQLiteException;
```

```
import android.database.sqlite.SQLiteOpenHelper;
import android.database.sqlite.SQLiteDatabase.CursorFactory;
import android.util.Log;

public class MyDBhelper extends SQLiteOpenHelper{
    private static final String CREATE_TABLE="create table "+
    Constants.TABLE_NAME+" ("+
    Constants.KEY_ID+" integer primary key autoincrement, "+
    Constants.TITLE_NAME+" text not null, "+
    Constants.CONTENT_NAME+" text not null, "+
    Constants.DATE_NAME+" long);";

    public MyDBhelper(Context context, String name, CursorFactory factory,
                      int version) {
        super(context, name, factory, version);
    }

    @Override
    public void onCreate(SQLiteDatabase db) {
        Log.v("MyDBhelper onCreate","Creating all the tables");
        try {
            db.execSQL(CREATE_TABLE);
        } catch(SQLiteException ex) {
            Log.v("Create table exception", ex.getMessage());
        }
    }

    @Override
    public void onUpgrade(SQLiteDatabase db, int oldVersion,
                          int newVersion) {
        Log.w("TaskDBAdapter", "Upgrading from version "+oldVersion
                               +" to "+newVersion
                               +", which will destroy all old data");
        db.execSQL("drop table if exists "+Constants.TABLE_NAME);
        onCreate(db);
    }
}
```

com.cookbook.data 包中的第三个文件是 Constants 类，如代码清单 11-11 所示。该类用于保存所有的 String 常量，因为它们在 MyDB 和 MyDBhelper 两个类中都要用到。

**代码清单 11-11　src/com/cookbook/data/Constants.java**

```
package com.cookbook.data;


public class Constants {
    public static final String DATABASE_NAME="datastorage";
    public static final int DATABASE_VERSION=1;
    public static final String TABLE_NAME="diaries";
    public static final String TITLE_NAME="title";
    public static final String CONTENT_NAME="content";
    public static final String DATE_NAME="recordDate";
    public static final String KEY_ID="_id";
}
```

## 技巧 108：使用独立的数据库包

本技巧使用上一个技巧建立的数据库包来演示 SQLite 数据存储。这里还将技巧 104 中的登录屏幕也融合进来，并允许创建和列出个人日记条目。首先，定义了用于创建日记条目的布局 XML 文件 **diary.xml**，如代码清单 11-12 所示，而屏幕输出则如图 11-3 所示。

**代码清单 11-12　res/layout/diary.xml**

```
<?xml version="1.0" encoding="utf-8"?>
<LinearLayout xmlns:android="http://schemas.android.com/apk/res/android"
    android:orientation="vertical"
    android:layout_width="match_parent"
    android:layout_height="match_parent"
    >
    <TextView
        android:layout_width="match_parent"
        android:layout_height="wrap_content"
        android:text="Diary Title"
    />
    <EditText
        android:id="@+id/diarydescriptiontext"
        android:layout_width="match_parent"
        android:layout_height="wrap_content"
    />
    <TextView
        android:layout_width="match_parent"
        android:layout_height="wrap_content"
        android:text="Content"
    />
    <EditText
        android:id="@+id/diarycontenttext"
        android:layout_width="match_parent"
        android:layout_height="200dp"
    />
    <Button
        android:id="@+id/submitbutton"
        android:layout_width="wrap_content"
        android:layout_height="wrap_content"
        android:text="submit"
        android:textSize="20dp"
    />
</LinearLayout>
```

图 11-3 日记条目创建界面

主 Activity 由 **Diary.java** 定义，如代码清单 11-13 所示。需要导入 com.cookbook.data 包，声明、初始化并打开 MyDB 对象以供使用。另外还显示了 **diary.xml** 定义的布局，并处理"submit"按钮按下的事件，将数据存储到数据库中。

代码清单 11-13 src/com/cookbook/datastorage/Diary.java

```
package com.cookbook.datastorage;

import android.app.Activity;
import android.content.Intent;
import android.os.Bundle;
import android.view.View;
import android.view.View.OnClickListener;
import android.widget.Button;
import android.widget.EditText;

import com.cookbook.data.MyDB;
public class Diary extends Activity {
    EditText titleET, contentET;
    Button submitBT;
    MyDB dba;

    @Override
    public void onCreate(Bundle savedInstanceState) {
        super.onCreate(savedInstanceState);
        setContentView(R.layout.diary);
        dba = new MyDB(this);
        dba.open();
        titleET = (EditText)findViewById(R.id.diarydescriptiontext);
        contentET = (EditText)findViewById(R.id.diarycontenttext);
        submitBT = (Button)findViewById(R.id.submitbutton);
        submitBT.setOnClickListener(new OnClickListener() {
            public void onClick(View v) {
                try {
                    saveItToDB();
                } catch (Exception e) {
                    e.printStackTrace();
                }
            }
        });
    }
    public void saveItToDB() {
        dba.insertDiary(titleET.getText().toString(),
                        contentET.getText().toString());
        dba.close();
        titleET.setText("");
        contentET.setText("");
        Intent i = new Intent(Diary.this, DisplayDiaries.class);
        startActivity(i);
    }
}
```

DataStorage.java 类的代码和代码清单 11-6 中的基本一样，只是把登录成功时启动的类由 MyPreferences.class 改为了 Diary.class：

```
Toast.makeText(DataStorage.this, "login passed!!",
               Toast.LENGTH_SHORT).show();
Intent i = new Intent(DataStorage.this, Diary.class);
startActivity(i);
```

最后，一定要更新 **AndroidManifest.xml** 文件以包含新建的 Activity，如代码清单 11-14 所示。

**代码清单 11-14　AndroidManifest.xml**

```
<?xml version="1.0" encoding="utf-8"?>
<manifest xmlns:android="http://schemas.android.com/apk/res/android"
      package="com.cookbook.datastorage"
      android:versionCode="1" android:versionName="1.0">
    <application android:icon="@drawable/icon"
                 android:label="@string/app_name">
        <activity android:name=".DataStorage"
                  android:label="@string/app_name">
            <intent-filter>
              <action android:name="android.intent.action.MAIN" />
              <category android:name="android.intent.category.LAUNCHER" />
            </intent-filter>
        </activity>
        <activity android:name=".MyPreferences" />
        <activity android:name=".Diary"/>
    </application>
    <uses-sdk android:minSdkVersion="7" />
</manifest>
```

至此，一个独立的数据库就集成进来了，条目列表的布局将在下一个技巧中进行讨论，以完成日记应用程序。

## 技巧 109：创建个人日记

本技巧利用 ListView 对象来显示 SQLite 数据表中的多条记录。我们把这些项目显示在一个垂直滚动列表中。ListView 需要一个数据适配器（data adapter）将底层数据的改变通知视图。需要创建两个 XML 文件：一个是 **diaries.xml**，定义了的 ListView 的内容，如代码清单 11-15 所示；另一个是 **diaryrow.xml**，定义了 ListView 内部的行的内容，如代码清单 11-16 所示。

**代码清单 11-15　res/layout/diaries.xml**

```
<?xml version="1.0" encoding="utf-8"?>
<LinearLayout xmlns:android="http://schemas.android.com/apk/res/android"
    android:orientation="vertical"
    android:layout_width="match_parent"
    android:layout_height="match_parent">
        <ListView
            android:layout_width="match_parent" android:dividerHeight="1px"
            android:layout_height="match_parent"
            android:id="list">
        </ListView>
</LinearLayout>
```

**代码清单 11-16　res/layout/diaryrow.xml**

```
<?xml version="1.0" encoding="utf-8"?>
<RelativeLayout android:layout_width="wrap_content"
    android:layout_height="wrap_content"
    android:layout_alignLeft="@+id/name" android:layout_below="@+id/name"
    xmlns:android="http://schemas.android.com/apk/res/android"
```

```
        android:padding="12dip">
        <TextView android:layout_width="wrap_content"
            android:layout_height="wrap_content" android:id="@+id/name"
            android:layout_marginRight="4dp" android:text="Diary Title"
            android:textStyle="bold" android:textSize="16dip" />
        <TextView android:id="@+id/datetext"
            android:layout_width="wrap_content"
            android:layout_height="wrap_content" android:text="Date Recorded"
            android:textSize="14dip" />
</RelativeLayout>
```

**DisplayDiaries.java** Activity 扩展了 ListActivity，在类内显示一个 ListView，其中定义了两个内部类：MyDiary 是一个数据类，用于保存日记条目的内容（标题、内容、日期）；DiaryAdapter 是一个 BaseAdapter 类，用于处理从数据库检索数据（使用 getData()）。下面的一些方法是从 BaseAdapter 派生而来，并被 ListView 调用的。

- getCount()：返回适配器中的项目数量。
- getItem()：返回指定的项目。
- getItemID()：返回项目的 ID（在本例中，没有项目 ID）。
- getView()：返回每个项目的视图。

注意，ListView 会调用 getView() 为每个项目绘制视图。要提升 UI 渲染的性能，从 getView() 返回的视图应当尽可能多地被回收重用。实现方法是通过创建一个 ViewHolder 类来保存视图。

调用 getView() 方法时，当前显示给用户的视图也会被传递过去，此时该视图会保存到 ViewHolder 中并打上标签。后面如果再用相同的视图调用 getView()，可以通过标签识别出已经存在 ViewHolder 中的视图。在本例中，可以在已有的视图上改变内容，而不用创建新视图。

主 Activity 如代码清单 11-17 所示，而日记条目在 ListView 上的显示结果见图 11-4。

**代码清单 11-17　src/com/cookbook/datastorage/DisplayDiaries.java**

```
package com.cookbook.datastorage;

import java.text.DateFormat;
import java.util.ArrayList;
import java.util.Date;

import android.app.ListActivity;
import android.content.Context;
import android.database.Cursor;
import android.os.Bundle;
import android.view.LayoutInflater;
import android.view.View;
import android.view.ViewGroup;
import android.widget.BaseAdapter;
import android.widget.TextView;

import com.cookbook.data.Constants;
import com.cookbook.data.MyDB;

public class DisplayDiaries extends ListActivity {
    MyDB dba;
    DiaryAdapter myAdapter;
```

```
private class MyDiary{
    public MyDiary(String t, String c, String r){
        title=t;
        content=c;
        recordDate=r;
    }
    public String title;
    public String content;
    public String recordDate;
}
@Override
protected void onCreate(Bundle savedInstanceState) {
    dba = new MyDB(this);
    dba.open();
    setContentView(R.layout.diaries);

    super.onCreate(savedInstanceState);
    myAdapter = new DiaryAdapter(this);
    this.setListAdapter(myAdapter);
}

private class DiaryAdapter extends BaseAdapter {
    private LayoutInflater mInflater;
    private ArrayList<MyDiary> diaries;
    public DiaryAdapter(Context context) {
        mInflater = LayoutInflater.from(context);
        diaries = new ArrayList<MyDiary>();
        getData();
    }
    public void getData(){
        Cursor c = dba.getDiaries();
        startManagingCursor(c);
        if(c.moveToFirst()){
            do{
                String title =
                    c.getString(c.getColumnIndex(Constants.TITLE_NAME));
                String content =
                 c.getString(c.getColumnIndex(Constants.CONTENT_NAME));
                DateFormat dateFormat =
                 DateFormat.getDateTimeInstance();
                String dateData = dateFormat.format(new
                 Date(c.getLong(c.getColumnIndex(
                                    Constants.DATE_NAME))).getTime());
                MyDiary temp = new MyDiary(title,content,dateData);
                    diaries.add(temp);
                } while(c.moveToNext());
        }
    }

    @Override
    public int getCount() {return diaries.size();}
    public MyDiary getItem(int i) {return diaries.get(i);}
    public long getItemId(int i) {return i;}
    public View getView(int arg0, View arg1, ViewGroup arg2) {
        final ViewHolder holder;
        View v = arg1;
        if ((v == null) || (v.getTag() == null)) {
            v = mInflater.inflate(R.layout.diaryrow, null);
            holder = new ViewHolder();
            holder.mTitle = (TextView)v.findViewById(R.id.name);
            holder.mDate = (TextView)v.findViewById(R.id.datetext);
            v.setTag(holder);
```

```
            } else {
                holder = (ViewHolder) v.getTag();
            }

            holder.mdiary = getItem(arg0);
            holder.mTitle.setText(holder.mdiary.title);
            holder.mDate.setText(holder.mdiary.recordDate);

            v.setTag(holder);

            return v;
        }

        public class ViewHolder {
            MyDiary mdiary;
            TextView mTitle;
            TextView mDate;
        }
    }
}
```

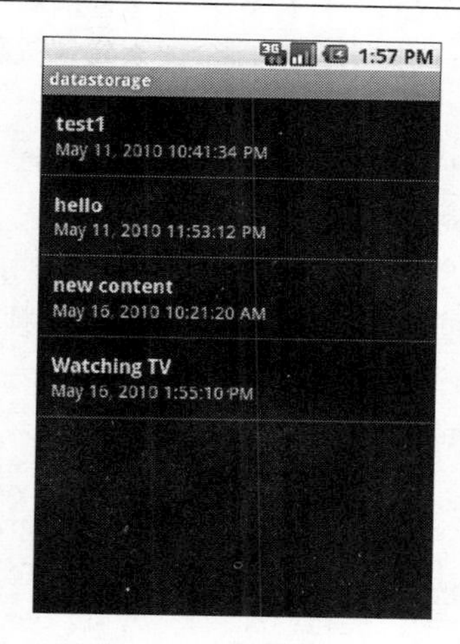


图 11-4 显示日记条目的 ListView

## 11.3 内容提供器

每个应用程序都有自己的沙箱（sandbox）[①]，不能访问其他应用程序的数据。若需要访问非自己的沙箱所提供的函数，应用程序必须预先在安装前显式地声明权限。Android 提供了名为 ContentProvider 的接口，来扮演应用程序间的桥梁，允许它们分享和改变彼此的数据。内容提供器使得应用层和数据层得以明确地分离。要使用它，需要在 **AndroidManifest.xml** 文件中设置权限。使用一个简单的 URI 模型即可访问它。

下面是 Android 系统中可以用到内容提供器的一些本地数据库。

- Browser：读取或修改书签、浏览器历史记录或 Web 搜索。

① 沙箱是一种按照安全策略限制程序行为的执行环境。——译者注

- CallLog：浏览或更新通话记录。
- Contacts：检索、修改或存储个人通讯录。通讯录信息被存储在 ContactsContract 对象下的一个三层数据模型表中。
  - ContactsContract.Data：包含各种个人数据。有一些预定义好的公用数据集，比如电话号码、邮箱地址。应用程序也可以自行定义表格的格式。
  - ContactsContract.RawContacts：包含一个与单个账号或个人相关联的数据对象的集合。
  - ContactsContract.Contacts：包含一个或多个 RawContracts 的聚集，有可能是描述同一个人。
- LiveFolder：一个特殊的文件夹，其内容由内容提供器提供。
- MediaStore：访问音频、视频和图像。
- Setting：浏览和检索蓝牙设置、铃声和其他设备偏好设置。
- SearchRecentSuggestions：配置后可与一个搜索建议提供器（search-suggestions provider）协作。
- SyncStateContract：查看内容提供器合约，把数据与一个数据数组账号相关联，想要以标准方法保存该数据的提供器可以使用它。
- UserDictionary：存储用户定义的单词，会被输入法用来预测用户输入。应用程序和输入法可以向字典中添加单词。单词可以带有相关的频率信息和区域信息。

要访问内容提供器，应用程序需要获取一个 ContentProvider 的实例，用于查询、插入、删除及更新内容提供器的数据，如下面的例子所示：

```
ContentResolver crInstance = getContentResolver(); //Get a ContentResolver instance
crInstance.query(People.CONTENT_URI, null, null, null, null); //Query contacts
ContentValues new_Values= new ContentValues();
crInstance.insert(People.CONTENT_URI, new_Values); //Insert new values
crInstance.delete(People_URI, null, null); //Delete all contacts

ContentValues update_Values= new ContentValues();
crInstance.update(People_URI, update_Value, null,null); //Update values
```

每个内容提供器需要包含一个 URI，用于注册及权限的访问。每个提供器的 URI 必须是独一无二的，并采用通用建议格式：

```
content://<package name>.provider.<custom ContentProvider name>/<DataPath>
```

为简明起见，URI 也可以被缩写成 content://com.cookbook.datastorage/diaries，下一个技巧中会用到这种格式。在 ContentProvider 接口中要用到 UriMatcher 类，以确保传递的 URI 是合适的。

## 技巧 110：创建自定义的内容提供器

了解了如何使用内容提供器之后，我们就可以试着往之前的技巧创建的日记项目中集成一个内容提供器。本技巧展示了如何将日记条目显示给其他选定的应用程序。自定义的内容提供器只需扩展 Android 的 ContentProvider 类，其中有 6 个方法，可以有选择地进行重写。

- query()：允许第三方应用程序检索内容。
- insert()：允许第三方应用程序插入内容。
- update()：允许第三方应用程序更新内容。
- delete()：允许第三方应用程序删除内容。
- getType()：允许第三方应用程序读取每个支持的 URI 结构。
- onCreate()：创建一个数据库实例以帮助检索内容。

例如，如果只允许其他应用程序从提供器中读取内容，则只需要重写 onCreate()和 query()两个方法。

代码清单 11-18 给出了一个自定义的 ContentProvider 类。它拥有一个被添加到 UriMatcher 的、基于 com.cookbook.datastorage 的 URI，以及名为 diaries 的数据库表。onCreate()方法使用代码清单 11-9 中的代码建立了一个 MyDB 对象，该对象负责数据库访问。query()方法检索日记数据库中的全部记录，这些记录是作为 uri 参数传递的。要选择范围更为具体的记录，可以在该方法上使用其他的参数。

**代码清单 11-18　src/com/cookbook/datastorage/DiaryContentProvider.java**

```
package com.cookbook.datastorage;

import android.content.ContentProvider;
import android.content.ContentValues;
import android.content.UriMatcher;
import android.database.Cursor;
import android.database.sqlite.SQLiteQueryBuilder;
import android.net.Uri;

import com.cookbook.data.Constants;
import com.cookbook.data.MyDB;

public class DiaryContentProvider extends ContentProvider {

    private MyDB dba;
    private static final UriMatcher sUriMatcher;
    //the code returned for URI match to components
    private static final int DIARIES=1;
    public static final String AUTHORITY = "com.cookbook.datastorage";
    static {
                sUriMatcher = new UriMatcher(UriMatcher.NO_MATCH);
                sUriMatcher.addURI(AUTHORITY, Constants.TABLE_NAME,
                                           DIARIES);
    }
    @Override
    public int delete(Uri uri, String selection, String[] selectionArgs) {
         return 0;
    }
    public String getType(Uri uri) {return null;}
    public Uri insert(Uri uri, ContentValues values) {return null;}
    public int update(Uri uri, ContentValues values, String selection,
```

```
        String[] selectionArgs) {return 0;}

    @Override
    public boolean onCreate() {
        dba = new MyDB(this.getContext());
        dba.open();
        return false;
    }

    @Override
    public Cursor query(Uri uri, String[] projection, String selection,
            String[] selectionArgs, String sortOrder) {
        Cursor c=null;
        switch (sUriMatcher.match(uri)) {
                case DIARIES:
                c = dba.getDiaries();
                break;
            default:
                throw new IllegalArgumentException(
                                                    "Unknown URI" + uri);
        }
        c.setNotificationUri(getContext().getContentResolver(), uri);
        return c;
    }
}
```

需要在 **AndroidManifest.xml** 中将提供器指定为可访问的，如代码清单 11-19 所示。

### 代码清单 11-19 AndroidManifest.xml

```
<?xml version="1.0" encoding="utf-8"?>
<manifest xmlns:android="http://schemas.android.com/apk/res/android"
      package="com.cookbook.datastorage"
      android:versionCode="1"
      android:versionName="1.0">
    <application android:icon="@drawable/icon"
                  android:label="@string/app_name">
      <activity android:name=".DataStorage"
                  android:label="@string/app_name">
              <intent-filter>
                <action android:name="android.intent.action.MAIN" />
                <category android:name="android.intent.category.LAUNCHER" />
              </intent-filter>
          </activity>
          <activity android:name=".MyPreferences" />
          <activity android:name=".Diary"/>
          <activity android:name=".DisplayDiaries"/>
          <provider android:name="DiaryContentProvider"
               android:authorities="com.cookbook.datastorage" />
     </application>
     <uses-sdk android:minSdkVersion="7" />
</manifest>
```

现在内容提供器已经准备好为其他应用程序所用了。为测试该内容提供器，我们创建名为 `DataStorageTester` 的新 Android 项目，其主 Activity 名为 `DataStorageTester`，在代码清单 11-20 中给出。创建一个 `ContentResolver` 的实例，用于从 `DataStorage` 内容提供器中查询数据。之后会返回一个 Cursor 对象，测试函数解析每个数据条目的第二列，并将其连接成一个 String，再使用 StringBuilder 对象将其显示在屏幕上。

**代码清单 11-20 src/com/cookbook/datastorage_tester/DataStorageTester.java**

```
package com.cookbook.datastorage_tester;

import android.app.Activity;
import android.content.ContentResolver;
import android.database.Cursor;
import android.net.Uri;
import android.os.Bundle;
import android.widget.TextView;

public class DataStorageTester extends Activity {
    TextView tv;

    @Override
    public void onCreate(Bundle savedInstanceState) {
        super.onCreate(savedInstanceState);
        setContentView(R.layout.main);
        tv = (TextView) findViewById(R.id.output);
        String myUri = "content://com.cookbook.datastorage/diaries";
        Uri CONTENT_URI = Uri.parse(myUri);
        //Get ContentResolver instance
        ContentResolver crInstance = getContentResolver();
        Cursor c = crInstance.query(CONTENT_URI, null, null, null, null);
        startManagingCursor(c);
        StringBuilder sb = new StringBuilder();
        if(c.moveToFirst()){
            do{
                sb.append(c.getString(1)).append("\n");

            }while(c.moveToNext());
        }
        tv.setText(sb.toString());
    }
}
```

在 **main.xml** 布局文件中，需要为用于输出的 TextView 的视图添加一个 ID，如代码清单 11-21 所示。

**代码清单 11-21 res/layout/main.xml**

```
<?xml version="1.0" encoding="utf-8"?>
<LinearLayout xmlns:android="http://schemas.android.com/apk/res/android"
    android:orientation="vertical"
    android:layout_width="match_parent"
    android:layout_height="match_parent"
    >
<TextView
    android:id="@+id/output"
    android:layout_width="match_parent"
    android:layout_height="wrap_content"
    android:text="@string/hello"
    />
</LinearLayout>
```

运行测试函数后会显示条目的标题，如图 11-5 所示。

图 11-5　另一个日记应用程序使用内容提供器的查询结果

## 11.4　文件的保存和载入

除了前面提及的几种 Android 特有的数据存储方法外，我们还可以使用标准的 Java 包 `java.io.File`。该包提供对平面文件的操作方法，如 `FileInputStream`、`FileOutput Stream`、`InputStream` 和 `OutputStream`。下面是从文件中读取内容和向文件写入内容的例子：

```
FileInputStream fis = openFileInput("myfile.txt");
FileOutputStream fos = openFileOutput("myfile.txt",
                                      Context.MODE_WORLD_WRITEABLE);
```

另一个例子是将摄像头捕捉的位图图片保存为 PNG 文件，如下：

```
Bitmap takenPicture;
FileOutputStream out = openFileOutput("mypic.png",
                                      Context.MODE_WORLD_WRITEABLE);
takenPicture.compress(CompressFormat.PNG, 100, out);
out.flush();
out.close();
```

资源目录下的文件也可以打开。例如，要打开位于 **res/raw** 文件夹中的 **myrawfile.txt**，使用下面的代码：

```
InputStream is = this.getResource()
                    .openRawResource(R.raw.myrawfile.txt);
```

### 技巧 111：使用 AsyncTask 进行异步处理

为使应用程序的性能达到最佳，最好确保主线程不被阻塞。可以卸载（offload）任务，将可在后台运行的线程分离出来。本技巧使用 `AsyncTask` 类执行一些主线程以外的逻辑。

AsyncTask 类接受三个参数：Params、Progress 和 Result。并不是总要使用全部三个参数，有时候传递 Void 也是可以的。使用 AsyncTask 时有以下 4 个主要方法会被作为逻辑步骤执行。

- onPreExecute()：该方法运行于主线程之上，一般用于建立异步任务。
- doInBackground()：该方法是任务逻辑所在。方法使用了一个独立的线程，因此不会出现线程阻塞。方法还使用了 publishProgress()方法将更新传回给主线程。
- onProgressUpdate()：该方法在 publishProgress()方法执行期间被调用，用于对主线程进行视觉上的更新。
- onPostExcute()：该方法会在 doInBackground()方法执行完毕后立即调用，并带有从后者传递过来的一个参数。

注意，在使用 AsyncTask 类时，任何使用过的线程都会在当前视图被移除或销毁之时销毁。

代码清单 11-22 中的代码会接受一个句子，并在其中查找“meow”一词。执行过程中，进度条会实时更新。

**代码清单 11-22 src/com/cookbook/async/MainActivity.java**

```
package com.cookbook.async;

import java.util.regex.Matcher;
import java.util.regex.Pattern;

import android.app.Activity;
import android.os.AsyncTask;
import android.os.Bundle;
import android.view.View;
import android.widget.ProgressBar;
import android.widget.TextView;

public class MainActivity extends Activity {

  TextView mainTextView;
  ProgressBar mainProgress;

  @Override
  protected void onCreate(Bundle savedInstanceState) {
    super.onCreate(savedInstanceState);
    setContentView(R.layout.activity_main);
    mainTextView = (TextView) findViewById(R.id.maintextview);
    mainProgress = (ProgressBar) findViewById(R.id.mainprogress);
  }

  private class MyAsyncTask extends AsyncTask<String, Integer, String> {
    @Override
    protected String doInBackground(String... parameter) {
      String result = "";

      Pattern pattern = Pattern.compile("meow");
      Matcher matcher = pattern.matcher(parameter[0]);

      int count = 0;
      while (matcher.find()){
        count++;
```

```
        try {
          Thread.sleep(100);
        } catch (InterruptedException e) {
          // Remember to error handle
        }
        publishProgress(count + 20);
      }

      result = "meow was found "+count+" times";

      return result;
    }

    @Override
    protected void onProgressUpdate(Integer... progress) {
      mainProgress.setProgress(progress[0]);
    }

    @Override
    protected void onPostExecute(String result) {
      mainTextView.setText(result);
    }
  }

  public void executeAsync(View view) {
    MyAsyncTask task = new MyAsyncTask();
    task.execute("Meow, meow, meow many times do you have meow?");
  }

}
```

# 第 12 章

# 基于位置的服务

基于位置的服务（Location-Based Service，LBS）催生了一些目前最流行的移动应用。位置信息可以被集成到许多功能上，比如因特网搜索、照相、游戏和社交网络。开发者可以利用可用的位置技术让他们的应用程序更具相关性和本地性。

本章首先介绍了获取设备位置的方法，然后是追踪、地理编码和地图绘制。此外，还有关于在地图上叠加标记和视图的技巧。

## 12.1 位置服务基础

要通过 Android 系统访问设备服务，应用程序需要以下这些组件。

- LocationManager：提供对 Android 系统定位服务访问的类。
- LocationListener：用于当位置改变时，从 LocationManager 接收通知的接口。
- Location：表示某个特定时刻确定的地理位置的类。

LocationManager 类需要通过名为 LOCATION_SERVICE 的 Android 系统服务进行初始化。该服务为应用程序提供设备的当前位置以及运动状况，并可以在设备进入或离开特定区域时发出警告。下面是一个初始化的例子：

```
LocationManager mLocationManager;
mLocationManager = (LocationManager)
          getSystemService(Context.LOCATION_SERVICE);
```

初始化 LocationManager 之后，需要选择一个位置提供器（location provider）。设备可能支持不同的位置技术，比如辅助全球定位系统（AGPS）、Wi-Fi 等，一般根据精度和电力方面的要求来确定合适的位置提供器。可以通过 android.location.Criteria 中定义的 Criteria 类实现这一点。该类使 Android 系统可以依据给定的需求，在可用的位置技术中寻找最佳者。下面是基于 criteria 选择位置提供器的一个例子：

```
Criteria criteria = new Criteria();
criteria.setAccuracy(Criteria.ACCURACY_FINE);
criteria.setPowerRequirement(Criteria.POWER_LOW);
String locationprovider =
              mLocationManager.getBestProvider(criteria, true);
```

还可以使用位置管理器的 `getProvider()` 方法来指定所用的位置估测技术。两种常用的提供器是基于卫星的 GPS（通过 `LocationManager.GPS_PROVIDER` 来指定）和蜂窝发射塔/Wi-Fi 识别（通过 `LocationManager.NETWORK_PROVIDER` 来指定）。前者更为精确，但当不能直接看到天空时（比如在室内），后者就能派上用场了。

除非另有说明，本章中的技巧都要用到下面两个支持文件。首先，主布局需要一个 `TextView`，如代码清单 12-1 所示，用于显示位置数据。

**代码清单 12-1　res/layout/main.xml**

```
<?xml version="1.0" encoding="utf-8"?>
<LinearLayout xmlns:android="http://schemas.android.com/apk/res/android"
    android:orientation="vertical"
    android:layout_width="match_parent"
    android:layout_height="match_parent"
    >
<TextView
    android:id="@+id/tv1"
    android:layout_width="match_parent"
    android:layout_height="wrap_content"
    android:text="@string/hello"
    />
</LinearLayout>
```

其次，需要在 **AndroidManifest.xml** 中将使用位置信息的权限授予用户，如代码清单 12-2 所示（对于本章的其他技巧，只需要改变包的名字即可）。对于更精确的定位，如 GPS，需要添加 ACCESS_FINE_LOCATION 权限。对于其他情况，则添加 ACCESS_COARSE_LOCATION 权限。应当注意，在 ACCESS_COARSE_LOCATION 下使用的那些传感器，在 ACCESS_FINE_LOCATION 下仍然可用。

**代码清单 12-2　AndroidManifest.xml**

```
<?xml version="1.0" encoding="utf-8"?>
<manifest xmlns:android="http://schemas.android.com/apk/res/android"
      package="com.cookbook.mylocationpackage"
      android:versionCode="1"
      android:versionName="1.0">
<uses-permission android:name="android.permission.ACCESS_FINE_LOCATION"/>

    <application android:icon="@drawable/icon"
                 android:label="@string/app_name">
        <activity android:name=".MyLocation"
                  android:label="@string/app_name">
            <intent-filter>
              <action android:name="android.intent.action.MAIN" />
              <category android:name="android.intent.category.LAUNCHER" />
            </intent-filter>
        </activity>
    </application>
    <uses-sdk android:minSdkVersion="4" />

</manifest>
```

## 技巧 112：检索最近保存的位置

由于估测位置可能需要一定的时间，我们可以调用 getLastKnownLocation()为给定的提供器检索上一次保存的位置信息。位置信息包含纬度、经度和协调世界时（Coordinated Universal Time，CUT）时间戳。有的提供器也许还能提供海拔高度、速度和方位信息（可通过在位置对象上调用 getAltitude()、getSpeed()、getBearing()来检索这些信息，还可使用 getExtras()检索卫星信息）。本技巧将显示纬度和经度信息。另一个可能用到的选项是 PASSIVE_PROVIDER，这是一个常量，是一个特殊的位置提供器，存储了对位置的上一次请求。主 Activity 在代码清单 12-3 中给出。

**代码清单 12-3 src/com/cookbook/lastlocation/MyLocation.java**

```
package com.cookbook.lastlocation;

import android.app.Activity;
import android.content.Context;
import android.location.Criteria;
import android.location.Location;
import android.location.LocationManager;
import android.os.Bundle;
import android.widget.TextView;

public class MyLocation extends Activity {
    LocationManager mLocationManager;
    TextView tv;

    @Override
    public void onCreate(Bundle savedInstanceState) {
        super.onCreate(savedInstanceState);
        setContentView(R.layout.main);
        tv = (TextView) findViewById(R.id.tv1);

        mLocationManager = (LocationManager)
                getSystemService(Context.LOCATION_SERVICE);

        Criteria criteria = new Criteria();
        criteria.setAccuracy(Criteria.ACCURACY_FINE);
        criteria.setPowerRequirement(Criteria.POWER_LOW);
        String locationprovider =
                mLocationManager.getBestProvider(criteria,true);
        Location mLocation =
               mLocationManager.getLastKnownLocation(locationprovider);

        tv.setText("Last location lat:" + mLocation.getLatitude()
                  + "long:" + mLocation.getLongitude());
    }
}
```

## 技巧 113：在位置改变时更新信息

LocationListener 接口用于在位置变化时接收通知。在初始化位置提供器后，需要调用位置管理器的 requestLocationUpdates()方法指定通知当前 Activity 位置变化的时间。时间的指定取决于下列参数。

- provider：应用程序使用的位置提供器。

- minTime：两次更新的最小时间间隔，以毫秒为单位（不过系统可能会增大该间隔以节约电力）。
- minDistance：两次更新的最小距离变化量，以米为单位。
- listener：用于接收更新的位置监听器。

可以重写位置监听器的 onLocationChanged()方法指定在新位置要进行的动作。代码清单 12-4 显示了如何把二者结合起来：最小时间间隔为 5 s，而距离变化要大于 2 m。在实际的应用中应当使用更大的值以节省电池电量。还要注意，不应在 onLocationChanged()方法中进行大任务量的处理工作，而要将数据赋值并传递给一个独立的线程。

**代码清单 12-4 src/com/cookbook/update_location/MyLocation.java**

```
package com.cookbook.update_location;

import android.app.Activity;
import android.content.Context;
import android.location.Criteria;
import android.location.Location;
import android.location.LocationListener;
import android.location.LocationManager;
import android.os.Bundle;
import android.widget.TextView;

public class MyLocation extends Activity implements LocationListener {
    LocationManager mLocationManager;
    TextView tv;
    Location mLocation;

    @Override
    public void onCreate(Bundle savedInstanceState) {
        super.onCreate(savedInstanceState);
        setContentView(R.layout.main);
        tv = (TextView) findViewById(R.id.tv1);

        mLocationManager = (LocationManager)
                getSystemService(Context.LOCATION_SERVICE);

        Criteria criteria = new Criteria();
        criteria.setAccuracy(Criteria.ACCURACY_FINE);
        criteria.setPowerRequirement(Criteria.POWER_LOW);
        String locationprovider =
                mLocationManager.getBestProvider(criteria,true);

        mLocation =
                mLocationManager.getLastKnownLocation(locationprovider);
        mLocationManager.requestLocationUpdates(
                    locationprovider, 5000, 2.0, this);
    }

    @Override
    public void onLocationChanged(Location location) {
        mLocation = location;
        showupdate();
    }
    // These methods are required
    public void onProviderDisabled(String arg0) {}
    public void onProviderEnabled(String provider) {}
    public void onStatusChanged(String a, int b, Bundle c) {}

    public void showupdate(){
        tv.setText("Last location lat:"+mLocation.getLatitude()
```

```
                        + "long:" + mLocation.getLongitude());
        }
    }
```

注意，除了在 Activity 一级实现 LocationListener 之外，还可以像下面那样在一个独立的内部类中声明它。这种声明可以很容易地添加到后续的任何技巧之中，提供一种对位置信息的更新机制：

```
        mLocationManager.requestLocationUpdates(
                    locationprovider, 5000, 2.0, myLocL);
    }

    private final LocationListener myLocL = new LocationListener(){
        @Override
        public void onLocationChanged(Location location){
            mLocation = location;
            showupdate();
        }

        // These methods are required
        public void onProviderDisabled(String arg0) {}
        public void onProviderEnabled(String provider) {}
        public void onStatusChanged(String a, int b, Bundle c) {}
    };
```

## 技巧 114：列出所有可用的提供器

本技巧能列出给定 Android 设备上各种可用的内容提供器。图 12-1 给出了一个输出结果示例，该结果可能根据设备的不同而不同。主 Activity 如代码清单 12-5 所示。要显示可用提供器的清单，可使用 getProviders(true) 方法。与前一个技巧形成对照的是，这里将 LocationListener 声明为一个匿名的内部类，而功能没有任何损失。

图 12-1 在一个真实的 Android 设备上输出所有可用的位置提供器及其最近保存的位置

**代码清单 12-5　src/com/cookbook/show_providers/MyLocation.java**

```
package com.cookbook.show_providers;

import java.util.List;

import android.app.Activity;
import android.content.Context;
import android.location.Criteria;
import android.location.Location;
import android.location.LocationListener;
import android.location.LocationManager;
import android.os.Bundle;
import android.widget.TextView;

public class MyLocation extends Activity {
    LocationManager mLocationManager;
    TextView tv;
    Location mLocation;

    @Override
    public void onCreate(Bundle savedInstanceState) {
        super.onCreate(savedInstanceState);
        setContentView(R.layout.main);
        tv = (TextView) findViewById(R.id.tv1);
        mLocationManager = (LocationManager)
                           getSystemService(Context.LOCATION_SERVICE);
        Criteria criteria = new Criteria();
        criteria.setAccuracy(Criteria.ACCURACY_FINE);
        criteria.setPowerRequirement(Criteria.POWER_LOW);
        String locationprovider =
                       mLocationManager.getBestProvider(criteria,true);

        List<String> providers = mLocationManager.getProviders(true);
        StringBuilder mSB = new StringBuilder("Providers:\n");
        for(int i = 0; i<providers.size(); i++) {
          mLocationManager.requestLocationUpdates(
            providers.get(i), 5000, 2.0f, new LocationListener(){

            // These methods are required
            public void onLocationChanged(Location location) {}
            public void onProviderDisabled(String arg0) {}
            public void onProviderEnabled(String provider) {}
            public void onStatusChanged(String a, int b, Bundle c) {}
          });
          mSB.append(providers.get(i)).append(": \n");
          mLocation =
              mLocationManager.getLastKnownLocation(providers.get(i));
          if(mLocation != null) {
              mSB.append(mLocation.getLatitude()).append(" , ");
              mSB.append(mLocation.getLongitude()).append("\n");
          } else {
              mSB.append("Location cannot be found");
          }
        }
        tv.setText(mSB.toString());
    }
}
```

## 技巧 115：将位置转化为地址（逆向地理编码）

Geocoder 类提供了从地址转化到经纬度坐标（即地理编码）以及从经纬度坐标转变到地址（即逆向地理编码）的方法。逆向地理编码也许只能生成部分的地址，比如城市和邮编，这取决于位置提供器可用的细节层次。

本技巧使用逆向地理编码，根据设备的位置获得一个地址，并将其显示到屏幕上，如图 12-2 所示。初始化 Geocoder 实例时需要提供一个上下文。如果所用的区域设置与系统区域设置不同的话，则还要提供区域设置信息，本例将其显式地设置为 Locale.ENGLISH。随后，getFromLocation() 方法给出了与提供的位置周围区域有关的地址列表。本例中，返回结果的最大数量被限定为 1（即只返回程序认为可能性最大的地址）。

图 12-2　逆向地理编码示例，将经、纬度坐标转换为地址

地理编码器（geocoder）返回一个 android.location.Address 对象的列表。向地址的转化有赖于一个并不包含在核心 Android 框架中的后端服务。比如，Google 地图 API 就提供了一个客户端地理编码器服务。然而，如果目标设备上没有此类服务，转化过程会返回一个空列表。作为一个字符串列表的地址会被逐行转储到一个 String 对象中，从而显示到屏幕上。主 Activity 如代码清单 12-6 所示。

**代码清单 12-6　src/com/cookbook/rev_geocoding/MyLocation.java**

```
package com.cookbook.rev_geocoding;

import java.io.IOException;
import java.util.List;
import java.util.Locale;

import android.app.Activity;
import android.content.Context;
import android.location.Address;
import android.location.Criteria;
import android.location.Geocoder;
import android.location.Location;
import android.location.LocationListener;
import android.location.LocationManager;
import android.os.Bundle;
import android.util.Log;
import android.widget.TextView;

public class MyLocation extends Activity {
    LocationManager mLocationManager;
    Location mLocation;
    TextView tv;
```

```
    @Override
    public void onCreate(Bundle savedInstanceState) {
        super.onCreate(savedInstanceState);

        setContentView(R.layout.main);
        tv = (TextView) findViewById(R.id.tv1);

        mLocationManager = (LocationManager)
                getSystemService(Context.LOCATION_SERVICE);

        Criteria criteria = new Criteria();
        criteria.setAccuracy(Criteria.ACCURACY_FINE);
        criteria.setPowerRequirement(Criteria.POWER_LOW);
        String locationprovider =
                mLocationManager.getBestProvider(criteria,true);

        mLocation =
                mLocationManager.getLastKnownLocation(locationprovider);

        List<Address> addresses;
        try {
          Geocoder mGC = new Geocoder(this, Locale.ENGLISH);
          addresses = mGC.getFromLocation(mLocation.getLatitude(),
                                       mLocation.getLongitude(), 1);
          if(addresses != null) {
            Address currentAddr = addresses.get(0);
            StringBuilder mSB = new StringBuilder("Address:\n");
            for(int i=0; i<currentAddr.getMaxAddressLineIndex(); i++) {
              mSB.append(currentAddr.getAddressLine(i)).append("\n");
            }

            tv.setText(mSB.toString());
          }
        } catch(IOException e) {
            tv.setText(e.getMessage());
        }
    }
}
```

## 技巧 116：将地址转化为位置（地理编码）

本技巧展示了如何将地址转化为经纬度坐标，即地理编码过程。该过程与上个技巧中的逆向地理编码过程几乎是相同的，只是要使用 getFromLocationName() 方法来代替之前的 getFromLocation() 方法。代码清单 12-7 在 String myAddress 中指定了一个地址，将其转化为位置，然后在屏幕上显示出来，如图 12-3 所示。

**代码清单 12-7　src/com/cookbook/geocoding/MyLocation.java**

```
package com.cookbook.geocoding;

import java.io.IOException;
import java.util.List;
import java.util.Locale;
```

```
import android.app.Activity;
import android.content.Context;
import android.location.Address;
import android.location.Criteria;
import android.location.Geocoder;
import android.location.Location;
import android.location.LocationListener;
import android.location.LocationManager;
import android.os.Bundle;
import android.widget.TextView;

public class MyLocation extends Activity {
    LocationManager mLocationManager;
    Location mLocation;
    TextView tv;

    @Override
    public void onCreate(Bundle savedInstanceState) {
        super.onCreate(savedInstanceState);

        setContentView(R.layout.main);
        tv = (TextView) findViewById(R.id.tv1);

        mLocationManager = (LocationManager)
                    getSystemService(Context.LOCATION_SERVICE);

        Criteria criteria = new Criteria();
        criteria.setAccuracy(Criteria.ACCURACY_FINE);
        criteria.setPowerRequirement(Criteria.POWER_LOW);
        String locationprovider =
                    mLocationManager.getBestProvider(criteria,true);

        mLocation =
                    mLocationManager.getLastKnownLocation(locationprovider);

        List<Address> addresses;

        String myAddress="Seattle,WA";
        Geocoder gc = new Geocoder(this);
        try {
            addresses = gc.getFromLocationName(myAddress, 1);
            if(addresses != null) {
                Address x = addresses.get(0);
                StringBuilder mSB = new StringBuilder("Address:\n");

                mSB.append("latitude: ").append(x.getLatitude());
                mSB.append("\nlongitude: ").append(x.getLongitude());
                tv.setText(mSB.toString());
            }
        } catch(IOException e) {
            tv.setText(e.getMessage());
        }
    }
}
```

图 12-3　地理编码示例，将地址字符串转换为经纬度坐标

## 12.2　使用 Google 地图

在 Android 系统中使用 Google 地图的方法有两种：通过浏览器的用户访问，以及通过 Google 地图 API 的应用程序访问。MapView 类是对 Google 地图 API 的一个封装。要使用 MapView 和 Google 地图的第 1 版，需要进行如下设置工作。

（1）下载并安装 Google API 的 SDK，具体步骤如下。

- 在 Eclipse 中使用 Android SDK 和 AVD manager 下载 Google API。
- 在要使用 API 的项目上右击，并选择 **Properties**。
- 选择 **Android**，再选择 **Google API**，从而为本项目启用这个 SDK。

（2）为使用 Google 地图服务，需获取一个有效的地图 API 密钥，具体步骤如下（参见 http://code.google.com/android/add-ons/google-apis/mapkey.html）。

- 使用 keytool 命令，为密钥 alias_name 生成一个 MD5 证书指纹：

```
> keytool -list -alias alias_name -keystore my.keystore
> result:(Certificate fingerprint (MD5):
        94:1E:43:49:87:73:BB:E6:A6:88:D7:20:F1:8E:B5)
```

- 使用 MD5 密钥库为 Google 地图服务进行签名，网址为 http://code.google.com/android/maps-api-signup.html。

（3）在 **AndroidManifest.xml** 文件中包含<uses-library android:name="com.google.android.maps" />，通知 Android 系统应用程序会使用来自 Google API SDK 的 com.google.android.maps 库。

（4）向 **AndroidManifest.xml** 文件中添加 android.permission.INTERNET 权限，这样应用程序就可以使用因特网接收来自 Google 地图服务的数据。

（5）在布局 XML 文件中包含一个 MapView。

更具体地说，使用 Google 地图的 Activity 需要两个支持文件。首先，**AndroidManifest.xml** 文件需要包含合适的地图库及权限，如代码清单 12-8 所示。

**代码清单 12-8　AndroidManifest.xml**

```
<?xml version="1.0" encoding="utf-8"?>
<manifest xmlns:android="http://schemas.android.com/apk/res/android"
      package="com.cookbook.using_gmaps"
      android:versionCode="1"
      android:versionName="1.0">
    <application android:icon="@drawable/icon"
                  android:label="@string/app_name">
        <activity android:name=".MyLocation"
                  android:label="@string/app_name">
            <intent-filter>
              <action android:name="android.intent.action.MAIN" />
              <category android:name="android.intent.category.LAUNCHER" />
            </intent-filter>
        </activity>
    <uses-library android:name="com.google.android.maps" />
    </application>
    <uses-sdk android:minSdkVersion="4" />
<uses-permission android:name="android.permission.INTERNET" />
<uses-permission android:name="android.permission.ACCESS_FINE_LOCATION"/>
</manifest>
```

其次，布局 XML 文件中需要声明合适的 MapView，以显示 Google 地图，如代码清单 12-9 所示。还可以通过声明 clickable 元素声明用户是否可以与地图进行交互，默认值为 false。后面的技巧会用到这一点。

**代码清单 12-9　res/layout/main.xml**

```
<?xml version="1.0" encoding="utf-8"?>
<LinearLayout xmlns:android="http://schemas.android.com/apk/res/android"
    android:orientation="vertical"
    android:layout_width="match_parent"
    android:layout_height="match_parent"
    >
<TextView
    android:id="@+id/tv1"
    android:layout_width="match_parent"
    android:layout_height="wrap_content"
    android:text="@string/hello"
    />
<com.google.android.maps.MapView
    android:id="@+id/map1"
    android:layout_width="match_parent"
    android:layout_height="match_parent"
    android:clickable="true"
    android:apiKey="0ZDUMMY13442HjX491CODE44MSsJzfDVlIQ"
    />
</LinearLayout>
```

注意，对于 Google 地图 API 版本第 2 版，有下列变化。

- 对 API 密钥的获取通过 Google API 控制台来完成（https://code.google.com/apis/console/）。
- 现在需要下列权限：

```
android.permission.INTERNET
android.permission.ACCESS_NETWORK_STATE
android.permission.WRITE_EXTERNAL_STORAGE
com.google.android.providers.gsf.permission.READ_GSERVICES
```

- 需要 OpenGL ES 版本 2，可以通过包含如下的`<uses-feature>`元素获取：

```
<uses-feature> element:
    <uses-feature
      android:glEsVersion="0x00020000"
      android:required="true"/>
```

- 在主布局 XML 文件中要加入下面的 Fragment：

```
<fragment xmlns:android="http://schemas.android.com/apk/res/android"
  android:id="@+id/map"
  android:layout_width="match_parent"
  android:layout_height="match_parent"
  android:name="com.google.android.gms.maps.MapFragment"/>
```

- 要确保在 `onCreate()` 中对包含上述 Fragment 的 XML 文件使用了 `setContentView`。例如：

```
setContentView(R.layout.main);
```

要了解 Google 地图 Android API 第 2 版的更多信息，可访问 https://developers.google.com/maps/documentation/android/start。

## 技巧 117：向应用程序中添加 Google 地图

要显示 Google 地图，需要让主 Activity 扩展 `MapActivity`，如代码清单 12-10 所示。它还必须指向主布局 XML 文件中地图布局的 ID，本例中为 `map1`。注意，还需要实现 `isRouteDisplayed()` 方法。显示结果如图 12-4 所示。

**代码清单 12-10　src/com/cookbook/using_gmaps/MyLocation.java**

```
package com.cookbook.using_gmaps;

import android.content.Context;
import android.location.Criteria;
import android.location.Location;
import android.location.LocationManager;
import android.os.Bundle;
import android.widget.TextView;

import com.google.android.maps.MapActivity;
import com.google.android.maps.MapView;

public class MyLocation extends MapActivity {
    LocationManager mLocationManager;
```

```
    Location mLocation;
    TextView tv;

    @Override
    public void onCreate(Bundle savedInstanceState) {
        super.onCreate(savedInstanceState);

        setContentView(R.layout.main);
        MapView mapView = (MapView) findViewById(R.id.map1);
        tv = (TextView) findViewById(R.id.tv1);

        mLocationManager = (LocationManager)
                getSystemService(Context.LOCATION_SERVICE);
        Criteria criteria = new Criteria();
        criteria.setAccuracy(Criteria.ACCURACY_FINE);
        criteria.setPowerRequirement(Criteria.POWER_LOW);
        String locationprovider =
                mLocationManager.getBestProvider(criteria,true);

        mLocation =
                mLocationManager.getLastKnownLocation(locationprovider);

        tv.setText("Last location lat:" + mLocation.getLatitude()
                + "long:" + mLocation.getLongitude());
    }

    @Override
    protected boolean isRouteDisplayed() {
        // This method is required
        return false;
    }
}
```

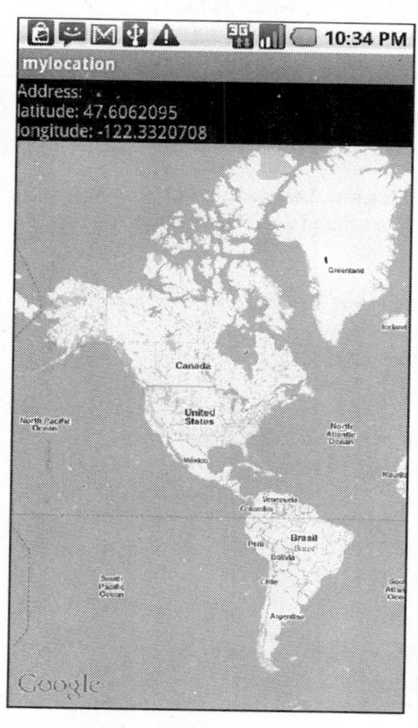


图 12-4 在应用程序内使用 Google 地图的例子

## 技巧 118：为地图添加标记

ItemizedOverlay 类提供了一种在 MapView 上绘制标记和覆盖物的方法。该类通过列表管理一个诸如一幅图像之类的 OverlayItem 元素的集合，并为每个元素处理绘制、放置、点击、焦点控制以及布局优化。我们要创建一个扩展了 ItemizedOverlay 的类并重写下列函数。

- addOverlay()：向 ArrayList 添加一个 OverlayItem。这将调用 populate()，读取项目并为绘制做准备。
- createItem()：被 populate()调用，检索给定的 OverlayItem。
- size()：返回 ArrayList 中的元素数量。
- onTap()：点击标记时的回调方法。

新创建的类在代码清单 12-11 中给出，运行结果则如图 12-5 所示。

**代码清单 12-11　src/com/cookbook/adding_markers/MyMarkerLayer.java**

```
package com.cookbook.adding_markers;

import java.util.ArrayList;

import android.app.AlertDialog;
import android.content.DialogInterface;
import android.graphics.drawable.Drawable;

import com.google.android.maps.ItemizedOverlay;
import com.google.android.maps.OverlayItem;

public class MyMarkerLayer extends ItemizedOverlay {

    private ArrayList<OverlayItem> mOverlays =
              new ArrayList<OverlayItem>();

    public MyMarkerLayer(Drawable defaultMarker) {
        super(boundCenterBottom(defaultMarker));
        populate();
    }
    public void addOverlayItem(OverlayItem overlay) {
        mOverlays.add(overlay);
        populate();
    }
    @Override
    protected OverlayItem createItem(int i) {
        return mOverlays.get(i);
    }
    @Override
    public int size() {
        return mOverlays.size();
    }
    @Override
    protected boolean onTap(int index) {
        AlertDialog.Builder dialog =
              new AlertDialog.Builder(MyLocation.mContext);
        dialog.setTitle(mOverlays.get(index).getTitle());
        dialog.setMessage(mOverlays.get(index).getSnippet());
```

```
            dialog.setPositiveButton("OK",
              new DialogInterface.OnClickListener() {
                public void onClick(DialogInterface dialog, int whichButton) {
                    dialog.cancel();
              }
            });
            dialog.setNegativeButton("Cancel",
              new DialogInterface.OnClickListener() {
                public void onClick(DialogInterface dialog, int whichButton) {
                    dialog.cancel();
              }
            });
            dialog.show();
            return super.onTap(index);
        }
    }
```

图 12-5　在地图上添加一个可点击的标记

下面对代码清单 12-11 中用粗体标出的代码展开几点说明。

- 声明了一个 OverlayItem 的容器 mOverlays，用于保存传递给覆盖物的所有项目。
- 在任何覆盖物项目被绘制之前，需要定义一个绑定点，所有覆盖的项目都通过它附加到地图上。为将该点指定为地图的底部中心，我们向类构造函数中添加了 boundCenterBottom。
- 重写了所需的方法，即 addOverlay()、createItem()、size()和 onTap()。在此，onTap()方法会在项目被点击时提供一个对话框。
- 在构造函数及 addOverlay()的末尾加上了 populate()方法，这样会告知 MyMarkerLayer 类要准备好全部的 OverItem 元素，并逐个绘制到地图上。

现在，可以将 ItemizedOverlay 添加到上一个技巧创建的 MapActivity 中。如代码清单 12-12 中粗体部分所示。

- 使用 mapView 的 getOverlays()方法检索已有的地图覆盖物项目。在函数的结尾处，标记层被添加到这个容器中。
- 定义一个 MyMarkerLayer 实例处理覆盖物项目。
- 检索地址对应的纬度和经度（以度为单位）。使用 GeoPoint 类定义目标点（point of interest）。Geopoint 接受的输入是以微度为单位的，因此需要把纬度和经度都乘以一百万（1E6）。
- 使用一个地图控制器来绘制指向 GeoPoint 的动画以及缩放视图。另外，使用 setBuiltInZoomControls()启用用户控制的缩放功能。
- 定义一个 OverlayItem 作为在目标 Geopoint 点处的消息。
- 使用 addOverlayItem()方法将项目添加到 MyMarkerLayer 上。该方法还会把已定义的 MyMarkerLayer 放到步骤 1 中已产生的覆盖物列表中。

**代码清单 12-12　src/com/cookbook/adding_markers/MyLocation.java**

```
package com.cookbook.adding_markers;

import java.io.IOException;
import java.util.List;

import android.content.Context;
import android.graphics.drawable.Drawable;
import android.location.Address;
import android.location.Geocoder;
import android.os.Bundle;
import android.widget.TextView;

import com.google.android.maps.GeoPoint;
import com.google.android.maps.MapActivity;
import com.google.android.maps.MapController;
import com.google.android.maps.MapView;
import com.google.android.maps.Overlay;

public class MyLocation extends MapActivity {
    TextView tv;
    List<Overlay> mapOverlays;
    MyMarkerLayer markerlayer;
    private MapController mc;
    public static Context mContext;

    @Override
    public void onCreate(Bundle savedInstanceState) {
        super.onCreate(savedInstanceState);
        mContext = this;
        setContentView(R.layout.main);
        MapView mapView = (MapView) findViewById(R.id.map1);
        tv = (TextView) findViewById(R.id.tv1);

        mapOverlays = mapView.getOverlays();
        Drawable drawable =
                  this.getResources().getDrawable(R.drawable.icon);
        markerlayer = new MyMarkerLayer(drawable);

        List<Address> addresses;
```

```
        String myAddress="1600 Amphitheatre Parkway, Mountain View, CA";

        int geolat = 0;
        int geolon = 0;

        Geocoder gc = new Geocoder(this);
        try {
            addresses = gc.getFromLocationName(myAddress, 1);
            if(addresses != null) {
              Address x = addresses.get(0);

              geolat = (int)(x.getLatitude()*1E6);
              geolon = (int)(x.getLongitude()*1E6);
            }
        } catch(IOException e) {
            tv.setText(e.getMessage());
        }

        mapView.setBuiltInZoomControls(true);
        GeoPoint point = new GeoPoint(geolat,geolon);
        mc = mapView.getController();
        mc.animateTo(point);
        mc.setZoom(3);

        OverlayItem overlayitem =
                new OverlayItem(point, "Google Campus", "I am at Google");
        markerlayer.addOverlayItem(overlayitem);
        mapOverlays.add(markerlayer);
    }

    @Override
    protected boolean isRouteDisplayed() { return false; }
}
```

## 技巧 119：向地图上添加视图

开发者可以向 MapView 中添加任意的 View 或 ViewGroup。本技巧显示了如何向地图中添加两个简单的元素，一个 TextView 和一个 Button。按钮被点击后，TextView 中的文本会改变。

将这两个视图添加到 MapView 上的方式是通过 LayoutParams 参数调用 addView() 方法。此处，元素的位置通过屏幕坐标对 $(x, y)$ 来指定，但开发者也可以给 LayoutParams 提供一个 GeoPoint 类作为替代。代码清单 12-13 给出了主 Activity，其中需要上一个技巧所定义的 MyMarkerLayer（见代码清单 12-11，只需修改其中的第一行，以反映包名的变化）。MapView 的显示结果如图 12-6 所示。

**代码清单 12-13 src/com/cookbook/mylocation/MyLocation.java**

```
package com.cookbook.mylocation;

import java.io.IOException;
import java.util.List;

import android.content.Context;
import android.content.Intent;
import android.graphics.Color;
import android.graphics.drawable.Drawable;
```

```
import android.location.Address;
import android.location.Geocoder;
import android.os.Bundle;
import android.view.View;
import android.view.View.OnClickListener;
import android.widget.Button;
import android.widget.TextView;
import com.google.android.maps.GeoPoint;
import com.google.android.maps.MapActivity;
import com.google.android.maps.MapController;
import com.google.android.maps.MapView;
import com.google.android.maps.Overlay;

public class MyLocation extends MapActivity {
    TextView tv;
    List<Overlay> mapOverlays;
    MyMarkerLayer markerlayer;
    private MapController mc;
    MapView.LayoutParams mScreenLayoutParams;
    public static Context mContext;

    @Override
    public void onCreate(Bundle savedInstanceState) {
        super.onCreate(savedInstanceState);
        mContext = this;
        setContentView(R.layout.main);

        MapView mapView = (MapView) findViewById(R.id.map1);
        mc = mapView.getController();
        tv = (TextView) findViewById(R.id.tv1);
        mapOverlays = mapView.getOverlays();
        Drawable drawable =
                   this.getResources().getDrawable(R.drawable.icon);
        markerlayer = new MyMarkerLayer(drawable);

        List<Address> addresses;
        String myAddress="1600 Amphitheatre Parkway, Mountain View, CA";

        int geolat = 0;
        int geolon = 0;
        Geocoder gc = new Geocoder(this);
        try {
          addresses = gc.getFromLocationName(myAddress, 1);
          if(addresses != null) {
              Address x = addresses.get(0);

              StringBuilder mSB = new StringBuilder("Address:\n");
              geolat =(int)(x.getLatitude()*1E6);
              geolon = (int)(x.getLongitude()*1E6);
              mSB.append("latitude: ").append(geolat).append("\n");
              mSB.append("longitude: ").append(geolon);
              tv.setText(mSB.toString());
          }
        } catch(IOException e) {
          tv.setText(e.getMessage());
        }

        int x = 50;
        int y = 50;
        mScreenLayoutParams =
            new MapView.LayoutParams(MapView.LayoutParams.WRAP_CONTENT,
                                     MapView.LayoutParams.WRAP_CONTENT,
                                     x,y,MapView.LayoutParams.LEFT);

        final TextView tv = new TextView(this);
        tv.setText("Adding View to Google Map");
```

```
        tv.setTextColor(Color.BLUE);
        tv.setTextSize(20);
        mapView.addView(tv, mScreenLayoutParams);

        x = 250;
        y = 250;
        mScreenLayoutParams =
            new MapView.LayoutParams(MapView.LayoutParams.WRAP_CONTENT,
                                     MapView.LayoutParams.WRAP_CONTENT,
                                     x,y,
                                     MapView.LayoutParams.BOTTOM_CENTER);

        Button clickMe = new Button(this);
        clickMe.setText("Click Me");
        clickMe.setOnClickListener(new OnClickListener() {
            public void onClick(View v) {
                tv.setTextColor(Color.RED);
                tv.setText("Let's play");
            }
        });

        mapView.addView(clickMe, mScreenLayoutParams);
    }

    @Override
    protected boolean isRouteDisplayed() { return false; }
}
```

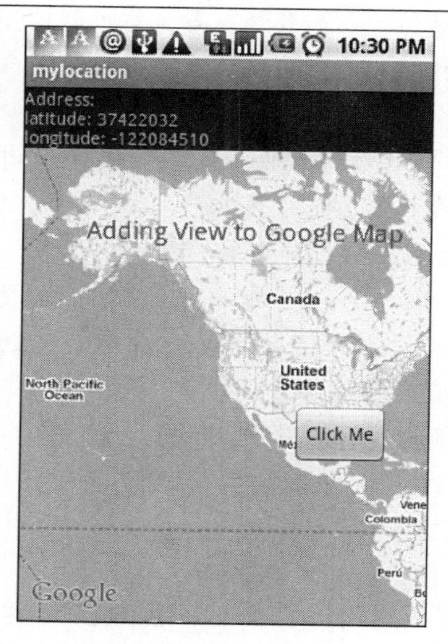


图 12-6 在地图上添加一个 `TextView` 和一个按钮

## 技巧 120：设置临近警告

`LocationManager` 提供了一种设置临近警告的方法，即当用户进入或离开指定的区域时触发警告。我们通过创建包含经纬度坐标及半径（以米为单位）的变量来指定区域。警报由一个 `PendingIntent` 规定，只要用户进入或离开指定的区域，该 `PendingIntent` 就会启动。还可以定义警报的失效时间，以毫秒为单位。代码清单 12-14 给出了一个实现示例。

**代码清单 12-14　创建一个不带失效时间的临近警告**

```
double mlatitude=35.41;
double mlongitude=139.46;

float mRadius=500f; // in meters

long expiration=-1; //-1 never expires or use milliseconds

Intent mIntent = new Intent("You entered the defined area");
PendingIntent mFireIntent
              = PendingIntent.getBroadCast(this, -1, mIntent, 0);

mLocationManager.addProximityAlert(mlatitude, mlongitude,
                                   mRadius, expiration, mFireIntent);
```

## 12.3　使用 Little Fluffy 位置库

在位置精度和保持合理的用电量之间做出权衡，有时候是件棘手的事情。如果用户为了获取地点修正而让 GPS 保持打开，将会耗掉可观的电量。如果应用程序只需要对用户位置的粗略的估计，则可以使用粗略定位（coarse location）来节约电池电量。然而，很多应用程序能同时从粗略定位和精确定位二者中获益，这需要开发者试着去计划如何处理对位置的获取方式，平衡不同方法（精确定位和定位粗略）的使用；还要考虑在应用程序移入后台后如何处理对位置的获取，使得电池电量的消耗不超过必要的范围。

Kenton Price of Little Fluffy Toys 公司编写了一个面向 Android 2.1 及以上版本的库，名为 Little Fluffy Location Library，其中考虑了上述问题。引用了该库的项目可以使用一个广播动作，后者包含一个 `LocationInfo` 对象，带有如下字段。

- `lastLocationUpdateTimestamp`：上次位置更新的时间，以毫秒为单位。
- `lastLocationBroadcastTimestamp`：上次位置更新广播的时间，以毫秒为单位。
- `lastLat`：上次更新的纬度，以度为单位。
- `lastLong`：上次更新的经度，以度为单位。
- `lastAccuracy`：上次更新的精确度，以米为单位。

此外，对象还包含以下工具方法。

- `refresh`：使用最新信息刷新所有字段。
- `anyLocationDataReceived`：判定上次重启之后是否接收到过新的位置数据。
- `anyLocationDataBroadcast`：判定上次重启之后是否有位置信息被广播出去。
- `hasLatestDataBeenBroadcast`：判定 `LocationInfo` 对象中包含的数据是否已经被广播过。
- `getTimestampAgeInSeconds`：返回距离上一次更新有多长时间，以秒为单位。

要使用 Little Fluffy 位置库，首先要到 http://code.google.com/p/little-fluffy-location-library/ 下载 **littlefluffylocationlibrary.jar** 文件，并通过将其复制到 **/libs** 目录将其包含到项目中。复制完成后，在文件上右击，选择 **Build Path→Add to Build Path**。

## 技巧 121：使用 Little Fluffy 位置库添加通知

要在项目中使用 Little Fluffy 位置库，需要向 manifest 文件中添加以下权限：

- ACCESS_FINE_LOCATION
- INTERNET
- RECEIVE_BOOT_COMPLETED
- ACCESS_COARSE_LOCATION
- ACCESS_FINE_LOCATION

还要向 manifest 文件中添加以下包引用：

- Android.hardware.location
- Android.hardware.location.gps

另外在 application 元素内还要加入下列元素：

```
<service android:name="com.littlefluffytoys.littlefluffylocationlibrary.
➥LocationBroadcastService" />
<receiver android:name="
➥com.littlefluffytoys.littlefluffylocationlibrary.StartupBroadcastReceiver
➥" android:exported="true">
  <intent-filter>
    <action android:name="android.intent.action.BOOT_COMPLETED" />
  </intent-filter>
</receiver>
<receiver
➥android:name="com.littlefluffytoys.littlefluffylocationlibrary.PassiveLocation
➥ChangedReceiver" android:exported="true" />
<receiver android:name=".FluffyBroadcastReceiver">
    <intent-filter>
        <action
➥android:name="com.cookbook.fluffylocation.littlefluffylocationlibrary.
➥LOCATION_CHANGED" android:exported="true"/>
    </intent-filter>
</receiver>
```

上述添加的元素会建立一个服务和一个接收器，用于从 Little Fluffy 位置库取得数据。其中最后一个 receiver 标签块会建立一个用于触发通知的广播接收器。代码清单 12-15 显示了在 manifest 中被引用为接收器的类。

**代码清单 12-15　/src/com/cookbook/fluffylocation/FluffyBroadcastReceiver.java**

```java
package com.cookbook.fluffylocation;

import android.app.Notification;
import android.app.NotificationManager;
import android.app.PendingIntent;
import android.content.BroadcastReceiver;
import android.content.Context;
import android.content.Intent;
import android.util.Log;

import com.littlefluffytoys.littlefluffylocationlibrary.LocationInfo;
import com.littlefluffytoys.littlefluffylocationlibrary.LocationLibraryConstants;
```

```
public class FluffyBroadcastReceiver extends BroadcastReceiver{
  @Override
    public void onReceive(Context context, Intent intent) {
      Log.d("LocationBroadcastReceiver", "onReceive: received location update");

      final LocationInfo locationInfo = (LocationInfo) intent
          .getSerializableExtra(LocationLibraryConstants
              .LOCATION_BROADCAST_EXTRA_LOCATIONINFO);

      // For API 16+ use Notification.Builder instead of Notification
      Notification notification = new Notification(R.drawable.ic_launcher,
          "Locaton updated " +
          locationInfo.getTimestampAgeInSeconds() +
          " seconds ago", System.currentTimeMillis());

      Intent contentIntent = new Intent(context, MainActivity.class);
      PendingIntent contentPendingIntent = PendingIntent.getActivity(context,
          0, contentIntent, PendingIntent.FLAG_UPDATE_CURRENT);

      notification.setLatestEventInfo(context, "Location update broadcast received",
          "Timestamped " +
          LocationInfo
              .formatTimeAndDay(locationInfo.lastLocationUpdateTimestamp, true),
                  contentPendingIntent);

      ((NotificationManager) context
          .getSystemService(Context.NOTIFICATION_SERVICE))
              .notify(1234, notification);
  }
}
```

代码清单 12-15 显示了对广播接收器的基本设置。其中对 Log.d()方法的使用意味着我们利用了日志技术来帮助调试应用程序。

还需要关注的是如何建立通知。notify()方法目前使用了 1234 这个值，这是一个与通知相绑定的 ID。如果已经存在全局通知端口（比如 NOTIFICATION_PORT），则可以用该端口来替换 1234。本例中的通知对于从 Gingerbread（2.3）到 Honeycomb（3.0）这些版本都是有效的。但从 API Level 16（Jelly Bean）版本开始，Notification 的构造函数就被废弃了。所以面向新近版本开发时，必须将 Notification 的构造函数转变为 Notification.Builder。

既然接收器已经就位了，下面就需要建立应用程序类。为此首先要确认已经给应用程序起了名字。在 manifest 文件中，application 元素应当含有如下属性：

```
android:name="com.cookbook.fluffylocation.FluffyApplication"
```

需要引用应用程序类的完整路径。在本例中使用 com.cookbook.fluffylocation.FluffyApplication。代码清单 12-16 给出了位于这一位置的类文件。

**代码清单 12-16　/src/com/cookbook/fluffylocation/FluffyApplication.java**

```
package com.cookbook.fluffylocation;

import com.littlefluffytoys.littlefluffylocationlibrary.LocationLibrary;

import android.app.Application;
import android.util.Log;
```

```
public class FluffyApplication extends Application {
  @Override
    public void onCreate() {
      super.onCreate();
      // Show debugging information
      Log.d("FluffyApplication", "onCreate()");

      LocationLibrary.showDebugOutput(true);

      // Default call would be the following:
      // LocationLibrary.initialiseLibrary(getBaseContext(),
      //       "com.cookbook.fluffylocation");

      // For testing, make request every 1 minute, and force a location update
      // if one hasn't happened in the last 2 minutes
      LocationLibrary.initializeLibrary(getBaseContext(),
             60 * 1000, 2 * 60 * 1000, "com.cookbook.fluffylocation");
    }
}
```

代码清单 12-16 给出了一个 onCreate() 方法，其中使用 Log.d() 方法设置了日志功能；还使用了 showDebugOutput(true) 来启用额外的调试信息。调试信息是很有用的，因为其中表明了库被装载的时间以及信息从应用程序传过来的时间。在注释中及 onCreate() 方法的尾部，我们都可以看到初始化 Little Fluffy 位置库的调用。使用这样的调用对于获取位置数据来说有些杀鸡用牛刀了，但出于开发方面的考虑，它能够每分钟都通知用户是否接收了新的位置数据。图 12-7 显示了 Little Fluffy 库的使用效果。

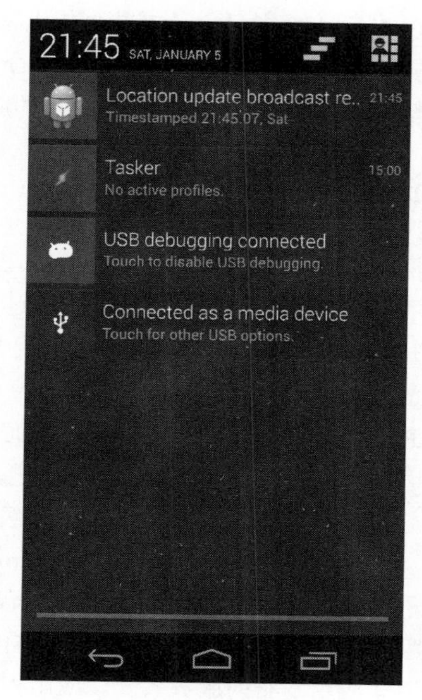


图 12-7 一条显示了从 Little Fluffy 位置库获取来的信息的通知

# 第 13 章

# 应用内计费

创建一个拥有追加销售（up-sell）能力的应用程序，允许用户购买物品或者添加功能，在应用程序的营销策略中扮演重要角色。多数用户可能熟悉“pay-to-win”策略和“礼物”系统。pay-to-win 系统通常允许用户免费下载应用程序，但用户可用的升级或游戏时间是受限的，除非花钱购买升级。礼物系统的工作原理相似，但用户不是花钱为自己升级，而是为其他人购买物品。

不久之前，Google 还没有在 Android 平台上为上述模型提供官方的支持系统，开发者需要建立自己的系统，或者集成第三方的附加销售服务或产品。不过如今，Google 已经为我们提供了一套非常健壮的集成计费系统。本章就将介绍如何对 Google 提供的应用内计费解决方案加以实现。

## Google Play 应用内计费

Google 提供了一套为应用程序添加应用内计费机制的 API。我们只能卖数字虚拟物品，没有任何一个 API 提供看得见摸得着的物理物品的销售功能。通过应用内购买方式出售的商品，要么为用户所拥有（比如进阶升级）或者为用户所消费（如游戏中的升级或钱币）。以应用内计费方式购买的东西是不能退换的。

如今有两个版本的计费 API 可用，不过版本 2 已经基本悬而不用，Google 极力促使用户升级到版本 3。尽管目前还不知道对版本 2 的支持何时会彻底取消，但新的开发者使用 API 时理应选择版本 3。2013 年初，Google 宣布将对 API 版本 3 进行升级，以支持订金功能，并添加对版本 2 引入的所有特性的支持。API 版本 2 的最低环境要求为 Android 1.6（API Level 4）以及 Google Play 3.5 版。API 版本 3 则最低要求 Android 2.2（API Level 8）以及 Google Play 3.9.16。

使用 Google 提供的任何版本的应用内计费 API 都需要遵守一个约定，即应用程序必须在 Google Play 商店中上架，并遵守应用发布的服务条款。应用程序还必须能够通过网络连接与 Google Play 服务器进行通信。

想通过 Google Play 使用应用内计费的开发者必须拥有一个商家账户（merchant account）。如果开发者已经在 Play 市场中创建了账户，则可登录开发者平台（https://play.google.com/apps/publish/），其中能找到在 Google Checkout 中建立商家账户的链接。该页面逐步显示了如何建立商家账户并将其与开发者账户连接。也可以到 Google Checkout Merchant 板块（www.google.com/wallet/merchants.html）去直接建立一个商家账户。在为一个应用程序测试应用内计费时，必须使用真实的信用卡，但产生的转款是可以退回的。

目前[①]Google 正在迁移开发者平台，如果找不到添加商家账户的链接，可以添加一个新应用程序，这样链接就应该会出现在 Price and Distribution 这部分下面。

## 技巧 122：安装 Google 的应用内计费服务

Google 提供了 Google Play 计费库（Billing Library）。该库包含了连接到 Google 的应用内计费服务所需的全部类和接口。可以从 Android SDK 安装它，位于 **SDK Manager** 中的 **Extras** 部分。图 13-1 显示了该库所在的位置。

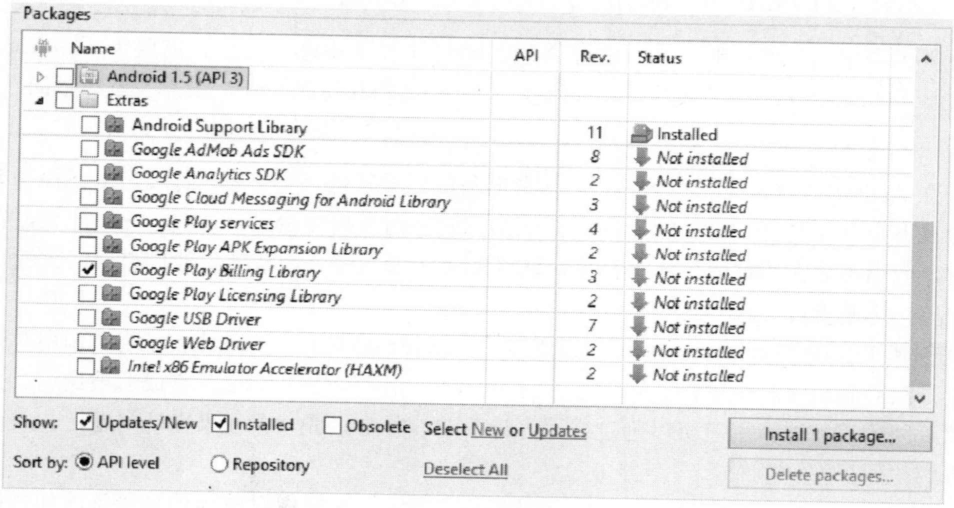

Packages

| Name | API | Rev. | Status |
|---|---|---|---|
| Android 1.5 (API 3) | | | |
| Extras | | | |
| Android Support Library | | 11 | Installed |
| Google AdMob Ads SDK | | 8 | Not installed |
| Google Analytics SDK | | 2 | Not installed |
| Google Cloud Messaging for Android Library | | 3 | Not installed |
| Google Play services | | 4 | Not installed |
| Google Play APK Expansion Library | | 2 | Not installed |
| ✔ Google Play Billing Library | | 3 | Not installed |
| Google Play Licensing Library | | 2 | Not installed |
| Google USB Driver | | 7 | Not installed |
| Google Web Driver | | 2 | Not installed |
| Intel x86 Emulator Accelerator (HAXM) | | 2 | Not installed |

Show: ✔ Updates/New ✔ Installed Obsolete Select New or Updates Install 1 package...

Sort by: ◉ API level ○ Repository Deselect All Delete packages...

图 13-1 安装 Google Play 计费库版本 3

安装 Google Play 计费库会在 SDK 安装目录下添加一些文件夹和文件，其中包括应用内计费实例应用程序，可供我们参考。这些添加的内容可以在文件系统中的 ***SDKInstallationDirectory*/extras/google/play_billing/in-app-billing-v03** 或者 ***SDKInstallationDirectory*/google-marker-billing/in-app-billing-v03** 文件夹中找到。其中包括 **IInAppBillingService.aidl**，任何要包含应用内计费功能的项目都需要引用这个文件。

在所需文件都添加进开发环境之后，还需要生成一个公钥。登录到开发者平台，创建一个

① 指本书写作时，本书原版的出版时间为 2013 年 6 月。——译者注

新的应用程序，为该程序起名，并点击 Prepare Store Listing 按钮。在新出现的页面的左侧会有若干标签，请定位到 Service & APIs 标签，将其中为应用程序生成的公钥复制下来。

下面我们来试验一下 Google 提供的示例应用程序 **TravialDrive**。创建一个新项目（使用默认选项，包括默认的 Activity 名称 **MainActivity**），并将示例应用中的内容复制到新建的应用程序中。下面对我们的应用程序做点小修改，让其中的类名与选定的包名一致，并修改 **AndroidManifest.xml**，让其中的内容也与包名匹配。

要为现有的应用程序添加应用内计费，需要把 **IInAppBillingService.aidl** 复制到项目的 **src** 目录下。注意，如果并未使用 Eclipse 作为集成开发环境，则要在 **src** 目录中创建如下路径，并将 **IInAppBillingService.aidl** 文件放入其中：

```
com/android/vending/billing
```

要验证安装的正确性，应对项目进行编译，并确保 **gen** 文件夹中含有 **IInAppBillingService.aidl** 文件。

## 技巧 123：为 Activity 添加应用内计费机制

要提供应用内计费机制，应用程序必须能够与计费服务进行通信。需要为 manifest XML 文件添加 `BILLING` 权限才能使功能可用。除应用所需的其他权限外，还需要添加下面一行：

```
<uses-permission android:name="android.permission.BILLING"/>
```

要在 Activity 和 Google Play 应用内计费服务间建立连接，需创建一个 `IabHelper` 对象。将当前上下文以及在开发者平台为应用程序生成的公钥一起传给 `IabHelper`。注意，在使用公钥时，应当考虑在运行时生成字符串。这样能阻止用户将公钥替换为他们自己的公钥，从而欺骗服务，逃避支付。

在创建了 `IabHelper` 之后，通过调用它的 `startSetup()` 方法绑定服务。调用时要传递另一个方法 `OnIabSetupFinishedListener()`，后者会在异步设置完成后被调用。会有一个对象返回给该方法，用它可以判定应用内计费的设置是否成功。如果出现问题，相关信息会在对象中传回。

Activity 关闭时，要解除与应用内计费服务的绑定，这样做对整体的系统资源和性能是有好处的。可以通过调用 `IabHelper` 对象的 `dispose()` 方法实现。

代码清单 13-1 给出了通过 Google Play 建立应用内计费机制的一份范例代码。

**代码清单 13-1 应用内计费范例**

```
IabHelper mHelper;

@Override
public void onCreate(Bundle savedInstanceState) {
  super.onCreate(savedInstanceState);
  setContentView(R.layout.activity_main);

  // Consider building the public key at run-time
```

```
    String base64EncodedPublicKey = "YourGeneratedPublicKey";

    mHelper = new IabHelper(this, base64EncodedPublicKey);

    mHelper.startSetup(new IabHelper.OnIabSetupFinishedListener() {
      public void onIabSetupFinished(IabResult result) {
        if (!result.isSuccess()) {
          // Replace Toast with error-handling logic
          Toast.makeText(context, "iab fail: "+result, Toast.LENGTH_LONG).show();
          return;
        }

        // iab successful, handle success logic here
      }
    });
  }
  @Override
  public void onDestroy() {
    if (mHelper != null) mHelper.dispose();
    mHelper = null;
  }
```

## 技巧 124：列出应用内可购买的项目清单

要想让用户在应用内进行购买，先要让他们知道有哪些可买的东西。可以在开发者平台上设置可购买的项目。每个项目创建时都带有一个项目编号或者 SKU[①]，价格则可定在 0.99 美元～200 美元之间。只要至少有一个项目可供用户购买，就可以编程请求 Google Play 通过应用内计费服务列出项目清单。

要在 Google Play 中为应用程序查询项目清单，可使用 `queryInventoryAsync()` 方法，并根据返回的对象编程决定相应逻辑。要在代码清单 13-1 的基础上实现，就需要在 `onCreate()` 方法中添加对 `queryInventoryAsync()` 方法的调用，该调用在应用内计费设置完成时进行。在安装成功之后，添加下面一行代码：

```
mHelper.queryInventoryAsync(mCurrentInventoryListener);
```

代码清单 13-2 显示了如何建立在 `queryInventoryAsync()` 方法中用到的 `Listener`。该监听器用来监听从 Google Play 服务传送回来的项目清单。

**代码清单 13-2 创建一个针对项目清单结果的监听器**

```
IabHelper.QueryInventoryFinishedListener mGotInventoryListener = new
IabHelper.QueryInventoryFinishedListener() {
  public void onQueryInventoryFinished(IabResult result, Inventory inventory) {
      if (result.isFailure()) {
        Toast.makeText(context, "inventory fail: "+result, Toast.LENGTH_LONG).show();
        return;
      }
      // Inventory has been returned, create logic with it
```

① SKU，即 Stock Keeping Unit，原意为库存量单位，即存进出计量的单位（可以是件、盒、托盘等），是物流管理的一种方法。后来被引申为产品统一编号的简称，即每种产品均对应有唯一的 SKU 号。此处当取引申义。——译者注

```
        // Do UI updating logic here
    }
};
```

要让用户可以为程序购买某个项目，应使用 launchPurchaseFlow()方法。该方法接受 5 个参数：Activity、产品 ID（String）、请求代码值（int）、要通知的监听器（OnIabPurchaseFinishedListener），以及 payload（String）。Google 建议使用 payload 来存储客户识别信息，用于购买时的验证，尽管它可能是任意随机生成的字符串。对 launchPurchaseFlow()方法的调用形式如下所示，可以把它放到一个触发事件中（比如按钮的点击事件）：

```
mHelper.launchPurchaseFlow(this, YOUR_SKU, 12345,
    mPurchaseFinishedListener, "R4nd0mb17+0hs7r1nGz/");
```

一旦订购成功，就会返回一个 Purchase 对象。处理它的办法跟处理 queryInventoryAsync()方法的办法差不多，在 Listener 中为返回的 Purchase 对象设置处理逻辑。代码清单 13-3 给出了一个实现示例。

**代码清单 13-3 完成购买**

```
IabHelper.OnIabPurchaseFinishedListener
    mPurchaseFinishedListener = new IabHelper.OnIabPurchaseFinishedListener() {
  public void onIabPurchaseFinished(IabResult result, Purchase purchase) {
    if (result.isFailure()) {
      Toast.makeText(context, "Purchase failed: "+result, Toast.LENGTH_LONG).show();
      return;
    }

    if (purchase.getSku().equals(YOUR_SKU)) {
      // Do something with this item
    } else if (purchase.getSku().equals(ANOTHER_SKU)) {
      // Do something with this item
    }
  }
};
```

# 第 14 章

# 推送消息

推送（push）消息是一种通信方法。在该方法中，已连接的客户端会收到远程系统上事件的通知。与之相对的是拉取（pull）消息，客户端需要每隔一段时间就查询一次远程系统，以获取消息。推送的消息由远程系统自身触发，客户端不需要请求状态更新。Android 通过 Google 云消息（Google Cloud Messaging，GCM）库来支持消息推送。GCM 在一切运行 API Level 8 或更高版本的 Androi 设备上都是可用的，这意味着当前在用的大部分设备都已包括在内。本章展示了如何与 GCM 集成，以及如何发送和接收消息。

## 14.1 Google 云消息设置

Google 云消息要依赖于 Google Play 商店以及在设备上已登录的 Google 用户账号。要发送信息，还需要一个 API 密钥。该密钥与 Google 账号相绑定，账号稍后可以用于在 Google Play 商店发布应用程序，因此一定要首先建立一个账号。

### 技巧 125：准备 Google 云消息

首先，必须获取一个 API 密钥。要做到这一点，需登录 Google 开发者账号并前往 https://code.google.com/apis/console。需要建立一个新的 API 项目来使用 GCM。如果是首次建立 API 项目，点击 **Create Project** 按钮；否则，点击窗口左上角的下拉按钮并选择 **Create**。无论使用了上述哪种方法，下面都要输入项目名称，比如 **cookbook**。项目创建后，会显示一个类似图 14-1 的界面。

要注意图中的两项内容。首先，有一个巨大的可用 API 的列表。要在其中找到 GCM，需要向下滚动相当的距离。其次，网页的 URL 会变成类似 **https://code.google.com/apis/console/b/0/?pli=1#project:123456** 的形式，记住 **#project** 后面的数字，它是独一无二的项目代号，稍后会扮演发送者 ID 的角色。即使设备上带有推送消息通道的应用程序不止一个，发送者 ID 也能确保 GCM 将消息发送给正确的应用。

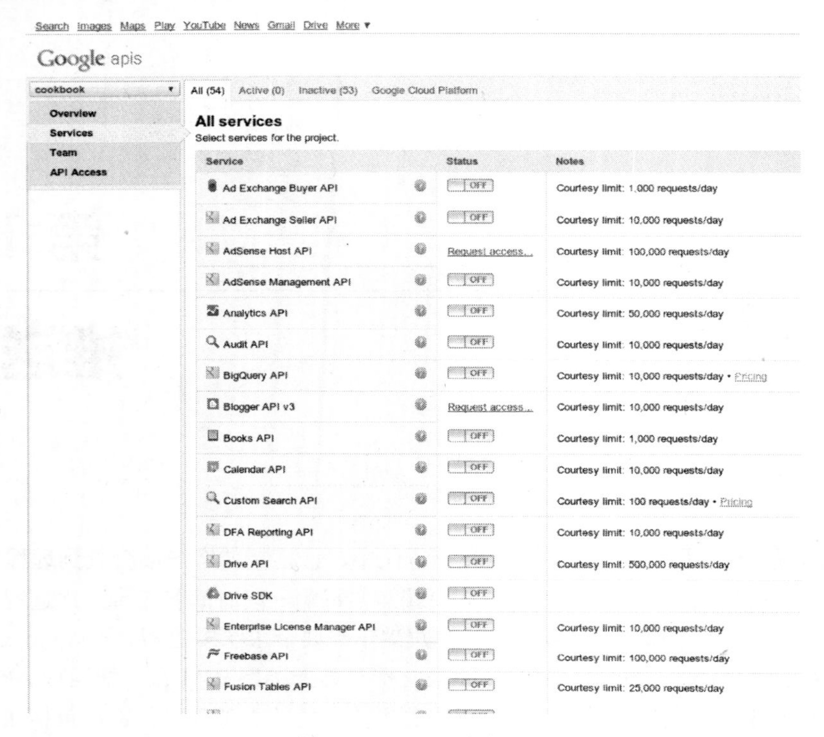


图 14-1　API 服务概览

向下滚动，找到 Google Cloud Messaging，将其右面的开关按钮变为 ON。同意下一个页面中的服务条款。为了获得 API 密钥，要通过左边的菜单跳转到 API Access 页面，该页面看上去会像图 14-2 那样。

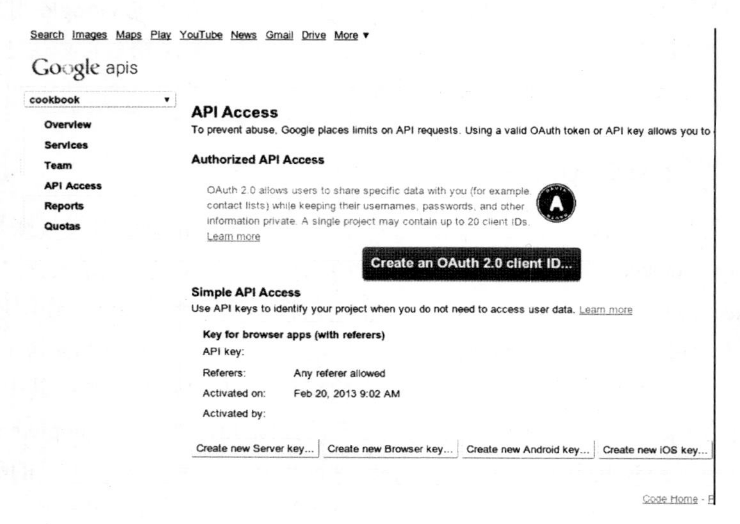


图 14-2　API Access 页面

点击 Create New Server Key...按钮，会弹出一个窗口，如果需要服务器 IP 地址，则在其中填写，将该处留空也没有问题。最后点击 Create，这样在 API Access 界面中就会出现一个新产生的服务器密钥。将该号码存入一个文本文件或记录下来，在稍后发送消息时要用到。

下一步，必须把附加库集成到项目中。打开 Android SDK Manager，进到 Extras 部分，勾选 Google Cloud Messaging for Android Libray 复选框。安装了 GCM 后，就能在 Android SDK 文件夹下找到**./extras/google/gcm** 目录。将 **gcm-client** 和 **gcm-server** 子目录下的所有**.jar** 文件都复制到项目的**/lib** 文件夹中。

# 14.2 发送和接收推送信息

多数情况下，推送通知会通过一个服务器系统发送，将后台某处发生的事件通知给用户或者应用程序。然而我们不急需用这种方式发送推送通知。因为本书是关于 Android 的，并且处理服务器本身是个复杂的话题，所以我们不使用这一方法。取而代之地，会从应用自身内部发送一条简短的 ping 命令到 GCM 端点。向设备自身发送推送通知似乎没有什么意义，但将推送通知从一部手机发送到另一部手机就很有吸引力了。向应用程序添加这一功能是轻而易举的。

## 技巧 126：准备 manifest

为了接收和发送消息，需要若干权限许可。首先，`android.permission.INTERNET` 对于一切数据传输操作都是必要的。还需要一个唤醒锁来确保消息被收到，即使在设备处于关闭或待机状态时也不例外，因此要加入 `android.permission.WAKE_LOCK`。GCM 有赖于 Google 账户，这意味着需要 `android.permission.GET_ACCOUNT` 权限来访问账户。为了能确实收到消息，由 GCM 库定义的名为 `google.android.c2dm.permission.RECEIVE` 的定制权限也必须被添加进来。（尽管 Google 云消息服务现在被称为 GCM，但它的旧名字 C2DM 仍会不时出现，可能会让人困惑。）

在上述权限之外，还需要创建一个权限，即 ***your.package.name*.C2D_MESSAGE**，其中 ***your.package.name*** 代表 manifest 标签中给出的包名。不过，只有目标环境为 Android 4.1 或更高版本时才需要该权限。

下一步，需要一个 `BroadcastReceiver` 类来接收消息，以及一个 `IntentService` 类来处理这些消息。该服务被声明为`<serviceandroid:name=".receive.GCMService"></service>`的形式。建议将服务名设为 `your.package.name.GCMService`，这样做 GCM 库会自动获得服务名。因为此处使用的服务名下面有个额外的子包，`BroadcastReceiver` 类确定服务名的部分必须要重写，稍后我们会看到这一点。

本例中的 `BroadcastReceiver` 类被命名为`.receive.GCMBroadcastReceiver`，并且需要它自己的权限 `com.google.android.c2dm.permission.SEND`。Intent filter 包含 `RECEIVE` 和 `SEND` 两个动作，以及一个 `your.app.package` 类别。

剩下的工作就是添加应用程序的 Activity 的其余部分以及其他的组件。本例的完整 **AndroidManifest.xml** 文件在代码清单 14-1 中给出。

**代码清单 14-1　AndroidManifest.xml**

```
<manifest xmlns:android="http://schemas.android.com/apk/res/android"
    package="cc.dividebyzero.android.cookbook.chapter14"
    android:versionCode="1"
    android:versionName="1.0">

    <uses-sdk android:minSdkVersion="8" android:targetSdkVersion="15" />

    <permission
android:name="cc.dividebyzero.android.cookbook.chapter14.permission.C2D_MESSAGE"
android:protectionLevel="signature" />
    <uses-permission
android:name="cc.dividebyzero.android.cookbook.chapter14.permission.C2D_MESSAGE"
/>

    <!-- App receives GCM messages -->
    <uses-permission android:name="com.google.android.c2dm.permission.RECEIVE" />
    <!-- GCM connects to Google Services -->
    <uses-permission android:name="android.permission.INTERNET" />
    <!-- GCM requires a Google account -->
    <uses-permission android:name="android.permission.GET_ACCOUNTS" />
    <!-- Keeps the processor from sleeping when a message is received -->
    <uses-permission android:name="android.permission.WAKE_LOCK" />
    <!-- Use this for sending out registration and other messages
         to a potential server -->
    <uses-permission android:name="android.permission.INTERNET"/>

    <application android:label="@string/app_name"
        android:icon="@drawable/ic_launcher"
        android:theme="@style/AppTheme"
        >
        <receiver
            android:name=".receive.GCMBroadcastReceiver"
            android:permission="com.google.android.c2dm.permission.SEND" >
          <intent-filter>
            <action android:name="com.google.android.c2dm.intent.RECEIVE" />
            <action android:name="com.google.android.c2dm.intent.REGISTRATION" />
            <category android:name="my_app_package" />
          </intent-filter>
        </receiver>

        <activity android:name=".Chapter14">
            <intent-filter>
                <action android:name="android.intent.action.MAIN" />

                <category android:name="android.intent.category.LAUNCHER" />
            </intent-filter>
        </activity>
        <activity android:name=".GCMPushReceiver"/>

        <service android:name=".receive.GCMService"></service>
    </application>

</manifest>
```

## 14.3　接收消息

要接收消息，需要做几件事情。首先，设备需要在 GCM 中注册。接下来必须把注册的 ID 传递给定制服务器，这样系统才知道向哪里发送消息。因为消息是从设备自身发出的，只要将

ID 存储在本地即可。之后，需要编写好对到来的消息做出实际反应的适当的代码。以上要求意味着有三样东西需要被添加到程序中：BroadcastReceiver 类、IntentService 类，以及主 Activity 中的范例文件注册代码。

## 技巧 127：添加 BroadcastReceiver 类

要使 BroadcastReceiver 类能够正常工作，必须扩展 com.google.android.gcm.GCMBroadcastReceiver 类。好消息是除了返回要启动的服务名外，我们没有太多的事情要做，因为剩下的过程都交由父类来处理了。因为服务位于一个子包中，必须返回 GCMService.class.getCanonicalName()。代码清单 14-2 给出了完整的实现。

**代码清单 14-2　GCMBroadcastReceiver.java**

```
public class GCMBroadcastReceiver extends
com.google.android.gcm.GCMBroadcastReceiver {

    @Override
    protected String getGCMIntentServiceClassName(Context context)
        return GCMService.class.getCanonicalName();
    }
}
```

## 技巧 128：添加 IntentService 类

IntentService 类必须扩展 GCMBaseIntentService 类，并实现其抽象方法。那些已注册和未注册的、可用于将注册的 ID 发送给后台的 event hook 必须得到处理。处理 onError 事件的方式就是将出现的错误记录到系统日志中。onMessage 事件被赋予了一个 Intent，该 Intent 在"msg"附加字段中存储了推送消息本身的实际的有效本体。本例中，设备可以以任何一种对应用程序有意义的方式响应到来的消息，比如，将到来的推送消息作为同步用户数据的唤醒信号。代码清单 14-3 给出了一个发送纯文本消息的例子，该消息会通过 Toast 显示出来。

**代码清单 14-3　GCMService**

```
public class GCMService extends GCMBaseIntentService {
    private static final String LOG_TAG = GCMService.class.getSimpleName();

    private Handler mToaster = new Handler(new Handler.Callback() {

        @Override
        public boolean handleMessage(Message msg) {
            Toast.makeText(
                    GCMService.this,
                    ((String) msg.obj),
                    Toast.LENGTH_SHORT
                    ).show();
            return true;
        }
    });

    @Override
```

```
    protected void onError(final Context ctx, final String errorMsg) {
        android.util.Log.v(LOG_TAG, "error registering device; " + errorMsg);
    }

    @Override
    protected void onMessage(final Context ctx, final Intent intent) {
        android.util.Log.v(LOG_TAG,
                "on Message, Intent="
                + intent.getExtras().toString()
                );
        Message msg = mToaster.obtainMessage(
                1,
                -1,
                -1,
                intent.getStringExtra("msg")
                );
        mToaster.sendMessage(msg);
    }

    @Override
    protected void onRegistered(Context ctx, String gcmRegistrationId) {
        android.util.Log.v(LOG_TAG,
                "onRegistered: gcmRegistrationId>>"
                + gcmRegistrationId + "<<"
                );
        sendRegistrationToServer(gcmRegistrationId);
    }

    @Override
    protected void onUnregistered(Context ctx, String gcmRegistrationId) {

        sendDeregistrationToServer(gcmRegistrationId);
    }

    private void sendRegistrationToServer(String gcmRegistrationId) {
        SharedPreferences.Editor editor = getSharedPreferences(
                AppConstants.SHARED_PREF,
                Context.MODE_PRIVATE
                ).edit();

        editor.putString(AppConstants.PREF_REGISTRATION_ID, gcmRegistrationId);
        editor.commit();
    }

    private void sendDeregistrationToServer(String gcmRegistrationId) {
        SharedPreferences.Editor editor = getSharedPreferences(
                AppConstants.SHARED_PREF,
                Context.MODE_PRIVATE
                ).edit();

        editor.clear();
        editor.commit();
    }

}
```

由于服务是由另一个线程启动的，到来的消息先被传送给一个 Handler，由该 Handler 来显示 `Toast`。因为 Handler 只能接受一种类型的信息，可以进行如下调用，通过将所有默认值以及文本信息作为字符串放入 `message.obj` 域来创建 Handler：

```
mToaster.obtainMessage(1, -1, -1, intent.getStringExtra("msg"));
```

在 onRegistered 方法中调用了 sendRegistrationToServer(gcmRegistrationId)；而在 onUnregistered 方法中则调用了 sendDeregistrationToServer(String gcmRegistrationId)。以上两个自定义的私有方法通常用于确保后台系统知道设备的 ID 以及其他附加信息，从而将设备与用户账户绑定。因为消息的发送和接收设备是同一个，本例没有使用网络通信，取而代之地，注册 ID 被存入了一个 **sharedpreferences** 文件。

## 技巧 129：注册设备

要实现消息接收，最后要做的就是在应用启动时注册设备。实现方法是在主 Activity 的 onCreate 方法中调用私有方法 registerGCM()，如代码清单 14-4 所示。

**代码清单 14-4 registerGCM()**

```
private void registerGCM() {

     GCMRegistrar.checkDevice(this);
     GCMRegistrar.checkManifest(this);
     final String regId = GCMRegistrar.getRegistrationId(this);
     if (regId.equals("")) {
        GCMRegistrar.register(this, getString(R.string.sender_id));
     } else {
        android.util.Log.v(LOG_TAG, "Already registered");
     }
}
```

GCM 会一次性地创建一个设备 ID，并将其安全地存储在设备上，因此最好在每次调用 GCMRegistrar.register(..)之前都检查 ID 是否存在。sender_id 类是在 GCM 注册时获取的。该 ID 被存储在**/res/values** 目录下名为 **sender_id.xml** 的附加文件中。在文件中，ID 被声明为一个字符串资源，因此它会被添加到 R.string 类中，如代码清单 14-5 所示。

对 checkDevice()和 checkManifest()的调用是强制性的，以确保设备和应用程序被正确地配置以使用 GCM。如果检查失败，就会抛出异常。

**代码清单 14-5 sender_id.xml**

```
<?xml version="1.0" encoding="utf-8"?>
<resources>
    <string name="sender_id">12345678</string>
</resources>
```

# 14.4 发送消息

向客户端发送消息是通过向 GCM 服务器传递目标 ID 以及消息有效本体来完成的。我们通常通过应用程序的后台系统实现这一功能。有若干基于常用的 Web 开发语言的库可以利用。这里我们将使用 Java 库将设备直接连接到 GCM 服务器。使用一个小型的 Activity 读取文本，并通过具有通信作用的 AsyncTask 将其移交给 GCM 服务器。这里要用到我们之前安装 GCM 附加包时得到的 **gcm-server.jar** 文件，因此要确保它已被放到应用程序的**/libs** 目录下。

## 技巧 130：发送文本消息

本例中，无论何时按下 Send 按钮，从布局的输入域中读取的简单文本消息都将被发送出去。代码清单 14-6 显示了我们用到的十分简单的布局。

**代码清单 14-6　gcm_acv.xml**

```
<?xml version="1.0" encoding="utf-8"?>
<LinearLayout xmlns:android="http://schemas.android.com/apk/res/android"
     android:layout_width="match_parent"
     android:layout_height="match_parent"
     android:orientation="vertical"
     >
     <EditText
          android:id="@+id/message"
          android:layout_height="wrap_content"
          android:layout_width="match_parent"
          />
     <Button
          android:id="@+id/message"
          android:layout_height="wrap_content"
          android:layout_width="match_parent"
          android:text="send"
          android:onClick="sendGCMMessage"
          />
</LinearLayout>
```

为了发送消息，我们在 sendGCMMessage()方法中创建并启动一个 AsyncTask 类。在 XML 布局中，sendGCMMessage()方法被定义为 onClick 事件的目标。消息字符串从文本域中读取，而目标 ID 来自 shared preferences（即服务存储注册设备 ID 的地方）。布局则在 onCreate 方法中被装载。完整的主 Activity 如代码清单 14-7 所示。

**代码清单 14-7　主 Activity**

```
public class GCMPushReceiver extends Activity{
     private static final String LOG_TAG = GCMPushReceiver.class.getSimpleName();
     private EditText mMessage;

     public void onCreate(Bundle savedState) {
          super.onCreate(savedState);
          setContentView(R.layout.gcm_acv);
          mMessage = (EditText) findViewById(R.id.message);
          registerGCM();
     }

     private void registerGCM() {

          GCMRegistrar.checkDevice(this);
          GCMRegistrar.checkManifest(this);
          final String regId = GCMRegistrar.getRegistrationId(this);
          if (regId.equals("")) {
               GCMRegistrar.register(this, getString(R.string.sender_id));
          } else {
               android.util.Log.v(LOG_TAG, "Already registered");
          }
     }
```

```
    public void sendGCMMessage(final View view) {
        final String message = mMessage.getText().toString();
        SendGCMTask sendTask = new SendGCMTask(getApplicationContext());
        SharedPreferences sp = getSharedPreferences(
                AppConstants.SHARED_PREF,
                Context.MODE_PRIVATE
                );

        final String targetId = sp.getString(
                AppConstants.PREF_REGISTRATION_ID,
                null
                );
        sendTask.execute(message, targetId);
    }
}
```

## 技巧 131：通过 AsyncTask 发送消息

使用 GCM 服务库的一个好处是，开发者不需要亲自处理诸如 HTTP 连接一类的东西。他们只需通过为 GCM 进行注册时得到的 API 密钥来初始化一个 `Sender` 对象。与发送器 ID 类似，API 密钥也被作为字符串资源存储在一个附加的 XML 文件中，如代码清单 14-8 所示。

**代码清单 14-8　api_key.xml**

```
<?xml version="1.0" encoding="utf-8"?>
<resources>
    <string name="api_key">12345678</string>
</resources>
```

消息本身是通过一个简单的构建器模式创建的，随后被传递给发送器，后者使用 `sender.sendNoRetry(gcmMessage, targetId)` 对其进行处理。这基本就是发送的全部过程了。另外还需要一些结果处理和范例文件代码，如代码清单 14-9 所示。

**代码清单 14-9　SendGCMTask.java**

```
public class SendGCMTask extends AsyncTask<String, Void, Boolean> {

    private static final int MAX_RETRY = 5;
    private static final String LOG_TAG = SendGCMTask.class.getSimpleName();
    private Context mContext;
    private String mApiKey;

    public SendGCMTask(final Context context) {
        mContext = context;
        mApiKey = mContext.getString(R.string.api_key);
    }

    @Override
    protected Boolean doInBackground(String . . . params) {
        final String message = params[0];
        final String targetId = params[1];
        android.util.Log.v(LOG_TAG,
                "message>>" + message + "<< "
                +"targetId>>"+ targetId + "<<"
                );

        Sender sender = new Sender(mApiKey);
```

```
        Message gcmMessage = new Message.Builder()
                .addData("msg", message)
                .build();

        Result result;

        try {
            result = sender.sendNoRetry(gcmMessage, targetId);

            if (result.getMessageId() != null) {
                String canonicalRegId = result.getCanonicalRegistrationId();

                SharedPreferences.Editor editor = mContext
                        .getSharedPreferences(
                                AppConstants.SHARED_PREF,
                                Context.MODE_PRIVATE
                        ).edit();
                String error = result.getErrorCodeName();

                if (canonicalRegId != null) {
                    // Same device has more than one registration ID: update
                    // database
                    editor.putString(
                            AppConstants.PREF_REGISTRATION_ID,
                            canonicalRegId
                            );                      editor.commit();
                } else if (error != null
                        && error.equals(Constants.ERROR_NOT_REGISTERED)) {
                    // Application has been removed from device: unregister
                    // database
                    editor.clear();
                    editor.commit();
                }
            }

            return true;
        } catch (IOException e) {
            // TODO autogenerated catch block
            e.printStackTrace();
        }

        return false;
    }
}
```

代码中涉及的上下文由构造函数给出，且只需要从资源文件中读取 API 密钥。消息字符串和目标设备 ID 从 Activity 传递到了执行方法中。相应的回调参见代码清单 14-7。

`Message.Builder.addData(..)`方法用于设置消息本体。其中用来设置字符串消息的密钥与代码清单 14-3 中用来从到来的推送消息中检索字符串的服务中所用的密钥是同一个。

发送方法的执行结果是：程序会通过设置相关的字段对产生的错误进行标记，并检查判定`.getErrorCodeName()`的值与 `ERROR_NOT_REGISTERED` 的值是否一致。如果一致，意味着设备将不能再（或始终就不能）用于接收推送消息，其 ID 应当从数据库中被移除。通过清除 shared preference 来完成这一工作。如果 `result.getCanonicalRegistrationId()`的值非空，意味着设备被重复注册，canonical ID 应当被用于发送消息。此时，应当用新的设备 ID 来更新 shared preference。

# 第 15 章
# 原生 Android 开发

本章给出了两种将原生 C 代码集成进 Android 应用程序的不同策略。一种是使用 Java 原生接口（Java Native Interface，JNI）用 C 语言编写封装函数，再用 Java 访问 C 语言代码库。另一种策略是利用原生 Activity，这类 Activity 允许应用程序不包含任何 Java 代码。

## Android 原生组件

当某个高计算密度的函数是 Android 应用程序非常关键的部分时，为提高效率，将密集计算的部分改用原生的 C 或 C++代码实现或许是值得的。有现成的 Android 原生开发包（Native Development Kit，NDK），可帮助开发原生组件。NDK 是伴随 Android SDK 发布的，其中包含一些库，可用于生成 C/C++库。设置和生成 Android 本地组件的步骤如下。

（1）从 http://developer.android.com/sdk/ndk/下载 Android NDK，其中包含详细的使用说明文档。

（2）安装 Eclipse C/C++ Development Tooling（CDT）。

（3）从 https://dl-ssl.google.com/android/eclipse/下载 Eclipse NDK 插件。

（4）通过下面的途径设置 NDK 路径：**Eclipse→Window→Preferences→Android→NDK→set path to NDK**。

（5）利用通常的方法创建一个 Android 项目。

（6）在创建好的项目上右击，并选择 **Android Tools→Add Native Support**。

（7）为原生库命名。

（8）点击 **Finish**。此时/**jni** 文件夹、make 文件 **Android.mk** 以及一个存根 **cpp** 文件都被创建。

（9）运行 **Project→Build Project** 编译 C 和 Java 文件。

使用 Eclipse IDE，在项目构建时原生库就能与应用程序正确绑定。

## 技巧 132：使用 Java 原生接口

本技巧使用一段 C 程序创建一个阶乘函数，而后使用一个 Java 编写的 Activity 调用 C 库函数（即阶乘函数），并将结果显示到屏幕上。首先，我们在代码清单 15-1 中给出 C 程序。

**代码清单 15-1 jni/cookbook.c**

```
#include <string.h>
#include <jni.h>

jint factorial(jint n){
     if(n == 1){
       return 1;
     }
     return factorial(n-1)*n;
}

jint Java_com_cookbook_advance_ndk_ndk_factorial( JNIEnv* env,
                                                  jobject this, jint n ) {
     return factorial(n);
}
```

在这个 C 程序中，用到了一种特殊数据类型 jint，是 C/C++中定义的 Java 类型。这提供了将原生类型传递给 Java 的方法。如果有必要从 Java 向 C 返回值，就需要进行类型转换。表 15-1 总结了 Java 和原生（C/C++）描述间的类型对应关系。

**表 15-1 Java 和原生语言间的类型对应关系**

| C/C++中的类型 | 原 生 类 型 | 描 述 |
|---|---|---|
| jboolean | unsigned char | 8 位无符号整型数 |
| jbyte | signed char | 8 位有符号整型数 |
| jchar | unsigned short | 16 位无符号整型数 |
| jshort | short | 16 位有符号整型数 |
| jint | long | 32 位有符号整型数 |
| jfloat | float | 32 位浮点数 |
| jlong | long long _int64 | 64 位有符号整型数 |
| jdouble | double | 64 位双精度型数 |

C 程序中包括两个函数。第一个是 factorial 函数，进行实际的计算。第二个函数名为 Java_com_cookbook_advance_ndk_ndk_factorial，将会在 Java 类中调用。该函数的名称应当始终定义为 JAVA_CLASSNAME_METHOD[①]的格式，因为它要作为接口使用。

第二个函数带有三个参数：一个 JNIEnv 指针，一个 jobject 指针，以及一个由 Java 方法声明的 Java 参数。JNIEnv 是一个 Java 原生接口指针，对于每个原生函数，它都作为一个参

① 即 JAVA_类名_方法名，且所有字母均大写。——译者注

数传递。这些函数被映射为 Java 方法，后者是一个包含了面向 Java 虚拟机（JVM）接口的结构，它包含了与 Java 虚拟机交互以及与 Java 对象协同所需的函数。在本例中，并不使用任何的 Java 函数，程序唯一需要传递的就是 Java 参数 `jint n`。

构建器（builder）使用的 make 文件如代码清单 15-2 所示，它应当被放到与 C 程序相同的路径中。该文件包含了供构建器使用的 `LOCAL_PATH` 定义，以及对 `CLEAR_VARS` 的一个调用，该调用用于在每次构建前清除所有的 `LOCAL_*`[1]变量。接着，`LOCAL_MODULE` 被识别为自定义库 `ndkcookbook` 的名称，用于识别将要生成的源代码文件。在上述声明之后，make 文件引用了 `BUILD_SHARED_LIBRARY`。这就是用来生成一个简单程序的通用 make 文件。关于 make 文件格式的更详细的信息可参见 NDK 的 **docs/**目录下的 **ANDROID-MK.HTML** 文件。

**代码清单 15-2　jni/Android.mk**

```
LOCAL_PATH := $(call my-dir)

include $(CLEAR_VARS)

LOCAL_MODULE     := ndkcookbook
LOCAL_SRC_FILES := cookbook.c

include $(BUILD_SHARED_LIBRARY)
```

下一步就是生成原生库了。对于 NDK-r4 及以上版本，只需调用项目的 NDK 根目录下提供的 `ndk-build` 脚本，即可使用相关的 make 文件来生成原生库。对于更早的版本，则需要执行 `make APP=NAME_OF_APPLICATION` 命令。在库生成以后，会创建一个包含 **libndkcookbook.so** 原生库的 **lib/**文件夹。对于 NDK-r4，该文件夹还含有两个有助于调试的 GDB 文件。

使用上述原生库的 Android Activity 会调用 `System.loadLibrary()` 来装载 **ndkcookbook** 库。之后需要对原生函数进行声明。上述过程在代码清单 15-3 中给出，而输出结果则如图 15-1 所示。

**代码清单 15-3　src/com/cookbook/advance/ndk/ndk.java**

```
package com.cookbook.advance.ndk;

import android.app.Activity;
import android.widget.TextView;
import android.os.Bundle;
public class ndk extends Activity {
    @Override
    public void onCreate(Bundle savedInstanceState) {
        super.onCreate(savedInstanceState);
        TextView tv = new TextView(this);
        tv.setText(" native calculation on factorial :"+factorial(30));
        setContentView(tv);
    }
    public static native int factorial(int n);
    static {
        System.loadLibrary("ndkcookbook");
    }
}
```

① 即所有以 LOCAL_开头的变量。——译者注

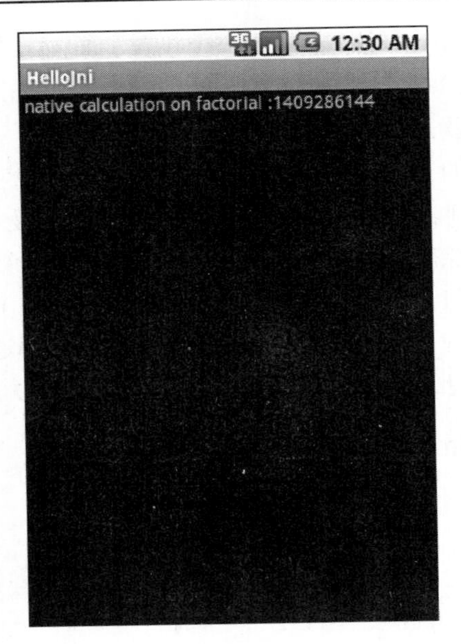


图 15-1　NDK 应用程序的输出

## 技巧 133：使用 NativeActivity

NativeActivity（意为原生 Activity）是处理 C 代码和 Android 框架之间通信的辅助类。有了它，就可以完全使用 C 来编写应用程序。创建一个使用原生 Activity 的项目的步骤如下。

（1）在 Eclipse 中使用通常的方法创建一个 Android 项目。

（2）在项目上右击，并选择 **Android Tools→Add NativeSupport**。

（3）为原生 Activity 起名。

（4）点击 **Finish**。这样就创建了 **/jni** 文件夹、make 文件 **Android.mk** 以及一个存根 **cpp** 文件。

（5）编辑 manifest，为其添加一个指向原生 Activity 的引用。

（6）如果需要其他库，编辑 **Android.mk** 文件添加。

（7）运行 **Project→Build Project** 来编译 C 和 Java 文件。

要声明本地的 Activity，只需将其加入 **AndroidManifest.xml** 文件中，如代码清单 15-4 所示。

**代码清单 15-4　AndroidManifest.xml**

```
<manifest xmlns:android="http://schemas.android.com/apk/res/android"
    package="com.cookbook.nativeactivitydemo"
    android:versionCode="1"
    android:versionName="1.0" >

    <uses-sdk
        android:minSdkVersion="14"
        android:targetSdkVersion="17" />

    <application
```

```
        android:allowBackup="true"
        android:icon="@drawable/ic_launcher"
        android:label="@string/app_name"
        android:theme="@style/AppTheme"
        android:hasCode="true"
        >

        <activity android:name="android.app.NativeActivity"
            android:label="@string/app_name"
            android:configChanges="orientation|keyboardHidden"
            >

          <meta-data android:name="android.app.lib_name"
            android:value="native-activity"
            />
            <intent-filter>
              <action android:name="android.intent.action.MAIN" />
              <category android:name="android.intent.category.LAUNCHER" />
            </intent-filter>
          </activity>

    </application>

</manifest>
```

Activity 标签的 android:name 属性必须设为 android.app.NativeActivity。metadata 标签告知系统要装载哪个库，其 android:name 属性必须要设为 android.app.lib_name。metadata 标签的 android:value 属性必须是文件名去掉 **lib** 前缀和 **.so** 后缀之后剩下的部分。文件名中不能带有空格，且字母必须全部小写。要确保原生的 **libs** 被正确地导出和编译，应该将 application 标签的属性 android:hasCode 设为 true。

有两种实现原生 Activity 的方法。第一种方法是直接使用 native_activity.h 头文件，在其中定义了实现 Activity 所需的全部结构体和回调函数。第二种，也是推荐的方法，是使用 android_native_app_glue.h。app glue 接口确保回调以一种不会阻塞 UI 线程的方式实现。原生应用程序仍旧在它们自己的虚拟机上运行，而所有向 Activity 的回调都在应用程序的主线程上执行。如果那些回调以一种会阻塞 UI 线程的方式执行，应用将会收到应用程序未响应（Application Not Responding）错误。最简单的解决办法就是使用 app glue 接口。app glue 会创建另一个线程，用于处理所有回调及输入事件，并将它们作为命令发送到代码的主函数中。

要在编译时带上 app glue，需要将其作为静态库添加到 **Android.mk** 文件中。在此，我们还会添加日志库及某些 OpenGL 库，因为原生 Activity 需要自己负责绘制自己的窗口，用来在屏幕上绘制的库首选 EGL。**Android.mk** 文件如代码清单 15-5 所示。

**代码清单 15-5 /jni/Android.mk**

```
LOCAL_PATH := $(call my-dir)
include $(CLEAR_VARS)
LOCAL_MODULE     := NativeActivityDemo
LOCAL_SRC_FILES  := NativeActivityDemo.cpp
LOCAL_LDLIBS     := -landroid -llog -lEGL -lGLESv1_CM
LOCAL_STATIC_LIBRARIES := android_native_app_glue
```

```
include $(BUILD_SHARED_LIBRARY)
$(call import-module,android/native_app_glue)
```

原生 Activity 需要拥有一个 void android_main(struct android_app* androidApp) 函数。该函数是启动 Activity 的主入口。它的工作机制与 Java 线程或其他常见的基于事件的系统的主函数非常类似，意味着其中需要一个 while(1) 循环来确保代码持续运行，直到被从外部停止。

android_app 结构体是一个辅助类，处理某些运行时所需的范例文件。它的 userData 域可以存放任何东西，在本例中它应当存储 Activity 的当前状态信息。虽然同一个 android_app 实例被传递给所有后续的函数，但总是可以在这里检索其状态。

还需要为结构体设置两个函数指针：一个用于处理输入事件（比如触摸或键盘事件），另一个用于处理 Activity 生命周期。因为它们在稍后会以静态的形式声明，这里只需要作为指针传递即可。android_main 函数如代码清单 15-6 所示。

**代码清单 15-6 android_main**

```
void android_main(struct android_app* androidApp) {

    struct activity_state activity;

    // Make sure glue isn't stripped
    app_dummy();

    memset(&activity, 0, sizeof(activity));
    androidApp->userData = &activity;
    androidApp->onAppCmd = handle_lifecycle_cmd;
    androidApp->onInputEvent = handle_input;
    activity.androidApp = androidApp;

    LOGI("starting");

    if (androidApp->savedState != NULL) {
        // We are starting with a previous saved state; restore from it
        activity.savedState = *(struct saved_state*)androidApp->savedState;
    }

    // Loop waiting for stuff to do

    while (1) {
        // Read all pending events
        int ident;
        int events;
        struct android_poll_source* source;

        // Wait for events
        while ((ident=ALooper_pollAll(-1, NULL, &events,(void**)&source)) >= 0) {

            // Process this event
            if (source != NULL) {
                source->process(androidApp, source);
            }

            // Check if the app has exited
```

```
            if (androidApp->destroyRequested != 0) {
                close_display(&activity);
                return;
            }
        }
    }
}
```

调用 app_dummy() 首先确保 app glue 没有从库中被剥离。应当始终在一个原生 Activity 中调用该函数。接着设定用于事件处理的指针，保存状态，然后开始事件循环。如果应用程序停止，android_app->destroyRequested 标志会被设置，此时就必须跳出循环。

Activity 的生命周期在 handle_lifecycle_cmd 函数中处理，该函数作为一个指针被传给 app glue。Activity 的每一个生命周期事件，都有对应的整型命令常量以及与窗口相关的用于获得焦点和调整大小的命令。handle_lifecycle_cmd 函数如代码清单 15-7 所示。

**代码清单 15-7 handle_lifecycle_cmd**

```
static void handle_lifecycle_cmd(struct android_app* app, int32_t cmd) {
    struct activity_state* activity = (struct activity_state*)app->userData;
    switch (cmd) {
        case APP_CMD_SAVE_STATE:
            // The system has asked us to save our current state. Do so.
            activity->androidApp->savedState = malloc(sizeof(struct saved_state));
            *((struct saved_state*)activity->androidApp->savedState) =
                activity->savedState;
            activity->androidApp->savedStateSize = sizeof(struct saved_state);
            break;
        case APP_CMD_INIT_WINDOW:
            // activity window shown, init display
            if (activity->androidApp->window != NULL) {
                init_display(activity);
                draw_frame(activity);
            }
            break;
        case APP_CMD_TERM_WINDOW:
            // activity window closed, stop EGL
            close_display(activity);
            break;
        case APP_CMD_INPUT_CHANGED:

            break;
        case APP_CMD_START:
            //activity onStart event
            LOGI("nativeActivity: onStart");
            android_app_pre_exec_cmd(app, cmd);
            break;
        case APP_CMD_RESUME:
            //activity onResume event
            LOGI("nativeActivity: onResume");
            android_app_pre_exec_cmd(app, cmd);
            break;
        case APP_CMD_PAUSE:
            //activity onPause event
            LOGI("nativeActivity: onPause");
            android_app_pre_exec_cmd(app, cmd);
```

```
            break;
        case APP_CMD_STOP:
            //activity onStop event
            LOGI("nativeActivity: onStop");
            android_app_pre_exec_cmd(app, cmd);
            break;
        case APP_CMD_DESTROY:
            //activity onDestroy event
            LOGI("nativeActivity: onDestroy");
            android_app_pre_exec_cmd(app, cmd);
            break;
    }
}
```

此处的 init 和 termination 事件只承担初始化 EGL 显示的作用。而所有其他的生命周期事件则只会通过调用 `LOGI("message")` 函数在系统日志中被记录下来。对 `android_app_pre_exec_cmd(app, cmd)` 的调用取代了 Java 代码中的一些强制调用（比如 `super.onCreate`、`super.onPause` 等）。

另一个需要对事件进行响应的函数是 `handle_input`。该函数的参数包括一个指向应用的指针以及一个指向事件的指针。事件类型通过调用 `AInputEvent_getType(event)` 来提取，返回值是一个整型常量。触摸事件的坐标通过调用 `AMotionEvent_getX(..)` 和 `AMotionEvent_getY(..)` 分别读取。以上内容如代码清单 15-8 所示。

**代码清单 15-8 handle_input**

```
static int32_t handle_input(struct android_app* app, AInputEvent* event) {
    struct activity_state* activity = (struct activity_state*)app->userData;
    if (AInputEvent_getType(event) == AINPUT_EVENT_TYPE_MOTION) {
        activity->savedState.x = AMotionEvent_getX(event, 0);
        activity->savedState.y = AMotionEvent_getY(event, 0);
        draw_frame(activity);
        return 1;
    }
    return 0;
}
```

显然，虽然 app glue 库代劳了一部分，更多的范例文件还是需要用 C 来编写。原生 Activity 的优点是能允许整个应用程序完全地用 C/C++来编写，对于大计算量或跨平台开发而言，这是一种理想的方式。

# 第 16 章

# 测试和调试

软件调试过程极有可能耗费等同于甚至超过开发过程的时长。掌握调试常见问题的各种方法能够节约大量的时间和精力。本章介绍调试 Android 应用程序的基本方法，并考察现有的多种调试工具。首先探讨的是常用的 Eclipse IDE 调试工具。接下来探讨的是 Android SDK 提供的工具。最后，介绍 Android 系统中可用的工具。应用程序各有不同，因此要根据应用程序自身的特点来选取适当的调试方法。

## 16.1 Android 测试项目

Android 测试框架提供给程序员一组可用于应用程序测试的工具。Android 测试套件是基于 JUnit 打造的。熟悉 JUnit 的人应该知道，它可以用于测试未扩展 Android 组件的项目。一个包含了 Android 开发工具包（ADT）的 Eclipse 安装就已然拥有创建测试套件所需的全部工具了。如果没有 Eclipse 或 ADT，可以使用 `android` 命令行工具来创建测试。

测试项目被创建为独立的项目，这类项目使用 instrumentation 来连接被测试的应用程序。

### 技巧 134：创建测试项目

强烈建议用 Eclipse 来创建测试项目，因为它内建了建立一个测试所需的全部工具。

要在 Eclipse 下创建测试项目，选择 **File→New→Other** 菜单项。在弹出的向导对话框中，选择 **Android** 部分并点选其中的 **Android Test Project**，再点击 Next 按钮。图 16-1 显示了 New Project 向导。

在下一个对话框的 Project Name 字段填入一个与测试用途相匹配的名称。在待测试的应用程序名后面加上 Test 一词不失为一种简单明了的命名法。更复杂地，也可以用应用程序执行的测试类型以及面向的项目来起名。我们这个项目就叫做 **HelloWorldTest**。你可以（在 Location 一栏）修改项目将要被存放的位置。在此我们使用默认设置，并单击 Next 按钮。图 16-2 显示了这一步对应的对话框。

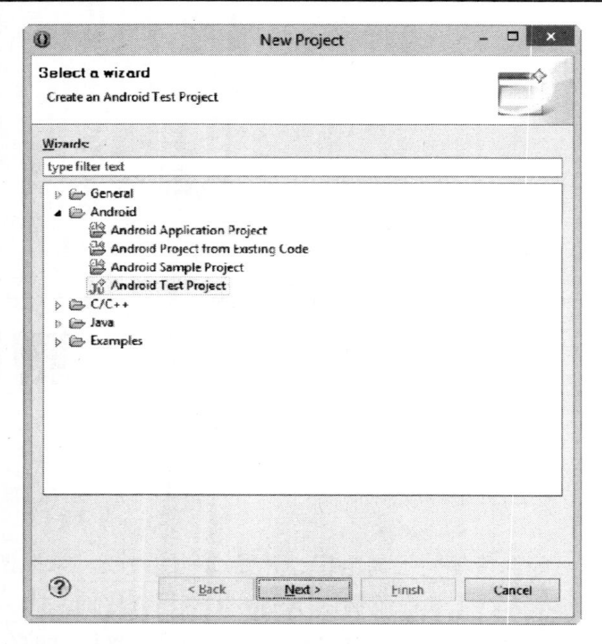


图 16-1　New Project 向导

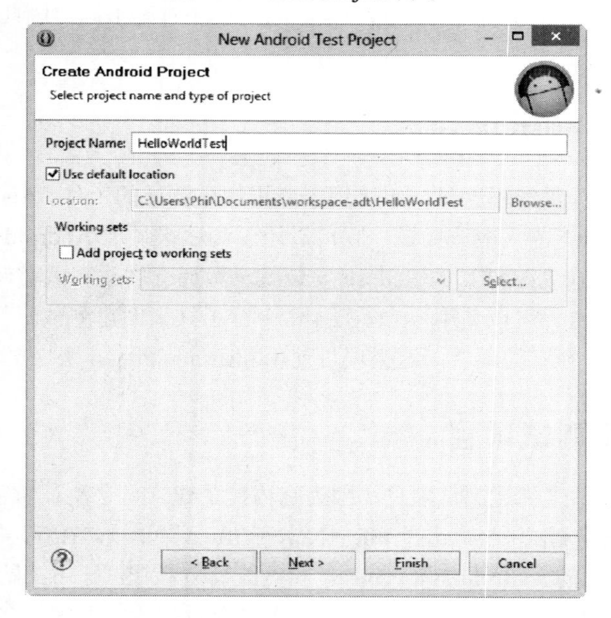


图 16-2　为测试项目起名

向导的下一步会让我们选择本测试项目实际要测试的目标项目。如果当前打开了多个项目，要确保选中了正确的一个。图 16-3 所示的例子中将 **MainActivity** 作为目标，它是目前 Eclipse 工作区中唯一打开的项目。

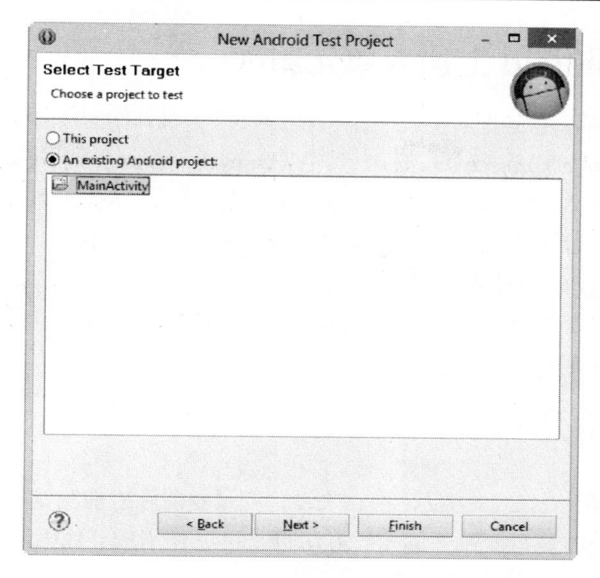


图 16-3 选择目标项目

要挑选特定的构建目标 SDK 级别，单击 Next 按钮，从新弹出的列表中进行选择。图 16-4 显示了向导在构建目标这一步的画面。如果测试并非面向特定的构建级别，直接点击 Finish 按钮即可。

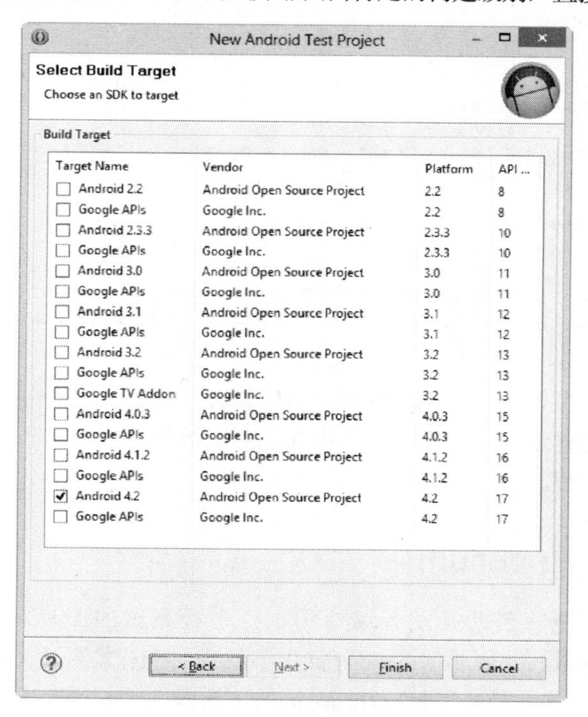


图 16-4 选择测试面向的目标 API 级别

## 技巧 135：在 Android 上加入单元测试

创建了测试项目之后，就该向其中加入要运行的测试了，具体做法是向测试包中添加类。测试包位于 **src/*com.yourtargetproject.projectname*.test**，其中 ***com.yourtargetproject.projectname*** 对应目标项目的包名。在父类中，放入 `ActivityInstrumentationTestCase2<MainActivity>`，其中 *`MainActivity`* 是目标项目中要用到的 Activity 的名字。`ActivityInstrumentationTestCase2` 是用于测试单个 Activity 的一种测试类。

创建好类后，在其中使用一条 import 语句引用要测试的 Activity，并添加一个测试用例构造函数。构造函数应当像如下这样建立：

```
public HelloWorldTest(Class<MainActivity> activityClass) {
  super(activityClass);
}
```

每次测试都要用到 `setUp()` 方法，用它来设置变量及清除以前测试的遗留。对于熟悉 JUnit 测试的人而言值得一提的是，`tearDown()` 方法也是可以使用的。代码清单 16-1 给出了创建和使用 `setUp()` 方法的一个基本代码构架。

**代码清单 16-1　测试要用到的维护方法**

```
protected void setUp() throws Exception {
  super.setUp();

  setActivityInitialTouchMode(false);

  mActivity = getActivity();

  // Test something in the activity by using mActivity.findViewById()

}
protected void tearDown() throws Exception {
  super.tearDown();
}
```

注意，要想使用 `setUp()` 方法，必须先使用 `setActivityInitialTouchMode(false)` 方法。

添加好测试用的值后，只要在测试项目上右击，并选择 **Run As→Android JUnit Test**，即可运行测试。这将会打开 JUnit 视图，在其中显示测试效果，包括每次运行的情况、出现的错误、故障，以及一份摘要。

## 技巧 136：使用 Robotium

Robotium 是一种能够帮助开发者编写和执行多种测试的工具。要使用它，首先要从 http://code.google.com/p/robotium/downloads/list 上下载它的 **.jar** 文件，并将其包含到测试项目中（在笔者写作时，最新的 **.jar** 版本为 **robotium-solo-3.6.jar**）。

让项目包含 Robotium 的方法是：在项目的根目录下创建一个名为 **libs** 的文件夹，并将

**robotium-solo-3.6.jar** 放入其中。在 Eclipse 中，要刷新一下项目，确保改动已经生效。转到 **libs** 文件夹并在 **robotium-solo-3.6.jar** 文件上右击。在弹出的菜单中选择 **Build Path→Add to Build Path**。

如果**.jar** 文件被存储在了另外的位置，那么打开测试项目的属性页面，选择 **Java Build Path**，再单击 **Add (external) Jar**。这样就能够查找**.jar** 文件并将其包含到项目中。

添加了**.jar** 文件后，打开测试项目所用的类文件，在其中加入下面的 import 语句：

```
import com.jayway.android.robotium.solo.Solo;
```

然后再添加一个新变量：

```
private Solo solo;
```

这个 solo 对象是与 Robotium 交互的主要途径，其中包含测试所需的全部方法。接下来向 setUp()方法中加入如下语句：

```
solo = new Solo(getInstrumentation(), getActivity());
```

这样就完成了 solo 对象的初始化，可以在自定义的函数中用它来进行测试。

下面是一个来自 Robotium 的范例测试项目的示例函数，展示了如何使用 solo 对象执行测试：

```
public void testMenuSave() throws Exception {

  solo.sendKey(Solo.MENU);
  solo.clickOnText("More");
  solo.clickOnText("Prefs");
  solo.clickOnText("Edit File Extensions");
  Assert.assertTrue(solo.searchText("rtf"));

  solo.clickOnText("txt");
  solo.clearEditText(2);
  solo.enterText(2, "robotium");
  solo.clickOnButton("Save");
  solo.goBack();
  solo.clickOnText("Edit File Extensions");
  Assert.assertTrue(solo.searchText("application/robotium"));
}
```

关于 Robotium 的更多信息以及更深入的教程，可以通过 http://code.google.com/p/robotium/wiki/RobotiumTutorials 获得。

## 16.2 Eclipse 内建测试工具

带有 ADT 插件的 Eclipse IDE 是一个对用户友好的开发环境，其中包含所见即所得（WYSIWYG)的用户界面以及将资源布局文件转换为生成Android可执行程序所需成分的工具。下一个技巧将给出配置过程的按部就班的引导。我们以一个特定的 Eclipse 版本（写作本书时为 3.7 Indigo）来说明，不过大多数步骤在不同的 Eclipse 版本之间没有差别。

### 技巧 137：指定运行配置

每个应用程序的运行配置都是一个单独的设定档，它告诉 Eclipse 如何运行项目和启动

Activity，以及是将应用程序安装到模拟器上还是安装到联机设备上。在每个应用程序首次创建时，ADT 会自动为其创建运行配置，而我们也可以按照本技巧中描述的方法自定义运行配置。

要创建新的运行配置或编辑已有的配置，可在 Eclipse 里选择 **Run→Run Configurations...**（或者 **Debug Configurations...**），打开如图 16-5 所示的 Run Configurations 窗口。在这个窗口中，有三个与应用程序测试有关的选项卡，包含了我们关心的设定项。

- Android：指定要运行的项目和 Activity。
- Target：选择要运行应用程序的虚拟设备。对于模拟器环境，这里可指定启动参数（如网速和延时），这样就能在模拟得更为真实的无线连接条件下测试应用程序的行为。开发者还可以选择在每次启动模拟器时清除其中存储的数据。
- Common：指定运行配置存储的位置，以及是否将配置显示在 Favourite 菜单中。

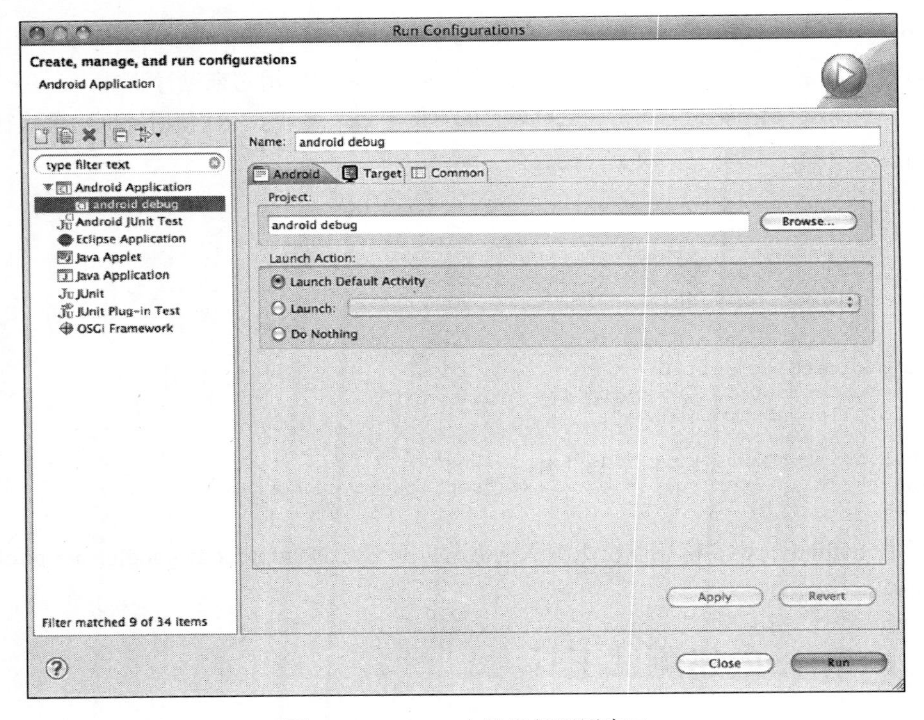


图 16-5　Eclipse 中的运行配置窗口

将以上项目设置妥当后，只需点击 Run 按钮，即可让应用程序在目标设备上开始运行。如果没有真实的 Android 设备连接到主机上，或者原本就选定了虚拟设备作为目标，就会启动模拟器来运行应用程序。

## 技巧 138：使用 DDMS

应用程序开始在目标设备上运行之后，可以打开 Dalvik 调试监控服务器（Dalvik Debug

Monitoring Server，DDMS）检查设备的状态，如图 16-6 所示。DDMS 可以通过命令行运行，也可以在 Eclipse 中选择 **Window→Open Perspective→DDMS** 菜单项运行。

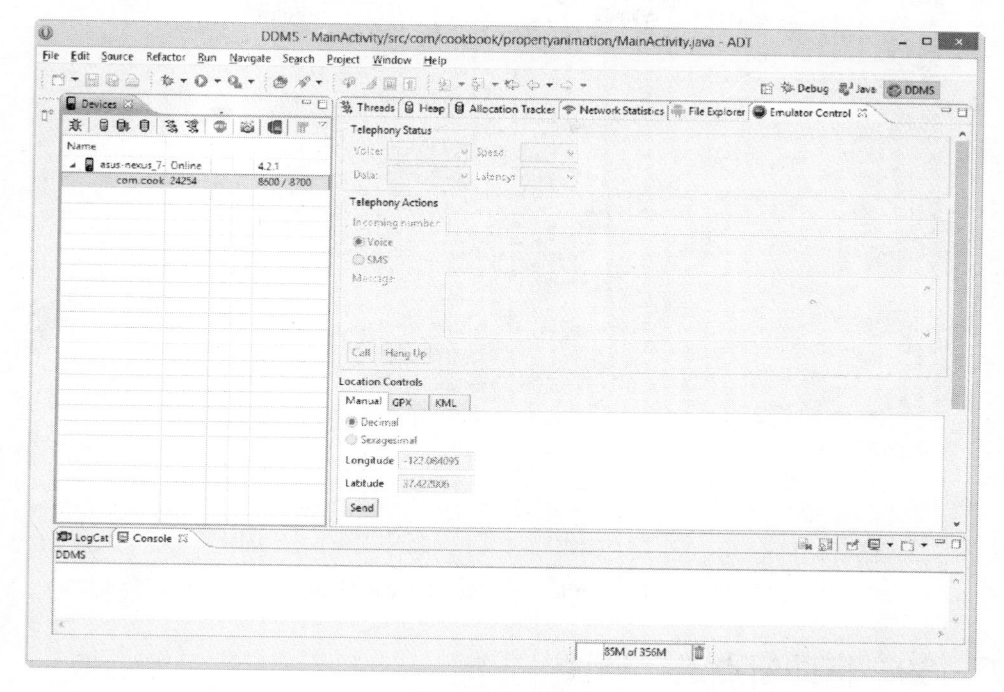


图 16-6 DDMS 控制面板

DDMS 环境中包含三个面板，提供了不同种类的调试数据。

- Devices（设备）：显示已连接的 Android 设备，包括模拟器和真实的 Android 设备。
- 底部面板：包含两个选项卡，即 LogCat 和 Console。LogCat 面板实时显示了设备的所有日志数据，其中包含系统日志消息和用户生成的日志消息。这些消息是通过在应用程序中使用 Log 类获取的。Console 选项卡对 Eclipse 用户来讲可能不陌生，其中显示了编译和运行时的 SystemOut 消息以及某些错误。
- 右上部面板：包含 6 个选项卡，即 Threads（线程）、Heap（堆）、Allocation Tracker（内存分配跟踪）、Network Statistics（网络数据统计）、File Explorer（文件浏览器）和 Emulator Control（模拟器控制）。它们大部分用于分析进程和网络带宽。模拟器控制选项卡包含控制语音、数据格式、网速、延时等的选项。另外还含有创建伪电话呼叫（用于测试）的选项，以及 GPS“欺骗”选项，帮助在模拟器上测试定位功能。点击设备选项卡中的设备会使得右上部的这些面板反映出所选的设备/模拟器的当前运行值，如图 16-7 所示。

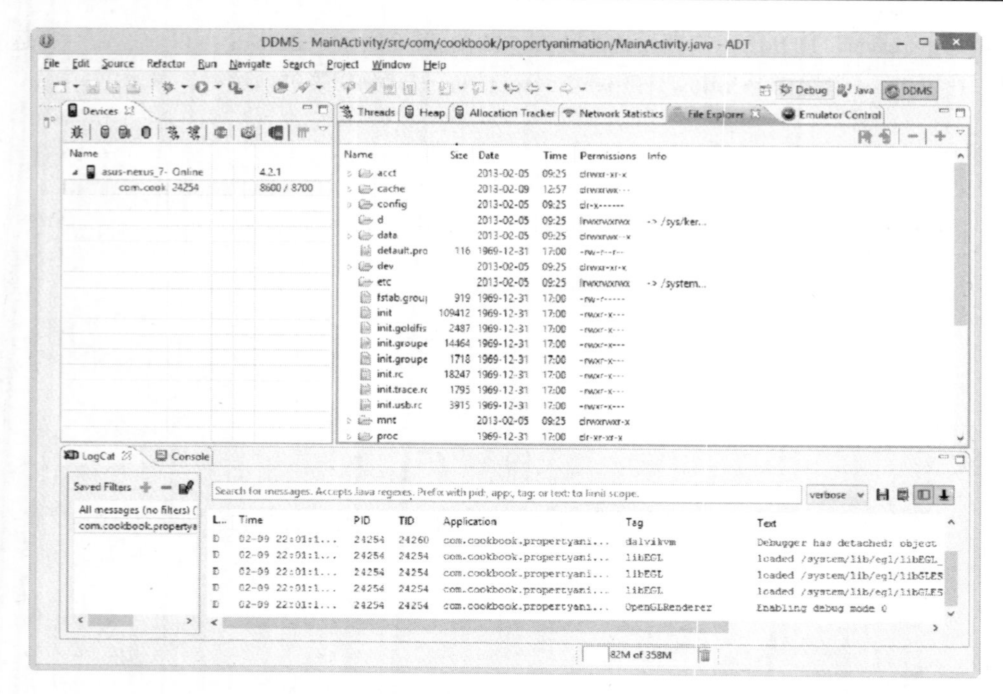


图 16-7　DDMS 控制面板，其中 File Explorer、Logcat 和 Devices 面板处于活动状态

## 技巧 139：借助断点进行调试

开发者还可以在调试模式下运行程序，并通过插入断点在运行时暂时“冻结”应用程序。首先当然需要在调试模式下启动程序，这样会显示如图 16-8 所示的对话框。如果选择“Yes”，就能切换到如图 16-9 所示的 Debug perspective（调试界面）。

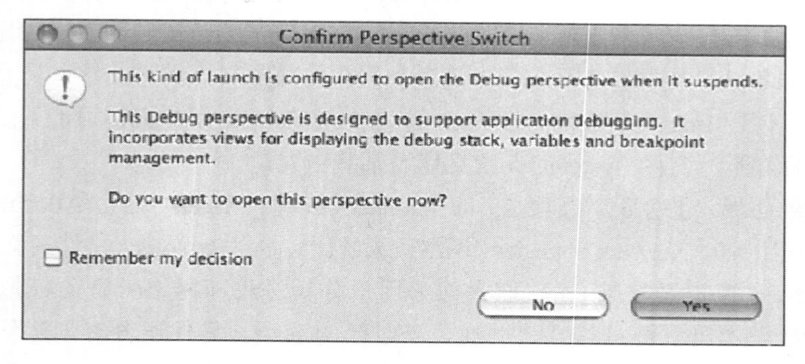


图 16-8　确认切换到调试界面的对话框

调试界面包括一个显示源代码文件的窗口以及其他一些窗口，包括 Variables（变量）、Breakpoints（断点）、Outline（大纲）等。在想要中断代码执行的行左边的边缘处双击鼠标，可以设置或取消断点。设置了断点的行上会出现一个蓝色的圆圈。

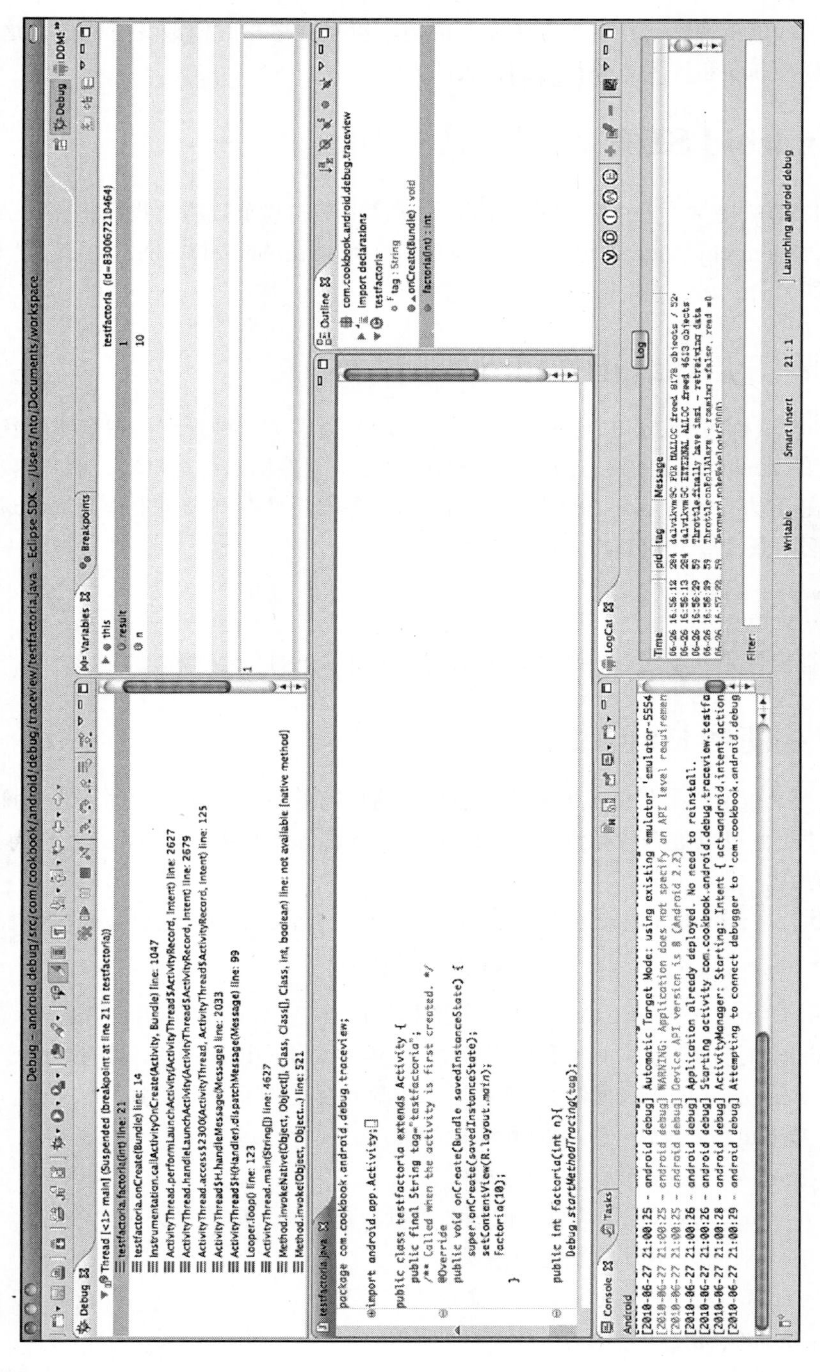

图 16-9 Eclipse 中的调试界面

对于嵌入式编程人员来说，使用断点是一种标准的调试方法。能够让程序停止在某条指令上、对函数进行单步调试、查看内存中变量的值，以及在运行时修改变量的值，这些功能构成了一种找出复杂 bug 和程序意外行为的强大的方法。

## 16.3 Android SDK 调试工具

Android SDK 提供了多个独立的调试工具。接下来的几个技巧中我们将探讨 Android 调试桥、LogCat、Hierarchy Viewer 和 TraceView 工具。可以在 Android SDK 安装位置下的 **tools/**目录中找到它们。

### 技巧 140：开启和终止 Android 调试桥

Android 调试桥（Android Debug Bridge，ADB）提供了一种管理模拟器实例状态或管理通过 USB 连接的 Android 设备状态的方法。ADB 由三个组件构成：客户端、服务器和守护进程。客户端组件在开发机上由 **adb** shell 脚本初始化。服务器组件在开发机上作为后台进程运行。可以分别用下面两条命令开启和终止服务器组件：

```
> adb start-server
> adb kill-server
```

守护进程组件是一个运行在服务器或 Android 设备上的后台进程。

### 技巧 141：使用 LogCat

LogCat 是 Android 提供的实时日志工具，它将所有的系统和应用程序日志数据收集到循环缓冲区中，使它们可以被查看和过滤。可以将其作为独立的工具来访问，或者作为 DDMS 工具的一部分来使用。

可以在执行了 **adb** shell 登录到设备后，在设备上使用 LogCat，或者通过 **adb** 执行 `logcat` 命令：

```
> [adb] logcat [<option>] ... [<filter-spec>] ...
```

所有使用了 `android.util.Log` 类的消息都有与之关联的标签和优先级。标签应当是有意义的，与 Activity 的功能有关。标签和优先级使日志数据易于阅读和过滤。可用的标签有以下几个。

- `V`：Verbose（优先级最低），会在日志中显示尽可能充分的信息。
- `D`：Debug，显示错误、信息和变量的值。
- `I`：Info，只显示信息，比如连接状态。
- `W`：Waring，显示警告消息，这些消息虽不是错误，但值得关注。
- `E`：Error，显示运行时发生的错误。
- `F`：Fatal，仅当程序发生崩溃时显示相关信息。
- `S`：Silent（优先级最高），不显示任何消息。

`logcat` 数据中包含的信息量很大，应当使用过滤器避免信息过量，这可以通过为 `logcat`

命令指定 tag:priority 参数实现：

```
> adb logcat ActivityManager:V *:S
```

该命令让 ActivityManager 的 verbose（V）数据得以显示，而其他日志命令都不予显示（S）。

Android 日志使用了循环缓冲区系统。默认情况下，所有信息都被记录到主日志缓冲区。还有其他两种缓冲区：一种包含与射频（radio）和通话（telephony）有关的消息，另一种则包含与事件相关的消息。可以使用-b 开关来启用不同的缓冲区：

```
> adb logcat -b events
```

该命令使缓冲区显示与事件相关的消息：

```
I/menu_opened( 135): 0
I/notification_cancel( 74): [com.android.phone,1,0]
I/am_finish_activity( 74):
[1128378040,38,com.android.contacts/.DialtactsActivity,app-request]
I/am_pause_activity( 74):
[1128378040,com.android.contacts/.DialtactsActivity]
I/am_on_paused_called( 135): com.android.contacts.RecentCallsListActivity
I/am_on_paused_called( 135): com.android.contacts.DialtactsActivity
I/am_resume_activity( 74): [1127710848,2,com.android.launcher/.Launcher]
I/am_on_resume_called( 135): com.android.launcher.Launcher
I/am_destroy_activity( 74):
[1128378040,38,com.android.contacts/.DialtactsActivity]
I/power_sleep_requested( 74): 0
I/power_screen_state( 74): [0,1,468,1]
I/power_screen_broadcast_send( 74): 1
I/screen_toggled( 74): 0
I/am_pause_activity( 74): [1127710848,com.android.launcher/.Launcher]
```

另外一个例子：

```
> adb logcat -b radio
```

该命令将显示与射频/通话有关的消息：

```
D/RILJ ( 132): [2981]< GPRS_REGISTRATION_STATE {1, null, null, 2}
D/RILJ ( 132): [2982]< REGISTRATION_STATE {1, null, null, 2, null, null,
null, null, null, null, null, null, null, null}
D/RILJ ( 132): [2983]< QUERY_NETWORK_SELECTION_MODE {0}
D/GSM ( 132): Poll ServiceState done: oldSS=[0 home T - Mobile T - Mobile
31026 Unknown CSS not supported -1 -1RoamInd: -1DefRoamInd: -1]
newSS=[0 home T - Mobile T - Mobile 31026 Unknown CSS not supported -1 -
1RoamInd: -1DefRoamInd: -1] oldGprs=0 newGprs=0 oldType=EDGE newType=EDGE
D/RILJ ( 132): [UNSL]< UNSOL_NITZ_TIME_RECEIVED 10/06/26,21:49:56-28,1
I/GSM ( 132): NITZ: 10/06/26,21:49:56-28,1,237945599 start=237945602
delay=3
D/RILJ ( 132): [UNSL]< UNSOL_RESPONSE_NETWORK_STATE_CHANGED
D/RILJ ( 132): [2984]> OPERATOR
D/RILJ ( 132): [2985]> GPRS_REGISTRATION_STATE
D/RILJ ( 132): [2984]< OPERATOR {T - Mobile, T - Mobile, 31026}
D/RILJ ( 132): [2986]> REGISTRATION_STATE
D/RILJ ( 132): [2987]> QUERY_NETWORK_SELECTION_MODE
D/RILJ ( 132): [2985]< GPRS_REGISTRATION_STATE {1, null, null, 2}
D/RILJ ( 132): [2986]< REGISTRATION_STATE {1, null, null, 2, null, null,
null, null, null, null, null, null, null, null}
D/RILJ ( 132): [2987]< QUERY_NETWORK_SELECTION_MODE {0}
```

LogCat 在调试基于 Java 的 Android 应用程序时非常有用。然而，当应用程序含有原生组件时，追踪起来就不那么容易。此时，应当将原生组件日志记录到 `System.out` 或 `System.err`。默认情况下，Android 系统将 `stdout` 和 `stderr`（即 `System.out` 和 `System.err`）输出到 `/dev/null`。可以通过下面的 ADB 命令将输出重定向到某个日志文件：

```
> adb shell stop
> adb shell setprop log.redirect-stdio true
> adb shell start
```

上述命令将停止正在运行的模拟器或设备实例，然后使用 shell 命令 `setprop` 启用输出重定向，再重启实例。

## 技巧 142：使用 Hierachy Viewer

使用 Hierarchy Viewer 是一种有效的调试和理解用户界面的方法。它提供了布局中视图层次结构的可视化表示，还有对显示效果的放大查看工具（显示在 Pixel Perfect 窗口中）。请注意，目前 Hierarchy Viewer 虽然仍然可用，但它之前包含的一些特性已被迁移到了 Android Debug Monitor 中，后者还包含了 DDMS。

我们使用 **hierarchyviewer** 工具来调用 Hierarchy Viewer，运行后会启动如图 16-10 所示的界面。其中显示了一个当前连接到开发机的 Android 设备列表。选择其中一个设备后，将显示设备上运行的程序列表。接下来就可以从中选择想要调试或进行用户界面优化的程序了。

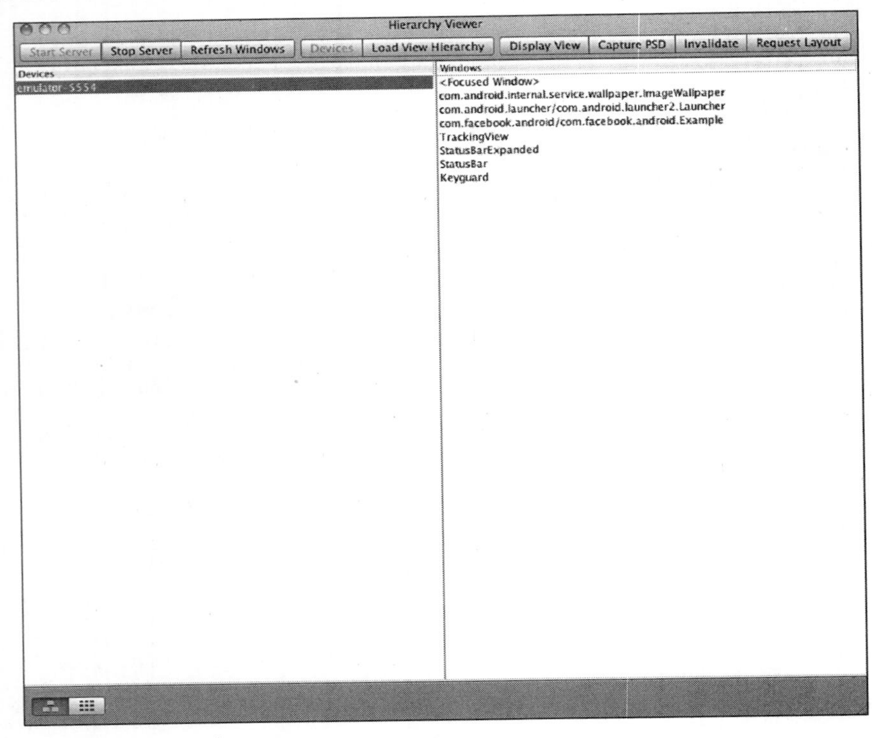


图 16-10 Hierarchy Viewer 工具

选择了程序之后，即可点击 Load View Hierarchy 查看由 Hierarchy Viewer 构建的视图树。其中包含以下 4 个视图。

- 树视图窗格：视图的层级图，位于窗体左部。
- 树概览窗格：层级图的缩略图。
- 属性视图窗格：选定视图的属性列表，位于窗体右上部。
- 布局视图窗格：布局的线框图，位于窗体右下部。

上述界面如图 16-11 所示。

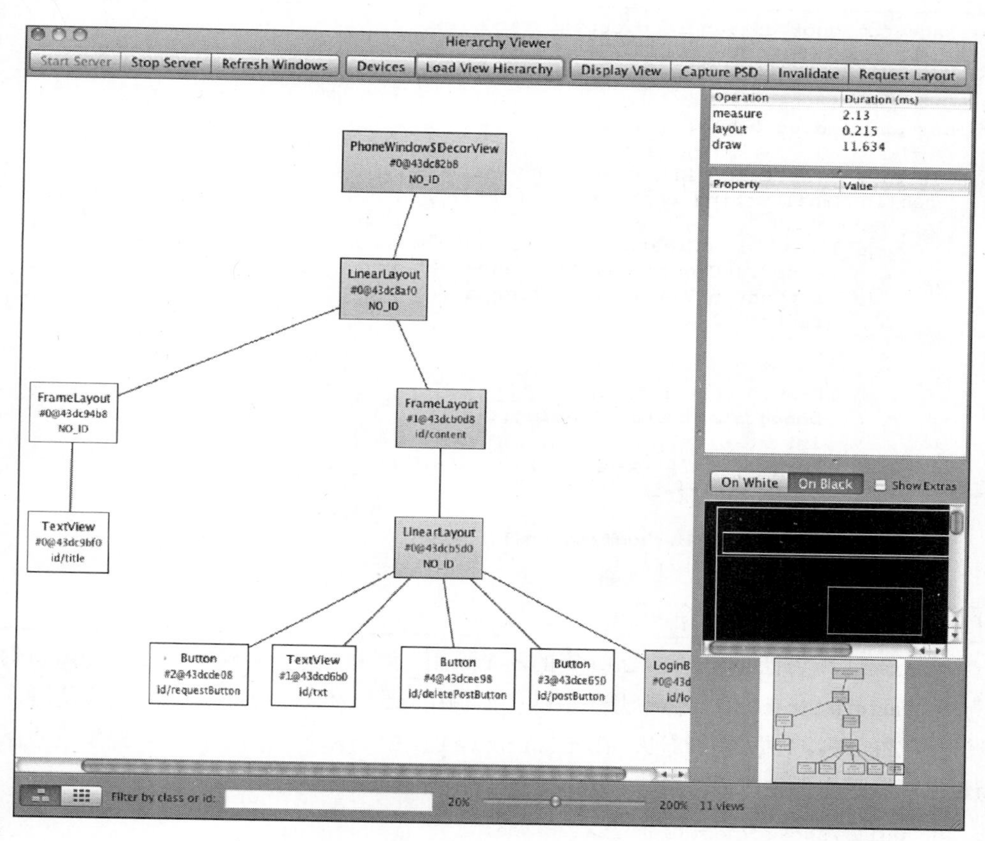


图 16-11　Hierarchy Viewer 工具中的布局视图

刚刚提到的 4 个视图显示关于视图层级的不同信息，有助于优化 UI。当选中了视图的某一节点，属性视图和线框视图就会更新。在 Android 系统中，对于应用程序产生的视图树有一定限制。树的深度不能大于 10，宽度不能超过 50。在 Android 1.5 及以前的版本中，超过这一限制会抛出栈溢出异常。尽管这一限制很宽松，但让视图树层次尽可能浅会使应用程序运行更快速流畅。可以合并视图或使用 `RelativeLayout` 代替 `LinearLayout` 来优化视图树，从而达到降低视图树层次的目的。

## 技巧143：使用TraceView

TraceView是一个用于优化程序性能的工具。要使用该工具，首先要在应用程序中实现Debug类。该工具会创建包含跟踪信息的日志文件，以便进行分析。本技巧指定了一个阶乘方法，以及一个调用阶乘方法的方法。主Activity由代码清单16-2给出。

**代码清单16-2 src/com/cookbook/android/debug/traceview/TestFactorial.java**

```
package com.cookbook.android.debug.traceview;

import android.app.Activity;
import android.os.Bundle;
import android.os.Debug;

public class TestFactorial extends Activity {
    public final String tag="testfactorial";
        @Override
        public void onCreate(Bundle savedInstanceState) {
            super.onCreate(savedInstanceState);
            setContentView(R.layout.main);
            factorial(10);
        }

        public int factorial(int n) {
            Debug.startMethodTracing(tag);
            int result=1;
            for(int i=1; i<=n; i++) {
                result*=i;
            }
            Debug.stopMethodTracing();
            return result;
        }
}
```

factorial()方法包含对Debug类的两个调用。调用startMethodTracing()时，会在名为**testfactorial.trace**的文件中开始追踪；而调用stopMethodTracing()方法时，系统会继续对生成的跟踪数据进行缓冲。在factorial(10)方法返回后，追踪文件生成并保存到**/sdcard/**中。追踪文件生成后，可以在开发机上使用下面的命令检索它：

```
> adb pull /sdcard/testfactorial.trace
```

接着可以使用Android SDK的**tools**文件夹下的**traceview**工具分析追踪文件：

```
> traceview testfactorial.trace
```

这一脚本命令运行后，会产生一个分析界面，如图16-12所示。

界面中包含Timeline（时间轴）面板和Profile（简况）面板。Timeline面板位于屏幕上半部，描述了每个线程和方法的起止时间。Profile面板位于屏幕下半部，提供了阶乘方法执行时发生情况的摘要。当Timeline面板中的游标移动时，面板上会随之显示对应的时间：追踪何时开始、方法何时被调用、追踪何时结束等。

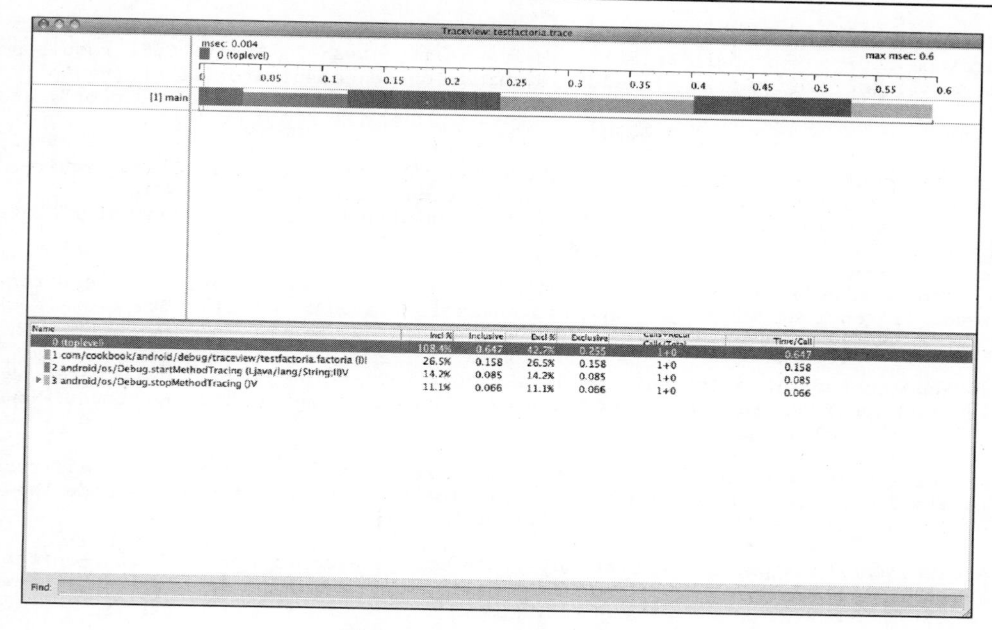


图 16-12 TraceView 分析界面

Profile 面板显示了阶乘方法中所有时间消耗情况的概要信息。除每段时间占总时间的百分比外，面板中还显示了 inclusive time 和 exclusive time。exclusive time 是执行方法本身所花费的时间，而 inclusive time 则是方法本身花费的时间加上由该方法所调用的其他方法消耗的时间。

***.trace** 文件由一个数据文件和一个关键字文件组成。数据文件用于保存追踪数据，关键字文件提供了从二进制标识符到线程名和方法名的映射。如果使用早期版本的 TraceView，需要手动将关键字文件和数据文件合并成一个 trace 文件。

利用 **dmtracedump** 工具可以基于 Android 追踪日志文件生成调用栈图。该工具要求安装第三方的 Graphviz dot 工具来创建图形化输出。

## 技巧 144：使用 lint

从版本 16 起，ADT 中包含了 **lint** 工具。该工具会自动在 Eclipse 中运行，其职责是检验代码，当代码可能不具备期望的功能或可能导致编译失败时，将其高亮显示。对于不使用 Eclipse 或者更想在命令行中为项目执行 **lint** 的开发者，可以在 SDK 安装目录的 **tools** 子目录上运行 **lint** 来达到目的。

下面是一个在命令行下运行 **lint** 的例子。注意，该例是在 Windows 的命令行下运行的，因此要迁移到其他平台上，可能要调整其中斜杠的方向。此外，在命令行中运行时，输出结果会被自动换行处理，可能与下面给出的效果完全一致：

```
lint \temp\PropertyAnimation
Scanning PropertyAnimation: . . . . . . . . . . . . . .
```

```
Scanning PropertyAnimation (Phase 2): . . . . . . . .
src\com\cookbook\propertyanimation\MainActivity.java:32: Error: Call requires API
level 11 (current min is 8): android.animation.ObjectAnimator#ofInt [NewApi]
        ValueAnimator va = ObjectAnimator.ofInt(btnShift, "backgroundColor", start,
end);
                                                   ~~~~~
src\com\cookbook\propertyanimation\MainActivity.java:32: Error: Class requires API
level 11 (current min is 8): android.animation.ValueAnimator [NewApi]
        ValueAnimator va = ObjectAnimator.ofInt(btnShift, "backgroundColor", start,
end);
        ~~~~~~~~~~~~~
src\com\cookbook\propertyanimation\MainActivity.java:33: Error: Call requires API
level 11 (current min is 8): android.animation.ValueAnimator#setDuration [NewApi]
        va.setDuration(750);
           ~~~~~~~~~~~
src\com\cookbook\propertyanimation\MainActivity.java:34: Error: Call requires API
level 11 (current min is 8): android.animation.ValueAnimator#setRepeatCount [NewApi]
        va.setRepeatCount(1);
           ~~~~~~~~~~~~~~
src\com\cookbook\propertyanimation\MainActivity.java:35: Error: Call requires API
level 11 (current min is 8): android.animation.ValueAnimator#setRepeatMode [NewApi]
        va.setRepeatMode(ValueAnimator.REVERSE);
           ~~~~~~~~~~~~~
src\com\cookbook\propertyanimation\MainActivity.java:36: Error: Call requires API
level 11 (current min is 8): android.animation.ValueAnimator#setEvaluator [NewApi]
        va.setEvaluator(new ArgbEvaluator());
           ~~~~~~~~~~~~
src\com\cookbook\propertyanimation\MainActivity.java:36: Error: Call requires API
level 11 (current min is 8): new android.animation.ArgbEvaluator [NewApi]
        va.setEvaluator(new ArgbEvaluator());
                        ~~~~~~~~~~~~~
src\com\cookbook\propertyanimation\MainActivity.java:37: Error: Call requires API
level 11 (current min is 8): android.animation.ValueAnimator#start [NewApi]
        va.start();
           ~~~~~
res\menu\activity_main.xml: Warning: The resource R.menu.activity_main appears to be
unused [UnusedResources]
res\values\strings.xml:5: Warning: The resource R.string.hello_world appears to be
unused [UnusedResources]
    <string name="hello_world">Hello world!</string>
            ~~~~~~~~~~~~~~~~~~
res\layout\activity_main.xml:13: Warning: [I18N] Hardcoded string "Rotate", should
use @string resource [HardcodedText]
        android:text="Rotate" />
        ~~~~~~~~~~~~~~~~~~~~~
res\layout\activity_main.xml:21: Warning: [I18N] Hardcoded string "Shift", should use
@string resource [HardcodedText]
        android:text="Shift" />
        ~~~~~~~~~~~~~~~~~~~~
res\layout\activity_main.xml:29: Warning: [I18N] Hardcoded string "Sling Shot",
should use @string resource [HardcodedText]
        android:text="Sling Shot" />
        ~~~~~~~~~~~~~~~~~~~~~~~~~
8 errors, 5 warnings
```

从输出可以看到，例子对应的项目有几处有待改正的错误，另有5条关于如何更好地遵循标准的警告或建议。

开发者可以通过传递`-disable`选项强制**lint**跳过某些需求检查，`-disable`后面应为希望跳过的错误或警告的类型。例如，如果开发者对可能存在的未被利用的资源、ID以及硬编码的文本等不感兴趣，可以将 **lint** 命令改为如下形式：

```
lint -disable UnusedResources,UnusedIds,HardcodedText \temp\PropertyAnimation
```

由于关闭了所有警告，所以命令只会返回需要被修正的错误。

# 16.4 Android 系统调试工具

Android 构建在 Linux 之上，因此很多 Linux 工具在 Android 中也可利用。例如，要显示当前正在运行的应用程序以及它们所使用的资源，可以使用 `top` 命令。当某个设备通过 USB 线连接到了主机上，或者模拟器正在运行时，可以在命令行下使用下面的命令：

```
> adb shell top
```

图 16-13 给出了该命令运行后的一个输出示例。

```
Terminal — adb — 202×56

User 2%, System 5%, IOW 0%, IRQ 0%
User 7 + Nice 0 + Sys 18 + Idle 292 + IOW 0 + IRQ 0 + SIRQ 1 = 318

  PID CPU% S  #THR     VSS     RSS PCY UID      Name
 8129   4% R     1    912K    396K  fg root     top
   40   0% S     8   7760K    300K  fg root     /sbin/adbd
  113   0% S    17 121204K  15756K  fg radio    com.android.phone
    4   0% S     1      0K      0K  fg root     events/0
    5   0% S     1      0K      0K  fg root     khelper
    6   0% S     1      0K      0K  fg root     suspend
    7   0% S     1      0K      0K  fg root     kblockd/0
    8   0% S     1      0K      0K  fg root     cqueue
    9   0% S     1      0K      0K  fg root     kseriod
   10   0% S     1      0K      0K  fg root     kmmcd
   11   0% S     1      0K      0K  fg root     pdflush
   12   0% S     1      0K      0K  fg root     pdflush
   13   0% S     1      0K      0K  fg root     kswapd0
   14   0% S     1      0K      0K  fg root     aio/0
   22   0% S     1      0K      0K  fg root     mtdblockd
   23   0% S     1      0K      0K  fg root     kstriped
   24   0% S     1      0K      0K  fg root     hid_compat
   25   0% S     1      0K      0K  fg root     rpciod/0
   26   0% S     1      0K      0K  fg root     mmcqd
   27   0% S     1    740K    316K  fg root     /system/bin/sh
   28   0% S     1    812K    292K  fg system   /system/bin/servicemanager
   29   0% S     3   3740K    396K  fg root     /system/bin/vold
   30   0% S     3   3720K    400K  fg root     /system/bin/netd
   31   0% S     1    668K    240K  fg root     /system/bin/debuggerd
   32   0% S     4   5392K    540K  fg radio    /system/bin/rild
   33   0% S     1  85100K  14192K  fg root     zygote
   34   0% S     5  21744K   1168K  fg media    /system/bin/mediaserver
   35   0% S     1    812K    340K  fg root     /system/bin/installd
   36   0% S     1   1620K    280K  fg keystore /system/bin/keystore
   37   0% S     1    740K    328K  fg root     /system/bin/sh
   38   0% S     1    840K    344K  fg root     /system/bin/qemud
   51   0% S     1    796K    312K  fg root     /system/bin/qemu-props
   59   0% S    47 157736K  26216K  fg system   system_server
  107   0% S     9 112220K  14944K  fg app_17   com.android.inputmethod.latin
  115   0% S     6 119100K  17712K  bg app_16   com.android.launcher
    1   0% S     1    312K    220K  fg root     /init
  167   0% S     6 111600K  14576K  bg system   com.android.settings
  183   0% S     6 106268K  14096K  bg app_30   com.android.alarmclock
  198   0% S     7 107172K  14780K  bg app_14   android.process.media
  208   0% S     8 107304K  13620K  bg app_22   com.android.defcontainer
  243   0% S     6 105436K  13436K  bg app_7    com.android.protips
  256   0% S     6 106592K  14084K  bg app_12   com.android.music
  264   0% S     6 105232K  13428K  bg app_8    com.svox.pico
  284   0% S     7 106704K  15888K  bg app_41   com.facebook.android
 2918   0% S     6 105864K  15340K  fg app_43   com.cookbook.android.debug.traceview
 8128   0% S     1    740K    328K  fg root     /system/bin/sh
  141   0% S     9 121044K  19768K  bg app_0    android.process.acore
    2   0% S     1      0K      0K  fg root     kthreadd
    3   0% S     1      0K      0K  fg root     ksoftirqd/0
```

图 16-13　top 命令的输出范例

top 命令还可以显示系统整体的 CPU 和内存使用的百分比。

另一个重要工具是 ps，它能够列出当前 Android 系统上运行的所有进程：

```
> adb shell ps
```

图 16-14 给出了该命令的一个输出示例。

```
Terminal — bash — 202×56

 Asset Allocations
    zip:/data/app/com.facebook.android-2.apk:/resources.arsc: 3K
nelsontos-MacBook-Pro:tools nto$ ./adb shell ps
USER     PID   PPID  VSIZE  RSS     WCHAN    PC         NAME
root      1     0     312    220   c009b74c 0000ca4c S /init
root      2     0     0      0     c004e72c 00000000 S kthreadd
root      3     2     0      0     c003fdc8 00000000 S ksoftirqd/0
root      4     2     0      0     c004b2c4 00000000 S events/0
root      5     2     0      0     c004b2c4 00000000 S khelper
root      6     2     0      0     c004b2c4 00000000 S suspend
root      7     2     0      0     c004b2c4 00000000 S kblockd/0
root      8     2     0      0     c004b2c4 00000000 S cqueue
root      9     2     0      0     c018179c 00000000 S kseriod
root      10    2     0      0     c004b2c4 00000000 S kmmcd
root      11    2     0      0     c006fc74 00000000 S pdflush
root      12    2     0      0     c006fc74 00000000 S pdflush
root      13    2     0      0     c00744e4 00000000 S kswapd0
root      14    2     0      0     c004b2c4 00000000 S aio/0
root      22    2     0      0     c017ef48 00000000 S mtdblockd
root      23    2     0      0     c004b2c4 00000000 S kstriped
root      24    2     0      0     c004b2c4 00000000 S hid_compat
root      25    2     0      0     c004b2c4 00000000 S rpciod/0
root      26    2     0      0     c019d16c 00000000 S mmcqd
root      27    1     740    316   c0158eb0 afd0d8ac S /system/bin/sh
system    28    1     812    292   c01a94a4 afd0db4c S /system/bin/servicemanager
root      29    1     3740   396   ffffffff afd0e1bc S /system/bin/vold
root      30    1     3720   400   ffffffff afd0e1bc S /system/bin/netd
root      31    1     668    240   c01b52b4 afd0c4dc S /system/bin/debuggerd
radio     32    1     5392   540   ffffffff afd0e1bc S /system/bin/rild
root      33    1     85100  14192 c009b74c afd0dc74 S zygote
media     34    1     22772  1280  ffffffff afd0db4c S /system/bin/mediaserver
root      35    1     812    340   c02181f4 afd0d8ac S /system/bin/installd
keystore  36    1     1620   280   c01b52b4 afd0c4dc S /system/bin/keystore
root      37    1     740    328   c003da38 afd0e7bc S /system/bin/sh
root      38    1     908    412   c00b8fec afd0e90c S /system/bin/qemud
root      40    1     3460   252   ffffffff 0000ecc4 S /sbin/adbd
root      51    37    796    312   c02181f4 afd0d8ac S /system/bin/qemu-props
system    59    33    157736 26712 ffffffff afd0db4c S system_server
app_17    107   33    110168 15436 ffffffff afd0eb08 S com.android.inputmethod.latin
radio     113   33    121204 16004 ffffffff afd0eb08 S com.android.phone
app_16    115   33    119100 17844 ffffffff afd0eb08 S com.android.launcher
app_0     141   33    121044 19996 ffffffff afd0eb08 S android.process.acore
system    167   33    111604 15024 ffffffff afd0eb08 S com.android.settings
app_30    183   33    106276 14432 ffffffff afd0eb08 S com.android.alarmclock
app_14    198   33    107172 15132 ffffffff afd0eb08 S android.process.media
app_22    208   33    107304 13908 ffffffff afd0eb08 S com.android.defcontainer
app_7     243   33    105440 13736 ffffffff afd0eb08 S com.android.protips
app_12    256   33    106596 14480 ffffffff afd0eb08 S com.android.music
app_8     264   33    105232 13772 ffffffff afd0eb08 S com.svox.pico
app_41    284   33    106704 16020 ffffffff afd0eb08 S com.facebook.android
app_43    2918  33    105864 15784 ffffffff afd0eb08 S com.cookbook.android.debug.traceview
root      13187 40    740    328   c003da38 afd0e7bc S /system/bin/sh
root      13188 13187 888    332   00000000 afd0d8ac R ps
nelsontos-MacBook-Pro:tools nto$
```

图 16-14　ps 命令的输出范例

输出中提供了每个正在运行的进程的进程 ID（PID）和用户 ID。使用 dumpsys 可以查看内存分配情况：

```
> adb shell dumpsys meminfo <package name>
```

图 16-15 给出了这条命令的输出示例。

上面介绍的几条命令不仅可以提供 Java 组件的信息，也能够提供原生组件的信息，这些信息对于优化和分析 NDK 应用程序是很有用的。除内存信息之外，还包括进程中使用了多少视图、多少 Activity 以及多少应用程序上下文等。

```
nelsontos-MacBook-Pro:tools nto$ ./adb shell dumpsys 284
Can't find service: 284
nelsontos-MacBook-Pro:tools nto$ ./adb shell dumpsys meminfo com.facebook.android
Applications Memory Usage (kB):
Uptime: 63486347 Realtime: 63486347

** MEMINFO in pid 284 [com.facebook.android] **
                    native   dalvik    other    total
            size:     3952     3719      N/A     7671
       allocated:     3325     2621      N/A     5946
            free:       42     1098      N/A     1140
           (Pss):      790     1978     1038     3806
  (shared dirty):     1532     4160     1060     6752
    (priv dirty):      684     1184      708     2576

 Objects
           Views:        0        ViewRoots:        0
     AppContexts:        0       Activities:        0
          Assets:        2    AssetManagers:        2
   Local Binders:        5    Proxy Binders:        9
Death Recipients:        0
 OpenSSL Sockets:        0

 SQL
            heap:        0       memoryUsed:        0
pageCacheOverflo:        0  largestMemAlloc:        0


 Asset Allocations
    zip:/data/app/com.facebook.android-2.apk:/resources.arsc: 3K
nelsontos-MacBook-Pro:tools nto$
```

图 16-15 dumpsys 命令的输出范例

## 技巧 145：设置 GDB 调试

GNU 项目调试器（GDB）是 Linux 中调试程序的常用工具。在 Android 中可以用 GDB 工具来调试原生库。在 NDK-r8 中，生成每个原生库时，还会生成 **gdbserver** 和 **gdb.setup**。可以用下列命令来安装 GDB：

```
> adb shell
> cd /data/
> mkdir myfolder
> exit
> adb push gdbserver /data/myfolder
```

要运行 GDB，可使用下面这条命令：

```
> adb shell /data/myfolder/gdbserver host:port <native program>
```

例如，当名为 myprogram 的程序在 IP 地址为 10.0.0.1 的 Android 设备上运行，且端口号为 1234 时，下面的命令就能够开启服务器：

```
> adb shell /data/myfolder/gdbserver 10.0.0.1:1234 myprogram
```

然后打开另一个终端窗口，在该程序上运行 GDB：

```
> gdb myprogram
 (gdb) set sysroot ../
 (gdb) set solib-search-path ../system/lib
 (gdb) target remote localhost:1234
```

在 GDB 提示符下，第一条命令设置了目标镜像的根目录，第二条命令设置了共享库的搜索路径，最后一条命令则设置了目标设备。远程目标 `localhost:1234` 运行起来后，就可以开始在 GDB 环境下进行调试了。关于 GDB 项目的更多信息参见 www.gnu.org/software/gdb/。

# 附录 A

# 使用 OpenIntents Sensor Simulator

OpenIntents Sensor Simulator（传感器模拟器）是一种可以随 Android 模拟器一起使用、辅助应用程序测试的工具。它可以模拟 GPS 坐标、罗盘方向、定向、压力、加速度及其他因素。它作为一个独立的 Java 应用程序运行，但需要一些编程工作才能与项目互动。

## A.1 建立 Sensor Simulator

首先要到 http://code.google.com/p/openintents/downloads/list?q=sensorsimulator 上下载 OpenIntents Sensor Simulator。将下载的文件（本书使用 **sensorsimulator-2.0-rc1.zip**）解压，并运行解压后的 **bin** 目录中的 **sensorsimulator.jar** 文件。

图 A-1 显示了 Sensor Simulator 运行时的情形。

下一步是启动模拟器，要么在控制台通过 AVD Manager 来启动，要么从 Eclipse 来启动。打开后，启动要用的模拟器，等待其引导完毕。

模拟器引导完毕后，需要通过 `adb` 命令来安装**.apk** 文件。在前面启动 **sensorsimulator.jar** 文件的那个 **bin** 目录中有两个**.apk** 文件：**SensorRecordFromDevice-2.0-rc1.apk** 和 **SensorSimulatorSettings-2.0-rc1.apk**。将 **SensorSimulatorSettings-2.0-rc1.apk** 文件复制到 Android 安装目录的 **platform-tools** 子目录下。复制好后，打开一个终端或命令行窗口，将当前路径切换为 **platform-tools** 目录。一种可能的路径如下：

```
/users/cookbook/Android/sdk/platform-tools/
```

切换到正确的目录之后，运行如下命令：

```
adb install SensorSimulatorSettings-2.0-rc1.apk
```

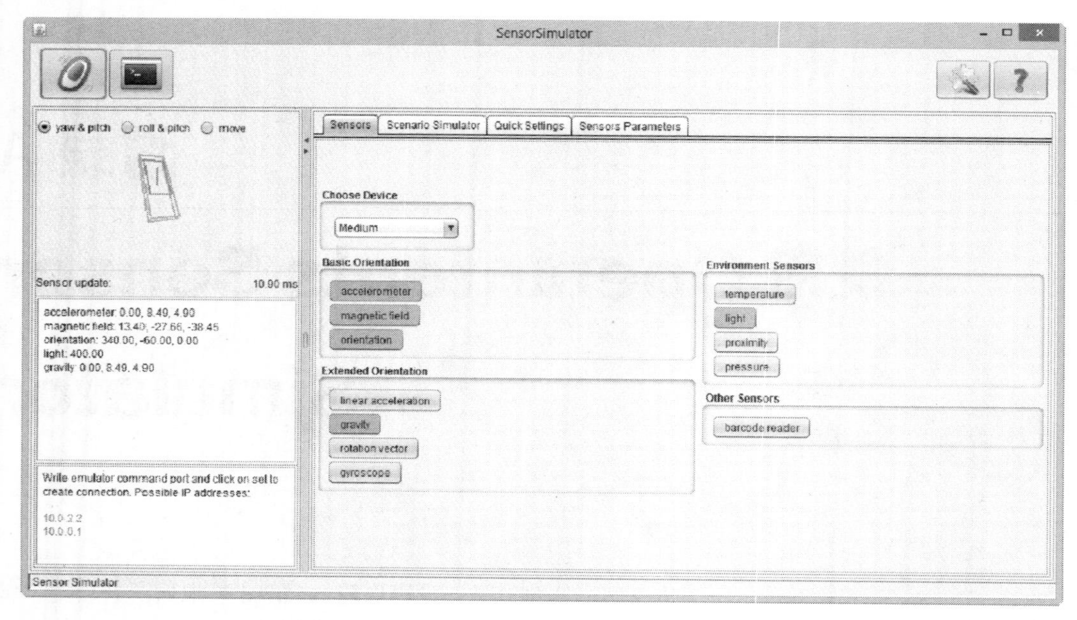


图 A-1　OpenIntentsSensor Simulator 2.0 RC1 版

注意，对于 Linux 和 Mac 用户，在运行命令时还要在 `adb` 前面加上`./`。

控制台或命令行窗口会显示几行信息，并以一条成功消息作为结束。如果没有收到成功消息，请确认模拟器正在运行，并再次尝试。成功安装了 app 后，它将作为已安装的应用程序呈现出来。

当 Android 应用程序被启动时，会显示 IP 地址和用于通信的套接字设置。IP 地址可以在左下角的桌面应用程序中找到。列出的可用 IP 地址不止一个是正常的。图 A-2 显示了 Android 应用程序启动时的初始设定屏幕。

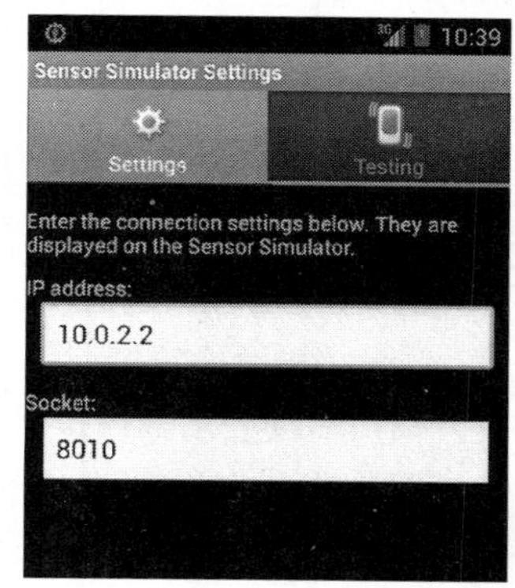


图 A-2　Sensor Simulator 的初始设定屏幕

要测试连接，点击 Android 应用程序的 Testing 选项卡。接着点击 Connect 按钮，并等待与桌面应用程序建立连接。连接建立后，可以变更桌面的设置，从而改变 Android 应用程序的对应值。图 A-3 显示了在桌面应用程序中所做的改动，图 A-4 则显示了 Android 应用程序中相应值的变化。

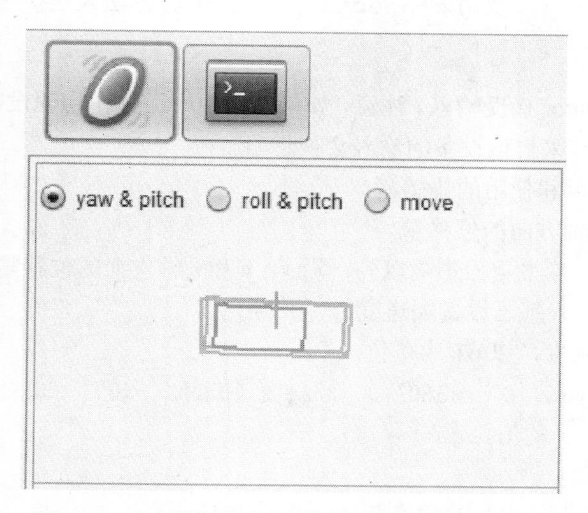


图 A-3 改变模拟器中的旋转矢量（rotation vector）

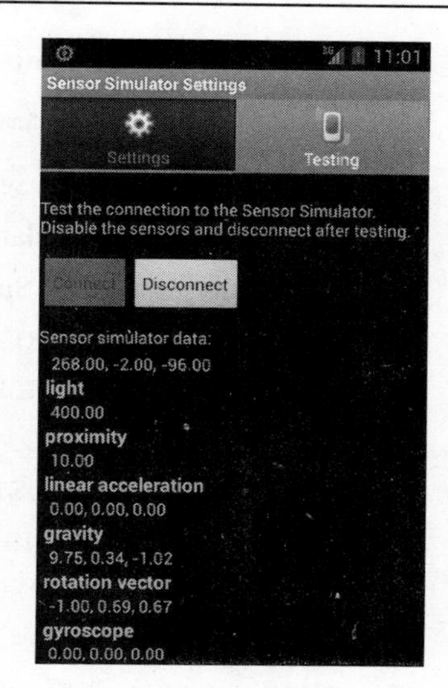


图 A-4 通过桌面应用程序发送的值对传感器数据进行修改后的结果

## A.2 将 Sensor Simulator 添加到应用程序中

我们还可以将 Sensor Simulator 包含到应用程序中，以在没有可用的物理设备时辅助测试工作。为此，要将 **sensorsimulator-lib-2.0-rc1.jar** 包含到项目中。该文件可以在 **sensorsimulator-2.0-rc1.zip** 解压后的 **lib** 文件夹中找到。将该文件复制到项目的 **libs** 文件夹中。如果使用 Eclipse，对文件夹刷新，就应当能看到该文件。在 **sensorsimulator-lib-2.0-rc1.ja** 文件上右击，在弹出的菜单中选择 **Build Path→Add to BuildPath**，这样就能将 Sensor Simulator 的**.jar** 文件自动引用并添加到项目的生成路径中。

由于要在 Android 应用程序和桌面应用程序之间通信，需要向 manifest XML 文件中添加如下权限：

```
<uses-permission android:name="android.permission.INTERNET"/>
```

接下来，在类文件中加入下列引用：

```
import org.openintents.sensorsimulator.hardware.Sensor;
import org.openintents.sensorsimulator.hardware.SensorEvent;
import org.openintents.sensorsimulator.hardware.SensorEventListener;
import org.openintents.sensorsimulator.hardware.SensorManagerSimulator;
```

要访问传感器，我们要在 `onCreate()` 方法中用 `SensorManagerSimulator` 取代 `SensorManager`。虽然这两个类都使用 `getSystemService()` 方法，但前者被修改为接受一个附加的参数。请使用下面的调用：

```
mSensorManager = SensorManagerSimulator.getSystemService(this, SENSOR_SERVICE);
```

由于使用了 SensorManagerSimulator，要用 connectSimulator()方法来连接它：

```
mSensorManager.connectSimulator();
```

现在应用程序被绑定为从 Sensor Simulator 接受输入。注意，如果应用程序运行在物理设备上，当应用程序与 Sensor Simulator 连线时，来自该设备的输入会被忽略。

下面列出可以通过 Sensor Simulator 映射和使用的传感器。

- TYPE_ACCELERATOR：*x*、*y* 和 *z* 轴方向的位移量。
- TYPE_LINEAR_ACCELERATION：除去重力影响以外，沿 *x*、*y* 和 *z* 轴方向的位移量。
- TYPE_GRAVITY：重力大小和方向（基于设备加速度）。
- TYPE_MAGNETIC_FIELD：磁场强度，以 uT 为单位。
- TYPE_ORIENTATION：偏航角（yaw，0° ～360° ）、倾斜度（pitch，-90° ～90° ）和旁向倾角（roll，-180° ～180° ）的值。
- TYPE_TEMPERATURE：环境温度。
- TYPE_LIGHT：光强，以国际单位制（SI）的勒克斯（lux）为单位。
- TYPE_PRESSURE：平均海平面压力，以百帕（hectopascal）为单位。
- TYPE_ROTATION_VECTOR：结合夹角和坐标轴二者而定义的方向。

# 附录 B

# 使用兼容包

兼容包（compatibility pack），或称支持库（support library），是为早期 Android 版本（支持的最低 API 级别为 4）添加支持的一组类。

## B.1 Android 支持包

支持库会向系统中添加下列包。

- android.support.v4.accessibilityservice
- android.support.v4.app
- android.support.v4.content
- android.support.v4.content.pm
- android.support.v4.database
- android.support.v4.net
- android.support.v4.os
- android.support.v4.util
- android.support.v4.view
- android.support.v4.view.accessibility
- android.support.v4.widget

表 B-1 显示了 android.support.v4.accessibilityservice 中包含的类。

表 B-1 android.support.v4.accessibilityservice

| 类型 | 名　称 | 描　述 |
|---|---|---|
| 类 | AccessbilityServiceInfoCompat | 使早于 API Level 4 的版本能够使用 AccessbilityService 的辅助类 |

表 B-2 显示了 android.support.v4.app 包中含有的接口、类和异常。

表 B-2 android.support.v4.app

| 类型 | 名 称 | 描 述 |
| --- | --- | --- |
| 接口 | FragmentManager.BackStackEntry | 表示 Fragment 返回栈的入口 |
| 接口 | FragmentManager.OnBackStackChangedListener | 监视返回栈的变化 |
| 接口 | LoaderManager.LoaderCallbacks<D> | 让客户端能够与管理器交互 |
| 类 | ActivityCompat | 用于访问在 API Level 4 以后添加的 Activity 类中的特性的辅助类 |
| 类 | DialogFragment | 添加 DialogFragment 的支持版本 |
| 类 | Fragment | 添加 Fragment 的支持版本 |
| 类 | Framgment.SavedState | 通过 FragmentManager.saveFragmentInstanceState 从 Fragment 返回的、被保存起来的状态信息 |
| 类 | FragmentActivity | 使用 Fragment（支持的版本）和 Loader API 调用的 Activity 的基类 |
| 类 | FragmentManager | 添加 FragmentManager 的支持版本 |
| 类 | FragmentPagerAdapter | 通过对每个页面使用一个 Fragment 进行页面管理的 PageAdapter |
| 类 | FragmentTabHost | 添加一个 TabHost，使得选项卡内容可以使用 Fragment |
| 类 | FragmentTransaction | 添加 FragmentTransaction 的支持版本 |
| 类 | ListFragment | 添加 ListFragment 的支持版本 |
| 类 | LoaderManager | 添加 LoaderManager 的支持版本 |
| 类 | NavUtils | 为更新后的 Android UI 漫游提供支持的辅助类 |
| 类 | NotificationCompat | 允许访问在 API Level 4 以后添加到 Notification 中的特性的辅助类 |
| 类 | NotificationCompat.Action | 与通知协同使用的 Action 的支持版本 |
| 类 | NotificationCompat.BigPictureStyle | 用于创建包含大图片的大幅面通知的辅助类 |
| 类 | NotificationCompat.BigTextStyle | 用于创建包含大量文本的大幅面通知的辅助类 |
| 类 | NotificationCompat.Builder | 用于创建 NotificationCompat 对象的构建器类 |
| 类 | NotificationCompat.InboxStyle | 用于创建包含列表（内容不能超过 5 个字符串）的大幅面通知的辅助类 |
| 类 | NotificationCompat.Style | 允许将丰富通知风格应用到 Notification.Builder 对象 |
| 类 | ServiceCompat | 用于访问服务特性的辅助类 |

续表

| 类型 | 名 称 | 描 述 |
| --- | --- | --- |
| 类 | ShareCompat | 用于在 Activity 间移动数据的辅助类 |
| 类 | ShareCompat.IntentBuilder | 通过建立 ACTION_SEND 和 ACTION_SEND_MULTIPLE Intent 及启动 Activity 与共享的 Intent 协同工作的辅助类 |
| 类 | ShareCompat.IntentReader | 用于从 ACTION_SEND Intent 读取数据的辅助类 |
| 类 | TaskStackBuilder | 创建在 Android 3.0 及以上版本中漫游时使用的返回栈 |
| 类 | TaskStackBuilderHoneycomb | 允许通过实现 TaskStackBuilder 访问 Honeycomb 的 API |
| 异常 | Fragment.InstantiationException | 在实例化期间如果发生错误，则由 instantiate (Context, String, Bundle) 抛出该异常 |

表 B-3 显示了 android.support.v4.content 包中含有的接口和类。

表 B-3 android.support.v4.content

| 类型 | 名 称 | 描 述 |
| --- | --- | --- |
| 接口 | Loader.OnLoadCompleteListener<D> | 判定某个 Loader 何时完成数据的装载 |
| 类 | AsyncTaskLoader<D> | AsyncTaskLoader 的支持版本 |
| 类 | ContextCompat | 使早于 API Level 4 的版本能够使用 Context 的辅助类 |
| 类 | CursorLoader | CursorLoader 的支持版本 |
| 类 | IntentCompat | 使早于 API Level 4 的版本能够使用 Intent 的辅助类 |
| 类 | Loader<D> | Loader 的支持版本 |
| 类 | Loader.ForceLoadContentObserver | 管理 ContentObserver 和 Loader 之间的连接的一个实现，用于在数据发生变化时重新装载数据 |
| 类 | LocalBroadcastMessage | 用于给对象注册和发送广播 Intent 的辅助类 |

表 B-4 显示了 android.support.v4.content.pm 包中含有的类。

表 B-4 android.support.v4.content.pm

| 类型 | 名 称 | 描 述 |
| --- | --- | --- |
| 类 | ActivityInfoCompat | 使早于 API Level 4 的版本能够使用 ActivityInfo 的辅助类 |

表 B-5 显示了 android.support.v4.database 包中含有的类。

表 B-5　android.support.v4.database

| 类型 | 名　称 | 描　述 |
| --- | --- | --- |
| 类 | DatabaseUtilsCompat | 使早于 API Level 4 的版本能够使用 DatabaseUtils 的辅助类 |

表 B-6 显示了 android.support.v4.net 包中含有的类。

表 B-6　android.support.v4.net

| 类型 | 名　称 | 描　述 |
| --- | --- | --- |
| 类 | ConnectivityManagerCompat | 使早于 API Level 16 的版本能够使用 Connectivity Manager 的辅助类 |
| 类 | TrafficStatsCompat | 使早于 API Level 14 的版本能够使用 TrafficStatsCompat 的辅助类 |
| 类 | TrafficStatsCompatIcs | TrafficStatsCompat 的一个实现版本，与 Ice Cream Sandwich API 共同使用 |

表 B-7 显示了 android.support.v4.os 包中含有的接口和类。

表 B-7　android.support.v4.os

| 类型 | 名　称 | 描　述 |
| --- | --- | --- |
| 接口 | ParcelableCompatCreatorCallbacks<T> | 使用 Parcelable 时用到的回调的支持版本 |
| 类 | ParcelableCompat | 使早于 API Level 4 的版本能够使用 Parcelable 的辅助类 |

表 B-8 显示了 android.support.v4.util 包中含有的类。

表 B-8　android.support.v4.util

| 类型 | 名　称 | 描　述 |
| --- | --- | --- |
| 类 | AtomicFile | AtomicFile 的支持版本 |
| 类 | LongSparseArray<E> | 一个将 long 型变量映射到对象的 SparseArray |
| 类 | LruCache<K,V> | LruCache 的支持版本 |
| 类 | SparseArrayCompat<E> | SparseArray 的 Honeycomb 版本，其中包含 removeAt()方法 |

表 B-9 显示了 android.support.v4.view 包中含有的接口和类。

表 B-9　android.support.v4.view

| 类型 | 名　称 | 描　述 |
| --- | --- | --- |
| 接口 | ViewPager.OnPageChangeListener | 响应页面上变化的接口回调 |

续表

| 类型 | 名　称 | 描　述 |
| --- | --- | --- |
| 接口 | ViewPager.PageTransformer | 当附加的页面发生滚动时调用的接口 |
| 类 | AccessibilityDelegateCompat | 使早于 API Level 4 的版本能够使用 AccessibilityDelegate 的辅助类 |
| 类 | GestureDetectorCompat | 检测手势和使用 MotionEvent 的事件 |
| 类 | KeyEventCompat | 使早于 API Level 4 的版本能够使用 KeyEvent 的辅助类 |
| 类 | MenuCompat | 使早于 API Level 4 的版本能够使用 Menu 的辅助类 |
| 类 | MenuItemCompat | 使早于 API Level 4 的版本能够使用 MenuItem 的辅助类 |
| 类 | PagerAdapter | 提供了用于填入 ViewPager 中包含的页面的适配器 |
| 类 | PagerTabStrip | 为 ViewPager 中包含的当前、后一个和前一个页面提供了一个可交互的指示器 |
| 类 | PagerTitleStrip | 为 ViewPager 中包含的当前、后一个和前一个页面提供了一个不可交互的指示器 |
| 类 | VelocityTrackerCompat | 使早于 API Level 4 的版本能够使用 Velocity 达式 Tracker 的辅助类 |
| 类 | ViewCompat | 使早于 API Level 4 的版本能够使用 View 的辅助类 |
| 类 | ViewCompatJB | 包含了特定于 Jelly Bean 版本的 View API 的访问 |
| 类 | ViewCompatJellyBeanMrl | 包含了对 Jelly Bean MR1 的 View API 的访问 |
| 类 | ViewConfigurationCompat | 使早于 API Level 4 的版本能够使用 ViewConfiguration 的辅助类 |
| 类 | ViewGroupCompat | 使早于 API Level 4 的版本能够使用 ViewGroup 的辅助类 |
| 类 | ViewPager | 包含了一个布局管理器，可在数据窗格间向左/向右翻转 |
| 类 | ViewPager.LayoutParams | 与添加到 ViewPager 上的视图协同使用的布局参数 |
| 类 | ViewPager.SavedState | ViewPager 的保存状态 |
| 类 | ViewPager.SimpleOnPageChangeListener | ViewPager.OnPageChangeListener 的一个实现，包含了其中方法的实现存根 |

表 B-10 显示了 android.support.v4.view.accessibility 包中含有的类。

表 B-10 android.support.v4. accessibility

| 类型 | 名 称 | 描 述 |
| --- | --- | --- |
| 类 | AccessibilityEventCompat | 使早于 API Level 4 的版本能够使用 AccessibilityEvent 的辅助类 |
| 类 | AccessibilityManagerCompat | 使早于 API Level 4 的版本能够使用 AccessibilityManager 的辅助类 |
| 类 | AccessibilityManagerCompat.AccessibilityStateChangeListenerCompat | 可访问性状态的监听器 |
| 类 | AccessibilityNodeInfoCompat | 使早于 API Level 4 的版本能够使用 AccessibilityNodeInfo 的辅助类 |
| 类 | AccessibilityNodeProviderCompat | 使早于 API Level 4 的版本能够使用 AccessibilityNodeProvider 的辅助类 |
| 类 | AccessibilityRecordCompat | 使早于 API Level 4 的版本能够使用 AccessibilityRecord 的辅助类 |

表 B-11 显示了 android.support.v4.widget 包中含有的接口和类。

表 B-11 android.support.v4.widget

| 类型 | 名 称 | 描 述 |
| --- | --- | --- |
| 接口 | SimpleCursorAdapter.CursorToStringConverter | 定义了如何将 Cursor 转换为 String 的一类接口（可被外部客户端使用） |
| 接口 | SimpleCursorAdapter.ViewBinder | 将来自 Cursor 的值绑定到视图的一类接口（可被 SimpleCursorAdapter 的外部客户端使用） |
| 类 | CursorAdapter | CursorAdapter 的支持版本 |
| 类 | EdgeEffectCompact | 使早于 API Level 4 的版本能够使用 EdgeEffect 的辅助类 |
| 类 | ResourceCursorAdapter | ResourceCursorAdapter 的支持版本 |
| 类 | SearchViewCompat | 使早于 API Level 4 的版本能够使用 SearchView 的辅助类 |
| 类 | SearchViewCompat.OnQueryTextListenerCompat | 针对查询中文本的改变的回调 |
| 类 | SimpleCursorAdapter | SimpleCursorAdapter 的支持版本 |

## B.2 为项目添加支持库

为项目添加支持库并不是件难事。首先要下载和安装支持库，然后将其添加到项目的构建路径中。

支持库通过 Android SDK Manager 下载，位于 **Extras** 文件夹的 **Android Support Library** 项。安装好之后，可以在 ***SDKInstallationFolder*/extras/android/support/*VersionNumber*/** 目录下找到所需的**.jar** 文件，其中的 ***SDKInstallationFolder*** 是指 Android SDK 的安装位置，而 ***VersionNumber*** 是指支持库的当前版本（在本书写作期间，最新的版本是 **v4**）。

要将**.jar** 添加到应用程序，需在项目中创建一个名为 **libs** 的文件夹，将**.jar** 文件复制到其中，并将**.jar** 文件添加到构建路径中。

在 Eclipse 中，刷新项目，打开 **libs** 文件夹，右击刚添加的**.jar** 文件，在弹出的菜单中选择 **Build Path→Add to Build Path**。之后，应当通过 import 语句将支持库中你所需的部分包含到代码中，并将所需的那些包含语句调整为支持的版本，从而完成对支持库的初始化。

# 附录 C

## 使用持续集成系统

持续集成（Continuous Integration，CI）系统通常作为敏捷开发或极限编程模型的一部分。下面是使用 CI 时通常遵循的工作流：

（1）开发者将工作代码签入到一个代码仓库中。

（2）签入的代码在构建服务器上被编译。

（3）执行测试（集成测试、单元测试，或二者兼有）。

（4）代码被部署到 staging 或 production 中。

有几种不同的完成项目构建过程的方式。当有重复性的构建任务时，使用构建系统有助于节约时间和减少人为错误。表 C-1 列出了一些有助于简化构建过程的构建工具。

表 C-1　Android 开发的 CI 过程中常用的构建工具

| 工　具 | 许　可　证 | 说　明 |
|---|---|---|
| Apache Ant（http://ant.apache.org/） | Apache 许可证 2.0 版 | 被 ADT 团队采用，是一种被广泛使用的、基于 Java 的构建系统。高度可定制化，在命令行下运行 |
| Apache Maven（http://maven.apache.org/） | Apache 许可证 2.0 版 | Maven 是一种用于项目管理的复杂的构建系统，能够将多个项目合并到同一个构建系统中 |

构建工具有助于管理构建过程。另外还存在一些完整的集成系统。虽然并不是必需的，但它们能通过将单元测试和集成测试同项目部署一起执行，从而大大地减少集成时间。它们能够在部署到产品服务器之前发现程序中的问题，从而避免带着错误发布后让用户灰心失望，甚至能预防可能的服务器故障停机，这足以证明这些工具的价值。表 C-2 列出了一些常见的 CI 系统。

表 C-2 常见的 CI 系统

| 系 统 | 许 可 证 | 说 明 |
|---|---|---|
| Apache Continuum（http://continuum.apache.org/） | Apache 许可证 2.0 版 | Continuum 是一种企业级的集成服务软件，提供了自动化构建（支持 Ant 和 Maven），发布管理，支持多种代码仓库系统[①]，包括 CVS、SVN、Git 和 ClearCase 等 |
| Jenkins（http://jenkins-ci.org/） | MIT 许可证 | Jenkins 是开源和基于 Java 构建的，是人们对于 CI 系统的一种流行选择。Jenkins 是能够帮助追踪诸如定时任务（cron job）等重复性过程的应用程序。可用于构建、测试和发布过程的监控和管理。在 Windows、Ubuntu/Debian、Red Hat/Fedora/CentOS、OS X、openSUSE、FreeBSD、OpenBSD、Solaris/OpenIndiana 和 Gentoo 等众多操作系统上都有可用版本。注意，Jenkins 原本是 Hudson 项目的一部分 |
| Hudson（http://hudson-ci.org/） | MIT 许可证 | 从 2012 年 1 月 24 日起，Hudson 从一个 Oracle 项目转而成为了 Eclipse Foundation 的一部分。它提供了许多与 Jenkins 相同的特性，包括与诸如 CVS、Subversion、Git、ClearCase 等流行的代码仓库系统集成，以及可通过插件来扩展功能等 |
| CruiseControl（http://cruisecontrol.sourceforge.net/） | BSD 风格 | 是一种模块化构建系统，能够与流行的构建系统（如 Ant 和 Maven）集成，支持集成和构建测试 |
| Bamboo（www.atlassian.com/software/bamboo/overview） | 商业软件，价格随使用情况的不同而变化 | 由 Atlassian 创建，能方便地与 JIRA 项目追踪系统集成，从而帮助推动项目的构建、测试和部署。该系统提供了基于需求（根据使用来计算收费）和基于下载（按年或月计费）两种系统版本。它为 Tomcat、JBoss 和 SSH/SCP 系统进行特性内建部署配置 |
| CircleCI（https://circleci.com/） | 商业软件，价格随用途和特性的不同而变化 | 一个新兴的 CI 系统，它提供了一组工具，用于连接 GitHub 仓库及在其上运行构建、测试和部署工作，从而减少管理和配置时间。如果想手动设置测试，可以参考 CircleCI 提供的文档，文档中还介绍了如何克服集成过程中常见的障碍 |

① code repository system，也称版本控制系统（Version Control System）。——译者注

# 附录 D

# Android 操作系统发布版本一览

这里总结了对开发者来说比较重要的 Android 操作系统的不同版本及其主要特性。

## D.1 Cupcake：Android OS 1.5，API Level 3，2009 年 4 月 30 日发布

- Linux 内核版本 2.6.27。
- 智能虚拟键盘，支持第三方键盘。
- AppWidget 框架。
- LiveFolder。
- 原始音频的录制和播放。
- 交互式 MIDI 播放引擎。
- 立体声蓝牙支持。
- 通过 RecognizerIntent（云服务）进行语音识别。
- 更快的 GPS 位置收集（使用 APGS）。

## D.2 Donut：Android OS 1.6，API Level 4，2009 年 9 月 15 日发布

- Linux 内核版本 2.6.29。
- 支持多种屏幕尺寸。
- 手势 API。
- 从文本转语音的引擎。
- 利用 SearchManager 来集成快速搜索框（Quick Search Box）。

## D.3 Eclair：Android OS 2.0，API Level 5，2009 年 10 月 26 日发布

Android OS 2.0.1，API Level 6，2009 年 12 月 3 日发布。

Android OS 2.1，API Level 7，2010 年 1 月 12 日发布。

- 能通过同步适配器 API 连接到任何后端。
- 应用程序可访问的嵌入式快捷通讯录（Embed Quick Contact）。
- 应用程序能够访问到不同设备的蓝牙连接。
- HTML5 支持。
- 通过 `MotionEvent` 类访问多点触控。
- 动态壁纸支持。

## D.4 Froyo：Android OS 2.2，API Level 8，2010 年 5 月 20 日发布

- Linux 内核版本 2.6.32。
- 启用了 Just-in-time（JIT）编译，提高了代码执行速度。
- 汽车和桌面 dock 主题。
- 对多点触控事件更好的定义。
- 云端到设备 API。
- 应用程序可以请求被安装到 SD 内存卡。
- 选定设备上的 Wi-Fi 共享支持。
- 视频和图像的缩略图工具。
- 键盘输入的多语言支持。
- Google 商店应用程序的错误报告。

## D.5 Gingerbread：Android OS 2.3，API Level 9，2010 年 12 月 6 日发布

Android OS 2.3.3，API Level 10，2011 年 2 月 9 日发布。

- Linux 内核版本 2.6.35。
- 支持更大的屏幕尺寸及分辨率（WXGA 或更高）。
- 对 SIP VoIP 因特网电话的原生支持。
- 键盘的改进。
- 增强的复制/粘贴功能。

- NFC 支持。
- 音频效果。
- 新的下载管理器。
- 支持设备上搭载多个摄像头。
- 支持 WebM/VP8 视频播放及 AAC 音频编码。
- 改进了电源管理功能，包括应用程序管理。
- 对于较新的设备，从 YAFFS 文件系统切换到 ext4 文件系统。
- 并发的垃圾收集机制，提升了性能。
- 为更多传感器（比如陀螺仪和气压计）提供原生支持。

## D.6 Honeycomb：Android OS 3.0，API Level 11，2011 年 2 月 22 日发布

Android OS 3.1，API Level 12，2011 年 5 月 10 日发布。

Android OS 3.2，API Level 13，2011 年 7 月 15 日发布。

- Linux 内核版本 2.6.36。
- 优化了对平板电脑的支持，以及新的“全息”用户界面。
- 添加了系统条（System Bar），可以快速访问通知、状态和软件导航按钮。
- 添加了 ActionBar，在屏幕顶部提供了对上下文选项、导航、微件及其他内容的访问。
- 可在系统条中使用“最近的应用”（Recent Apps），简化了多任务切换。
- 为大屏幕重新设计了键盘。
- 简化的复制/粘贴界面。
- 用多浏览器选项卡取代了以前的新建浏览器窗口，还有无痕浏览模式。
- 硬件加速。
- 支持多核处理器。
- 通过服务器名指示（Server Name Indication）增强了 HTTPS 栈。
- 可连接 USB 配件。
- 扩展了“最近的应用”列表。
- 可调整大小的主屏幕微件。
- 支持外部键盘和定点设备（pointing device）。
- 支持 FLAC 音频播放。
- 支持在屏幕关闭时保持 Wi-Fi 连接。
- 为未针对平板电脑屏幕分辨率进行优化的应用准备的兼容显示模式。
- 用户空间文件系统（Fliesystem in Userspace，FUSE，内核模块）[1]。

① FUSE 是类 Unix 操作系统提供的一种机制，使非特权用户在不修改内核代码的前提下，能够创建自己的文件系统。——译者注

## D.7 Ice Cream Sandwich：Android OS 4.0，API Level 14，2011 年 10 月 19 日发布

Android OS 4.0.3，API Level 15，2011 年 12 月 16 日发布。

- Linux 内核版本 3.0.1。
- 使从 Android 3.x 版本引入的软按钮在手机上可用。
- 可定制的桌面启动器（Launcher）。
- 集成了屏幕捕捉功能。
- 能够在锁屏时直接访问应用。
- 改进的复制/粘贴功能。
- 更好的语音集成，以及连续实时的从语音转文本“听写”。
- 人脸识别解锁功能。
- 新的选项卡式 Chrome 浏览器，最大支持 16 个选项卡，并能自动同步书签（包括用户添加的 Chrome 书签）。
- 新的 Roboto UI 字体族。
- 在设置中加入数据使用（Data Usage）部分，可以跟踪数据流量，当超过定额时关闭数据连接。
- 能够关闭在后台使用数据流量的应用。
- 更新了人际应用，带有社交集成、状态更新和高分辨率图像。
- 引入了 Android Beam，一种近场通信特性，能够在近距离快速交换书签、通讯录信息、方向信息、YouTube 视频和其他数据。
- 支持 WebP 图像。
- 对 UI 进行硬件加速。
- Wi-Fi 直连。
- 让通常的设备能支持 1080p 视频录像。
- Android VPN 框架（AVF）和 TUN（不包含 TAP）[①]内核模块（在 4.0 版之前，VPN 软件只能在 root 过的设备上运行）。

## D.8 Jelly Bean：Android OS 4.1，API Level 16，2012 年 7 月 9 日发布

Android OS 4.2，API Level 17，2012 年 11 月 13 日发布。

① TUN 和 TAP 是操作系统内核中的虚拟网络设备。TAP 等同于一个以太网设备，它操作 OSI 参考模型中的第二层（数据链路层）上的数据包，比如以太网数据帧。TUN 则模拟了 OSI 第三层（网络层）设备，操作该层的数据包，比如 IP 数据包。——译者注

- Linux 内核版本 3.0.31。
- 为 Android 系统执行的所有绘图和动画操作进行垂直同步定时（vsync timing），并在图形管道中进行三重缓冲。
- 增强的可访问性。
- 双向的文本和其他语言支持。
- 可由用户安装的键盘映射。
- 可扩展的通知。
- 在特定应用中关闭通知的功能。
- 能对快捷方式和微件进行自动重排或调整大小，使新项目能添加到主屏幕上。
- 为 Android Beam 提供蓝牙数据连接。
- 离线语音听写。
- 屏幕较小的平板电脑改用用于手机的界面布局及主屏幕的扩展版本。
- 改进了语音搜索功能。
- 多声道音频。
- USB 音频。
- 连续无缝地播放音频。
- 预装 4.1 版系统的设备上，系统自带的浏览器被 Google Chrome 的移动版所取代。
- Google Now 搜索应用程序。
- 其他桌面启动器可以从应用抽屉（app drawer）中添加微件，而无需请求 root。
- Photo Sphere 全景图片。
- 改进了锁屏功能，包括对微件的支持，以及直接划动切换到相机功能。
- 通知电源管理。
- “白日梦”屏幕保护程序，在空闲或停驻状态下显示信息。
- 多用户账户（仅平板电脑上可用）。
- 支持基于 Miracast 的无线显示[①]。
- 为更多应用增加的多个扩展的通知及可操作通知（Actionable Notification）。
- SELinux。
- 始终保持开启状态的 VPN。
- 高端的 SMS 确认。
- Android 4.1 中为小型平板电脑提供的类似手机的桌面启动器被扩展到更大的平板电脑上。
- 将蓝牙游戏手柄加入支持的 HID 设备范畴。

① Miracast 技术简化了设备间的发现和连接设置过程，使用户可以迅速在设备间传输视频。——译者注